D0710213

About Island Press

Island Press is the only nonprofit organization in the United States whose principal purpose is the publication of books on environmental issues and natural resource management. We provide solutions-oriented information to professionals, public officials, business and community leaders, and concerned citizens who are shaping responses to environmental problems.

In 1994, Island Press celebrates its tenth anniversary as the leading provider of timely and practical books that take a multidisciplinary approach to critical environmental concerns. Our growing list of titles reflects our commitment to bringing the best of an expanding body of literature to the environmental community throughout North America and the world.

Support for Island Press is provided by The Geraldine R. Dodge Foundation, The Energy Foundation, The Ford Foundation, The George Gund Foundation, William and Flora Hewlett Foundation, The James Irvine Foundation, The John D. and Catherine T. MacArthur Foundation, The Andrew W. Mellon Foundation, The Joyce Mertz-Gilmore Foundation, The New-Land Foundation, The Pew Charitable Trusts, The Rockefeller Brothers Fund, The Tides Foundation, Turner Foundation, Inc., The Rockefeller Philanthropic Collaborative, Inc., and individual donors.

About AMERICAN FORESTS *and Forest Policy Center*

AMERICAN FORESTS, formerly the American Forestry Association, was founded in 1875 and is the oldest national citizens' conservation organization in the United States. It pioneered public education and citizen action establishing policies for sound, sustainable forest management. Today, AMERICAN FORESTS continues to pursue these goals through programs such as Global ReLeaf and seeks to educate and encourage individual and community action to protect and care for trees and forests, and thereby the environment as a whole, in the United States and worldwide.

The Forest Policy Center, a program of AMERICAN FORESTS, serves as a bridge between the scientific community and policymakers, providing timely, impartial synthesis of scientific information relating to current issues in the protection and sustainable management of forest ecosystems. A primary function of the Forest Policy Center is to identify current and emerging information needs, coordinate research among scientists with specific knowledge and expertise on the topic, and interpret the results to policymakers and the public. As part of AMERICAN FORESTS, the Center plays an important role reinforcing the scientific underpinning for AMERICAN FORESTS' action-oriented programs in policy, urban forestry, and international resource policy.

Remote Sensing and GIS in Ecosystem Management

Remote Sensing and GIS in Ecosystem Management

Edited by

V. Alaric Sample

American Forests
Forest Policy Center

ISLAND PRESS

Washington, D.C. • Covelo, California

Library of Congress Cataloging-in-Publication Data

Remote sensing and GIS in ecosystem management / edited by V. Alaric Sample.
p. cm.
Based on papers from a conference.
Includes bibliographical references and index.
ISBN 1-55963-284-4 (cloth). — ISBN 1-55963-285-2 (paper)
1. Forest ecology. 2. Ecosystem management. 3. Ecosystem management—United States—Case studies. 4. Information storage and retrieval systems—Ecosystem management. 5. Geographic information systems. 6. Ecosystem management—Remote sensing. I. Sample, V. Alaric.
QH541.5.F6R45 1994
634.9—dc20 94-17547
CIP

Printed on recycled, acid-free paper 

Manufactured in the United States of America
10 9 8 7 6 5 4 3 2 1

Contents

Acknowledgments

A great many individuals and organizations contributed to the development of this book, and to the workshop on which it is based. The initial inspiration came from the many policymakers in the U.S. Congress and elsewhere who have recently been confronted with a number of complex issues regarding the identification, characterization, and protection of the remaining areas of late-successional forest ecosystems. Pressed for decisions that will have tremendous social and economic—as well as ecological—implications, these policymakers have had to sort through information from a variety of experts. Much of this information has been in the form of large, complex displays of geographic data derived from remotely sensed imagery. Their uncertainty and hesitation over how they should interpret and use this new kind of information have been compounded when experts of apparently equal technical competence and authority seemed to derive different—even contradictory—information from the same basic remotely sensed data.

It is clear that this new technology—remote sensing and geographic information systems (GIS)—has a great deal to offer policymakers who are seeking to balance competing interests and still provide protection for forest areas of the highest ecological significance. However, these policymakers have seen their share of new technologies come and go, and important questions remain. Exactly what is the technology capable of telling them? And what is it *not* capable of telling them? To the extent that the current technology cannot provide the information that is most needed, what can be done to overcome these limitations? Is this just another example of technological promise running far ahead of performance, or has it proven its usefulness in advancing the effectiveness of forest ecosystem management, planning, and decision-making enough to justify substantial additional investment?

These questions have been articulated in many ways by many policymakers—not only legislators but decision-makers in the top levels of land and resource management organizations, both public and private. These were the people who motivated this effort, and I am grateful to each and every one of them for taking the time to explain their concerns and thoughtfully question the information presented to them.

I would like to acknowledge the special contribution to the conceptualization of the workshop made by John Teply of the Forest Service's Pacific Northwest Region, and authors Kass Green and Peter Morrison, each of them pioneers in applying remote sensing and GIS technology to the critical ecological information needs of the day. The initial idea was continuously refined and improved upon with the help of Tom Spies, of the Forest Service's Pacific Northwest Research Station, Jeff Olson, of the Bolle Institute for Sustainable Forestry at The Wilderness Society, and Jim Rochelle of Weyerhaeuser.

Major financial support for this effort was provided by Pew Charitable Trusts, and I would like to thank both Josh Reichert and Cecily Kihn of Pew for their personal support and assistance throughout. Charlie Van Sickle of the Forest Service's Southeastern Forest Experiment Station and Marvin Meier of the Forest Service Southern Region office in Atlanta also provided important financial assistance as well as technical guidance in the development and conduct of the workshop. Additional assistance from Pacific Meridian Resources, the James W. Sewall Company, and The Wilderness Society helped provide the critical mass of financial support necessary to see this project through to completion. This combination of support from charitable foundations, corporations, public agencies, and nonprofit organizations is itself an indication of the shared interest in advancing the ecological application of this important technology. It is also a reflection of the kind of cooperation, constructive dialogue, and mutual respect that the Forest Policy Center strives to facilitate in all of its efforts to promote the development of sound, progressive forest policies.

In the development of this book, Barbara Dean and Barbara Youngblood of Island Press provided invaluable guidance and editorial assistance, offered with almost divine patience and understanding. Sylvan Kaufman, a Fellow at the Forest Policy Center for 1993, made a substantial contribution to the book, transcribing major sections of the manuscript and working closely with the scientists and resource managers to develop workshop presentations into the chapters that follow.

Perhaps the greatest debt is owed to Susan Stedfast, Research Associate at the Forest Policy Center, for her indomitable good spirit in helping to organize and conduct the workshop, and then patiently shepherding more than three dozen authors and their manuscripts over a period of many months. Susan is as cheerfully persistent as she is organized, instilling far more than simple admiration among those who have awaited the publication of this volume, and among those for whom deadlines have never been more than rough approximations of when a work would be completed. The accuracy and completeness of this book are tributes to her diligent oversight in its compilation and editing.

Finally, I would like to acknowledge with thanks the contributions of each of the authors, and of their colleagues who provided peer review for each of the chapters that appear here. Ecologists, resource managers, and remote sensing/GIS applications specialists were brought together in a unique format that required active listening, interdisciplinary cooperation, and real-time revisions to prepared presentations. Authors, participants, and peer reviewers not only adapted well, but embraced these challenges with an enthusiasm that confirmed the need and value of such interaction. Ecosystem management at the landscape scale will require a far higher level of integration and coordination among disciplines than ever before. It is hoped that this will be the first of many efforts to facilitate such integration to promote the protection and sustainable management of forest ecosystems.

V. Alaric Sample
Washington, DC
January 4, 1994

Introduction

V. Alaric Sample

Ecologists and resource managers alike are being challenged by the need to greatly expand the scale on which they analyze and manage forest ecosystems. This need arises partly from the growing public concern over the loss of sensitive plant and animal species, in the United States and worldwide. The rate of loss is unprecedented in human history and is especially pronounced in species associated with late-successional forest ecosystems, which continue to dwindle as human population expands. At the same time, the current single-species approach to bringing sensitive species back from the brink of extinction is raising broad concerns due to its unpredictable, and sometimes extensive, economic and social impacts. But it is also of concern to ecologists who fear that it simply won't work, and that a more comprehensive approach is needed—one based on protecting entire plant and animal communities by using an ecosystem management approach at a landscape scale.

As ecologists grapple with the incipient science of ecosystem management, so resource managers are beginning to grapple with what amounts to a fundamental shift in the way they conceive of and manage forest resources. After more than a century as the guiding principal of forestry, "sustained-yield management" of a few commercial tree species is evolving and gradually being replaced by another principle: the protection and sustainable management of forest *ecosystems*, taking into consideration the full array of plant and animal species that occupy these complex natural communities.

Recent advances in remote sensing technology, and the processing of remotely sensed data through geographic information systems (GIS) to directly address specific analysis or management queries, present ecologists and resource managers with a tool of tremendous potential value—but only if they understand its capabilities sufficiently to capture that potential. Application of remote sensing/GIS technology has helped ecologists and resource managers recognize landscape-level ecological issues. However, use of this relatively new technology to support ecosystem-based landscape management is still in its early developmental stages. Remote sensing/GIS technology has advanced rapidly in recent years, but it is not clear that ecologists and resource managers are fully aware of its expanding capability to address their needs for landscape-scale spatial information. Nor is it clear that ecologists and resource managers have adequately identified and articulated their information needs to specialists in remote sensing and GIS applications who may be able to match those needs with new or emerging capabilities.

The central purpose of this volume, then, is to identify and articulate current and emerging information needs of ecologists and resource managers for use in policy development and decision-making regarding the management of forest ecosystems, and to explore the current and potential applications of remote sensing/GIS technology to address those information needs.

A second purpose is to provide a basic information resource to legislative and judicial policymakers who do not have a technical background in either remote sensing or resource management, but who are nonetheless called upon to make decisions and provide direction regarding the protection and sustainable management of forest ecosystems. Policymakers confronted with issues over land and resource allocation are often presented with visual representations developed with remote sensing and GIS technology, but many have little or no understanding of how the information is developed, what its limitations are, and how different representations can be derived from the same basic data.

This book represents an unusually high level of coordination and collaboration among leading landscape ecologists, resource managers, and remote sensing/GIS applications specialists, and lays a foundation for greater cooperation in the future. The book is organized into three major sections. Part I is a general description of the need for landscape-scale analysis to support forest ecosystem research and management, and current challenges in the development of remote sensing and GIS applications for this purpose. From an ecologist's perspective, Jerry Franklin discusses the need for a multiscale spatial and temporal approach to understanding and maintaining late-successional ecosystems, including old growth. An effective strategy for protecting biological diversity must look beyond forest reserves and consider the entire "landscape matrix," including lands that will continue to be managed primarily for economic purposes. Bill Gregg takes a resource manager's viewpoint and sees GIS technology as a catalyst for building constituencies for ecosystem management. He sees the increased use of remote sensing/GIS and communications technology as fostering a shared understanding of the complex tradeoffs involved in managing ecosystems to meet local and regional needs while furthering the goal of a sustainable biosphere. As a remote sensing/GIS applications specialist, Roger Hoffer asserts that the recent explosion of data available through these technologies poses new challenges for effectively utilizing the *data* to produce the *information* needed to facilitate ecosystem management. An important role of GIS applications specialists is to assist researchers and managers in understanding the choices they face in terms of the type and characteristics of the information desired, the accuracy and reliability of the data needed, and the economic tradeoffs of various sources of data and processing techniques.

Part II consists of case studies of four major forest regions of the United States—the Pacific Northwest, the Southwest, the southern Appalachians, and the northern Lake States—each with a different set of ecological and resource management concerns in a different social, economic, and cultural context. For each regional case, an ecologist and a resource manager each describe their specific information needs, and a remote sensing/GIS applications specialist describes the current and emerging technological capabilities for addressing those needs.

As home to some of the nation's largest remaining areas of native old-growth forest, the Pacific Northwest was the focus for several early efforts at ecosystem classification using remote sensing and GIS technology. James Rochelle, Peter Morrison, and Warren Cohen each examine aspects of the evolution and future of techniques used to compare historical and current extent, location, and condition of late-successional forests in western Oregon and western Washington. To Rochelle, the key to the development of an ecosystem management approach is the examination and interpretation of spatial patterns of forests as they exist now and as they could exist in the future. Historical trends toward greater fragmentation of these ecosystems, adds Morrison, could be redirected through the application of landscape approaches to forest management. Cohen looks ahead to emerging techniques for correlating the particular spatial and spectral properties of particular sources of remotely sensed data to specific biological attributes to improve the mapping of Pacific Northwest forests. Tom Spies articulates the fundamental change-to a multiscale spatial and temporal perspective—in both ecological analysis and resource management—needed to understand and maintain late-successional ecosystems, including old growth. Improvements are needed in our scientific understanding of ecological characteristics such as structure, composition, dynamics, and function. But to achieve ecosystem management on the ground, it is also important to identify and incorporate components in management activities such as objective setting, inventory, planning, implementation, and monitoring. Remote sensing and GIS techniques will continue to evolve and will be important tools for understanding the composition, dynamics, and interactions of landscape components in these unique and highly valued forest ecosystems.

Using Pacific Northwest forests as a case study, John Sessions, Sarah Crim, and Norm Johnson describe an important breakthrough in the development of spatial computer models to guide resource management decision-making. Most of the resource management decision models used in the past, such as the FORPLAN model developed by the USDA Forest Service to guide national forest planning, have used a linear-programming approach. Complex ecological interactions must be approximated and expressed as linear relationships, and then incorporated into the model as a series of constraints on the maximization of a particular objective, such as timber harvests or present net value (PNV). Such models can describe the optimal allocation of lands among a variety of land uses, but they offer little guidance as to exactly where on the landscape those uses will take place, or how different land uses on adjacent areas will relate to one another when viewed at the landscape scale. Sessions, Crim, and Johnson unveil what may become a model for planning and decision models of the future, overcoming the limitations of nonspatial models. Their prototype decision model is designed to allow scientists and resource managers to test a variety of ways to meet resource management objectives with a much clearer idea of the landscape-scale cumulative effects on aesthetics, water quality, biological diversity, and other important values.

The native aspen/pine forests of montane Arizona and New Mexico once formed an unbroken band along the mountains that ranged from the border of Colorado to the Chihuahua Mountains of northern Mexico. Today, the remaining forests provide important habitat for the northern goshawk, the Mexican spotted owl, and many other plant and animal species that characterize these upper-elevation late-successional forest ecosystems. Ecosystem classification using satellite imagery has helped to determine the spatial distribution of the remaining areas of old-growth forests, as well as the composition, structure, and condition of these forests. Margaret Moore describes how the determination of key ecological parameters using remote sensing and GIS technology can serve as an aid in the entire inventory, monitoring, and decision-making process. The controversy over the amount of late-successional habitat necessary to maintain viable populations of regional threatened and endangered species has increased as the amount of habitat has declined. Several old-growth inventories have facilitated protection efforts by using GIS technology to quickly identify "probable old-growth" using a combination of crown closure, cover type, size classes, and topography. Craig Allen and Jessica Gonzales provide more specific descriptions of efforts by the National Park Service and U.S. Forest Service, respectively, to determine the distribution and abundance of probable old-growth areas in the Jemez Mountains in northern New Mexico. The usefulness of the analysis forecosystem management and planning is discussed, along with some of the limitations and problems encountered in conducting the analysis. Doug Schleusner, who is also involved in the northern New Mexico project, takes the evaluation a step further, exploring the spatial information needs in national forest-level planning and the role that remote sensing and GIS technology might play in decisions made at that level.

The highlands of the southern Appalachian Mountains in Georgia, South Carolina, North Carolina, Tennessee, and Virginia contain some of the last vestiges of the vast native hardwood forests that once covered much of the eastern United States. The loss and fragmentation of these forests has been linked to declining populations of many species of songbirds that summer there and winter in the Amazon and other areas of the neotropics that are also under development pressures. Most of the eastern hardwood forests are in private ownership; these lands have been extensively logged, with the result that the majority of these forests are second or even third growth. The region's forests are under increasing pressure from rapid population growth, resulting in the development of many forest areas for retirement and seasonal second homes.

A large and fairly contiguous area of the southern Appalachians is under federal supervision as national parks or national forests. In 1989, the United Nations recognized the tremendous natural value of this area by designating it a World Biosphere Reserve. The Reserve is focused on the Great Smokies National Park. A larger "zone of cooperation" surrounds the park but the management of these lands, particularly national forest land, has been a source of continuing concern. Timber cutting, particularly clearcutting and the new roads that come along with it, are seen by many as threatening the ecological integrity of this especially valuable forest region.

In addition to the possible threat to songbird populations, there is considerable worry over what effects current forest management might have on the local black bear population. While much of the denning, dietary, and isolation requirements of the roughly 3000 bears in the region remain unclear, there is general agreement on some factors. These include large elevated tree cavities (associated largely with old-growth forests), an ample supply of acorns and other hard mast, and low road density. This last factor has become more important with the rise of illegal poaching of the black bear.

Ecosystem management in the South requires effective ecological land classification but, notes Charles Van Sickle, it must also recognize the social and economic context of these ecosystems and that this context varies widely across major physiographic regions. He describes in general terms the differences in approaches being used to analyze ecosystems in coastal plain, piedmont, mountains, and interior highlands. He provides a more specific description of a separate plan developed for an area with a special set of geographic and social features, but which spanned jurisdictional boundaries of three states and three federal administrative areas. David Cawrse discusses what can be done on federal lands in the area and describes the Forest Service's efforts to evaluate the quantity, quality, diversity, and connectedness of old-growth stands in advance of timber sale planning. Broader resource management plans are developed with a recognition of desired future conditions at three levels: stand, watershed, and region.

Scott Pearson tackles the problem of fragmentation at the landscape scale, recognizing the natural patterns of heterogeneity in southern Appalachian forests and the importance of this heterogeneity to ecological processes. Information on ecological processes, GIS-based maps of natural communities, and projections about activities that modify forests are needed in order to implement management strategies that will minimize forest fragmentation. Richard Flamm and Monica Turner have developed a GIS-based spatial model that will permit resource managers and planners to simulate the effects of alternative management scenarios on landscape structure. The model simulates not only the amount of land cover change that occurs on the landscape, but the spatial expression of this change, allowing issues such as biodiversity conservation, the importance of specific landscape elements to conservation goals, and long-term landscape integrity to be addressed. Of particular value in these regions will be the model's potential usefulness in enhancing coordination among different landowners in managing a regional landscape.

With the exception of a few limited areas in northern Minnesota and Wisconsin, the native boreal and northern hardwood forests that once clothed the upper Lake States were almost completely cleared away during the wave of European settlement and forest exploitation that crossed the area in the nineteenth century. However, some ecologists believe that, given time, the existing second-growth forest could continue its successional development and eventually recreate forests very similar to the old-growth forests found by the first European explorers and settlers. Several studies to improve the scientific basis for restoring old-growth characteristics are underway. Many have advocated that extensive areas of publicly owned forest land in this region be reserved from commercial timber harvesting in order to foster the development of these late-successional forest ecosystems. The goal is more than historical; it is an attempt to reestablish the species composition and diversity that existed in these relatively stable, highly developed forest ecosystems.

Thomas Crow provides a broad overview of the social and economic as well as ecological considerations that will guide the management of the northern Lake States forests. As the forests regenerated after initial development are reaching maturity, they are expected to play an important role in regional economic development. Nonetheless, there is a growing acceptance of more comprehensive ecosystem management strategies intended to protect a diversity of forest values. This approach requires different information for planning, analysis, and decision-making, as compared to the more narrow, traditional approaches to forest management. New applications of remote sensing/GIS technology are seen as critical to providing this information. David Mladenoff and George Host describe the special challenges the Lake States forests pose for ecologists and GIS specialists. These landscapes lack pronounced topography but have complex variations in soils and landforms that influence ecosystems. The region has a large number of tree species with a wide range of ecosystem characteristics, but with few differentiable spectral signatures. Among the specific information needs identified are cover mapping at various resolutions and temporal change detection, including succession and management effects and land use changes.

David Cleland, Tom Crow, J. Hart, and Joyce Thompson add that ecosystem management requires working definitions and inventories of ecosystems, as well as an understanding of ecological patterns and processes that vary at different spatial and temporal scales. There is a need to further develop GIS capabilities for arraying numerous elements of landscapes and evaluating features across a variety of scales. This kind of integration is crucial to resource managers in conducting spatial analyses of existing and desired future conditions, and analyzing management alternatives and effects. Remote sensing and GIS technology has been used to address the needs of both ecologists and resource managers, and Mark MacKenzie sees this technology as having provided a mechanism for intellectual exchange between these two groups. This integration is providing a more complete understanding of the system and effective ways to manage the resource. GIS analysis of presettlement spatial information with current land cover derived from remote sensing has been used to quantify the nature of past disturbance-induced land cover change. This may provide useful insights into ecosystem processes important to the restoration of late-successional forest ecosystems in the Lake States. When used with current and future spatial data sets, it may also be useful in detecting ecological changes due to succession, management, and disturbance.

Part III is a general description of the potential for further development (or declassification) of military and aerospace remote sensing/GIS technologies and their application to address the heretofore unmet needs for ecological information. Forrest Hall describes NASA's efforts to assist other agencies, universities, and the private sector in adapting NASA-developed technologies for broader scientific and commercial application. Hall also illustrates NASA's commitment to new environmental applications of space-based remote sensing technologies, using as an example the BOREAS project, an international scientific effort to analyze boreal forest ecosystems around the globe and assess their role in global climate change. The Department of Defense, notes Don Artis, is also beginning to explore opportunities to improve remote sensing and GIS capabilities to support ecosystem management by declassifying data obtained through a wide variety of surveillance satellites. Military remote sensing experts are still largely unaware of the particular needs and potential usefulness of their remotely sensed data for environmental and natural resource management purposes, and there is a new receptiveness to declassifying certain data that are no longer deemed sensitive for national security purposes. If these experts are made aware of these needs, there may be a significant potential for providing remote sensing information of types not currently available from commercial or civilian satellites.

Kass Green provides a summary and overview of the capabilities and limitations of remote sensing/GIS technology to address current ecological information needs, and the potential for broader application and use of GIS technology by landscape ecologists, resource managers, and conservationists. Green notes that, while the technology presents the tools for analysis, it does not guarantee the wisdom to use them effectively and wisely. She draws from several case examples to illustrate both the promise and the potential pitfalls of remote sensing/GIS technology, examining some of the fundamental assumptions of the technology and the sensitivity of analysis results to those assumptions.

Finally, Zane Cornett offers new insights into a somewhat less apparent value of remote sensing and GIS technology to ecosystem management—its use as a device for more effective public involvement in resource management decision-making by federal and state agencies and major private land managers. Many public concerns expressed during resource management planning are motivated by a sense of place, a concern over what the consequences will be for a particular location of special value or interest. As noted above, planning models used in the past have generally been nonspatial. Linear-programming models can specify the optimal number of acres to allocate among various land uses to maximize timber production, wildlife habitat, or present net value, but they are very limited when it comes to answering the question, "What changes can I expect to take place on this particular acre of land?" Remote sensing and GIS technology offers the first real opportunity to display detailed spatial information for a range of planning alternatives, and then make changes in land allocations in real time based on location-specific public concerns that may not have been expressed in earlier, more abstract stages of the planning process. More informed public choices regarding the protection and sustainable use of forest ecosystems are more likely to result in a consensus with both the breadth and depth necessary to survive the daily challenges of plan implementation.

Part I

Emerging Information Needs for the Protection and Sustainable Management of Forest Ecosystems

1

Developing Landscape-Scale Information to Meet Ecological, Economic, and Social Needs

William P. Gregg, Jr.

There is a growing consensus among managers and the public that natural resources should be managed according to principles of sustainability. Management practices must be ecologically sound, economically feasible, and socially and politically acceptable (Salwasser et al. 1992). Ecosystems are increasingly seen as fundamental units of interaction between people and Nature. Management goals must be holistic. They must reflect clear vision of the desired condition of the land to maintain aesthetics, soil processes, water quality and availability, atmospheric composition, native biological diversity, and areas of special sensitivity to disturbance. They must also reflect a clear vision for maintaining the health, economic prosperity, cultural diversity, and social well-being of the human communities of today and tomorrow. In addition, management practices to implement these goals must take into account the many uncertainties that affect natural and human systems, and the interacting effects of local decisions and regional and global conditions and trends. They must be based on an ethic of cooperation, in which people are seen as partners in planning and decision-making.

Ecosystem management is a process for implementing principles of sustainability. The process can apply to any geographic area where integrated approaches to achieve desired ecological conditions are practicable. However, ecosystem management most often applies to large geographic areas, which have been referred to as greater ecosystems (Grumbine 1990, Clark and Zaunbrecher 1987), regional ecosystems (Keystone Center 1991), regional landscapes (Noss 1983), and biogeocultural regions (U.S. MAB 1989). Although there is substantial recent literature on ecosystem management, a detailed definition has not yet become widely accepted. However, the following characteristics are frequently mentioned (Clark and Harvey 1988, Grumbine 1990, Noss

1990, 1992, Salwasser et al. 1992). Ecosystem management areas are typically open-ended. They are large enough to maintain viable populations of native species, including wide-ranging mammals. They accommodate natural disturbance regimes—that is, they can be expected to enable the long-term survival of native species under expected frequencies and intensities of fire, drought, temperature extremes, outbreaks of pests and diseases, and other stochastic events. Human use and occupancy occur at levels that do not cause ecological degradation. Management has a long time horizon, on the order of decades to centuries, that facilitates natural evolution of ecosystems and species. All organisms benefit from the management process. Charismatic, commercial, and other special-status species are considered in the context of the total management program. People are considered an integral part of the ecosystem. Management reflects understanding of the interactions of natural and human systems at many spatial and temporal scales.

Ecosystem management reflects and fosters an ethic of sustainability, and requires continuing consultation and cooperation among landowners and other stakeholders. As the number of stakeholders increases, the process becomes more effective in facilitating the complex tradeoffs that enable use of ecosystems to meet human needs in ways that sustain natural ecosystem functions and components in a changing environment.

Ecosystem management requires the creative balancing of four general types of needs:

- *Material and energy needs* refer to the use of ecosystems to produce materials, commodities, and energy required for human survival or that contribute to the material advancement of human societies. Timber production areas, croplands, subsistence use areas, water development projects, and mineral and energy development projects serve primarily these needs.
- *Social needs* refer to the use of ecosystems to provide the context for the social interactions of individuals and communities. Human settlements, recreation areas, and cultural landscapes exemplify uses that serve primarily social needs.
- *Spiritual needs* refer to uses that facilitate the aesthetic, meditative, symbolic or religious experience of individuals or cultural groups. Nature preserves, national parks, wilderness areas, religious sites, and areas of traditional cultural importance serve primarily these types of needs.
- *Informational needs* refer to uses that provide the knowledge and skills that support the theory and practice of ecosystem

> management through research, education, training, and demonstration. Biosphere reserves, long-term ecological research sites, and experimental forests exemplify areas where developing, sharing, and applying information are especially important.

Diverse, highly focused, and well-established constituencies of public agencies and private sector organizations facilitate, promote, or implement use and management of resources to meet material and energy needs (e.g., forest products and energy development industries), social needs (e.g., sporting associations, community organizations), and spiritual needs (e.g., wilderness and nature conservation organizations). However, constituencies for use and management of ecosystems to meet informational needs remain embryonic. Government agencies and academic institutions have established hundreds of research sites, and have identified numerous areas for obtaining information on natural and managed ecosystems to address particular management issues and research questions. However, only a few government programs (e.g., U.S. Man and the Biosphere Program, National Science Foundation's Long-Term Ecological Research Areas) and professional organizations (e.g., Association of Ecological Research Areas, Ecological Society of America's Sustainable Biosphere Initiative) actively promote and facilitate broad, interdisciplinary scientific uses of ecosystems that contribute to ecosystem management.

Existing landscape patterns reflect the interactions of particular governmental agencies responsible for regulating and managing uses relating to the needs of particular constituencies. Although the types and patterns of uses currently being established in many regions are becoming less effective in sustaining economic productivity and ecosystem functions, natural resource managers have found it difficult to achieve a broad acceptance of goals and practices to facilitate beneficial changes. An underlying problem is the lack of a sufficiently broad public understanding and ownership of the information used to formulate new goals and management practices. Although cooperation in developing information for ecosystem management is increasing, public constituencies for these activities have only recently begun to emerge. Yet, these constituencies will be essential for significant progress in implementing ecosystem management. In the complex political environments of many jurisdictions and competing interests that comprise the constituencies for ecosystem management, good information, widely shared, broadly understood, and generally accepted will be the most important enabling factor in the ecosystem management process.

Scientists and managers have a mutual interest in identifying representative ecological areas where long-term cooperative programs will be organized and implemented to develop the theory and practice of ecosystem management. The four case-study regions examined in the following chapters exemplify "landscapes for learning" where efforts have been initiated to build constituencies for such programs. Remote sensing and geographic information systems are basic technological tools in these programs. They offer the means for obtaining and organizing a large volume of scientific data, and a way to encourage consideration of ecosystems and landscapes as information resources. The technologies are becoming increasingly important in assessing the status of ecosystems and biological diversity (Scott et al. 1993). They are facilitating understanding and modeling of the temporal and spatial relationships among natural forces and human activities in creating contemporary landscapes, and thus improving the informational basis for establishing goals and strategies for managing these interactions. They offer managers useful tools in resolving conflicts among competing interests by strengthening common understanding of ecosystem conditions and trends. By facilitating communication among the stakeholders in ecosystem management, they can help reveal whether conflicts are due to lack of information or to differences in valuation of natural resources, and can play a catalytic role in developing the new cooperative mechanisms that will be needed to implement ecosystem management.

Addressing issues of scale is one of the greatest challenges facing contemporary ecology (Wiens 1992). Through cooperative programs that support ecosystem management, remote sensing and GIS technologies can help ecologists develop and test hypotheses regarding the effects of scale on ecological relationships and processes. They can link understanding of ecological processes and mechanisms from site-specific and organismal studies with ecosystem models at many scales, and can apply this knowledge in discovering and assessing possibilities for adaptively managing ecosystems and biological diversity in response to regional and global change.

Biological diversity, sustainable ecological systems, and global change are likely to be the greatest management challenges in the next century. These interrelated issues are the priorities of contemporary ecosystem science (Lubchenco et al. 1991, Huntley et al. 1992) and the focus of domestic and international cooperation to improve the scientific basis for ecosystem management (Comanor and Gregg 1993, UNESCO 1993). By bringing together resource managers, ecologists, and specialists in remote sensing and GIS technologies from four large ecological regions of North

America, the chapters that follow set a precedent for the broader exchange of experience that will be required to implement the goal of a sustainable biosphere.

References

Clark, T.W., and A.H. Harvey. 1988. Management of the Greater Yellowstone Ecosystem: an annotated bibliography. Northern Rockies Conservation Cooperative, Jackson Hole, Wyo. 51 pp.

Clark, T.W., and D. Zaunbrecher. 1987. The Greater Yellowstone Ecosystem: The ecosystem concept in natural resource policy and management. Renewable Resources Journal 5(3):8–16.

Comanor, P., and W.P. Gregg. 1993. Global change research in U.S. National Parks. George Wright Forum 9(3–4):67–74.

Grumbine, E. 1990. Protecting biological diversity through the greater ecosystem concept. Natural Areas Journal 10(3):114–120.

Huntley, B.J., et al. 1992. The sustainable biosphere: The global imperative. Ecology International 20:5–15.

Keystone Center. 1991. Biological diversity on federal lands: Report of a Keystone policy dialogue. The Keystone Center, Keystone, Colo. 96 pp.

Lubchenco, J., et al. 1991. The Sustainable Biosphere Initiative: An ecological research agenda. Ecology 72(2):371–412.

Noss, R.F. 1983. A regional landscape approach to maintain diversity. BioScience 33:700–706.

Noss, R.F. 1990. Issues of scale in conservation biology. In: P.L. Fiedler and S.K. Jain (eds.). Conservation biology: The theory and practice of nature conservation, preservation, and management. Chapman and Hall.

Noss, R.F. 1992. The Wildland Project: Plotting a North American wilderness recovery strategy. Wild Earth (Special Issue). Cenozoic Society, Inc., Canton, N.Y. 88 pp.

Salwasser, H., D.W. MacCleery, and T. Snellgrove. 1992. A perspective on people, wood, and environment in U.S. forest stewardship. George Wright Forum 8(4):21–43.

Scott, J.M., et al. 1993. Gap analysis: A geographic approach to protection of biological diversity. Wildlife Monographs 123:1–41.

UNESCO. 1993. MAB International Coordinating Council, 12th Session, 25–29 January 1993. Draft report. 19 pp.

U.S. MAB Directorate on Biosphere Reserves. 1989. Guidelines for the selection and management of biosphere reserves (draft). Available from U.S. MAB Secretariat, Department of State, Washington, D.C.

Wiens, J.A. 1992. Ecology 2000: An essay on future directions in ecology. Bull. Ecol. Soc. Amer. 73(3):155–164.

2

Developing Information Essential to Policy, Planning, and Management Decision-Making: The Promise of GIS

Jerry F. Franklin

Remote sensing and geographic information systems (GIS) together are crucial tools for the challenges resource managers face now and into the twenty-first century. Remotely sensed data—such as aerial photographs and satellite imagery—have been available for many years, and the last decade has witnessed a rapid evolution in the kinds of sensors used, the analytic approaches, and the methodologies for interpreting data.

GIS goes a giant step beyond our previous tools for tracking spatial data. Drawing on both traditionally gathered and remotely sensed data, it allows us as resource managers to develop, analyze, and display spatially explicit information and gives us the ability, for the first time, to deal with larger spatial scales such as drainages (≤5000 acres) and regional landscapes (e.g., several million acres). The lessons of forestry over the last decade have taught us that we must expand our view, that we must step back to see how things relate at larger levels.

GIS may be the most important technology resource managers have acquired in recent memory. Following is an exploration of five arenas in which GIS should be invaluable—(1) inventorying and monitoring, (2) management planning, (3) policy setting, (4) research, and (5) consensual decision-making—and, finally, the question of who will be responsible for overseeing the universe GIS spawns.

Inventorying and Monitoring

Despite all our efforts at inventorying and talk about monitoring, we still have only a fuzzy idea of what quantities of resources we have, where they are located, and whether they are growing, shrinking, or holding steady.

Take, as a prime example, the old-growth forests of the Pacific Northwest. In the late 1980s, we found ourselves trying to determine how much of this forest type we had and where. The Wilderness Society, spearheaded by Peter Morrison, used aerial photography and satellite imagery to help create the desired information base. Pacific Meridian Resources, contracting with the USDA Forest Service, developed a very large first-order approximation of old-growth forest acreage and layout on the national forests in which the northern spotted owl is resident.

Yet in 1991, when Congress charged the Scientific Panel on Late-Successional Forest Ecosystems (the "Gang of Four") with the task of identifying significant late-successional/old-growth forest on federal lands within the owl's range as the basis for establishing a reserve system, we found our information base to be sorely deficient despite all prior inventorying efforts (Johnson et al. 1991). To develop a best approximation, the Scientific Panel drew from a broad array of sources—in addition to the Forest Service, The Wilderness Society, the National Audubon Society, Bureau of Land Management, the Environmental Remote Sensing Applications Laboratory at Oregon State University, the U.S. Fish and Wildlife Service, and state wildlife agencies in Oregon, Washington, and California. Because the databases alone were inadequate, we brought in agency experts from the various forests and districts to assist with mapping and interpretation. The Scientific Panel's experience, then, provides an important cautionary note: despite its merits, remote sensing does have its limitations (see Chapter 4). The subsequent exercise carried out by the Forest Ecosystem Management Assessment Team for President Clinton (Thomas et al. 1993) demonstrated an improved GIS capability, however, when the full resources of the federal government are focused on this activity.

Of inventorying and monitoring, the latter is probably the major new thrust for resource managers. Without pointing fingers, we must all acknowledge we have done an abominable job of monitoring the condition of our resources, despite repeated promises by chiefs of the Forest Service, heads of the National Park Service, and others to do so. Talk has not metamorphosed into action—and that is going to have to change.

Monitoring is a critical issue for two reasons. First, public agencies are entering into many legal contracts or semilegal social contracts through court-ordered agreements, environmental impact decisions, plans, and so on, that depend on monitoring specific management practices. If agencies agree to monitor as part of a plan or agreement and then fail to do so, they are probably going to be sued. Second, and more fundamental, because all management prescriptions are, at best, working hypotheses whose outcomes are uncertain, we must monitor to gain the feedback necessary to make course corrections. Monitoring is essential to adaptive management.

Scientists have never simply conceived possible management approaches, then tested and proved them prior to implementation. Approaches have emerged over time, through experimentation, experience, and accident. For example, managing Douglas fir forests in the Pacific Northwest through patch clearcutting was an unproven concept when Leo Isaac and others proposed it in 1945. Many aspects didn't work out as expected, including the reliance on natural regeneration; tree planting was needed to assure prompt development of a new forest. Impacts on nongame wildlife and fish were not even considered. Monitoring helped correct some deficiencies in the original concept (reforestation), and lack of adequate monitoring masked other failures of this experiment. We are just beginning to acknowledge how little *proven* scientific information we base our management on, to recognize that we are always—and have always been—operating with significant levels of uncertainty. We know we need to monitor to check the effectiveness of our actions, and we know that monitoring must be spatially explicit; average values will not suffice.

Management Planning

GIS promises to be a major tool for management planning within administrative units and among ownerships because of the importance of larger spatial scales. No longer can we deal with one forest patch at a time, a 20- or 40-acre parcel, even a several-hundred-acre timber sale, without putting it into a larger context. Management planning without that context yields unwanted *and unexpected* down sides, such as cumulative effects or habitat fragmentation. Resource managers must think about patch size, about whether to disperse or aggregate activities, about edge densities, about connectivity in the landscape—which is far more than corridors: it is how the entire landscape matrix is configured.

Since about 1980, forest planning has been important in moving us toward building interdisciplinary teams and surfacing issues. However, the lack of spatially explicit information is one of the major reasons the first round of national forest plans failed. Models like FORPLAN, which lack such built-in information, simply mislead us (as Norm Johnson, FORPLAN's creator, would agree). Unrealistic timber targets have been made even more so because spatially explicit constraints could not be accounted for in existing models. For instance, the Scientific Panel on Late-Successional Forest Ecosystems had to tell Congress that the allowable sale quantities the Forest Service was projecting were unrealistic because spatial constraints such as adjacency requirements had not been considered. A subsequent study (Thomas et al. 1993) indicates that the situation was even worse than the Scientific Panel proposed.

GIS is powerful when dealing with one ownership over a smaller spatial scale (e.g., a forest or district) and, as is the case for most of the United States, with mixed ownerships over larger spatial scales (e.g., major drainages). For example, in Washington state we are talking about GIS analyses at the drainage level, which necessitates that state, federal, and private interests sort out their respective roles and deal with issues, like the northern spotted owl, that transcend ownership. We no longer see landowners suggesting they have the right to operate independently; indeed, private owners are accepting responsibility for integrating their activities with those of public owners. In the riparian zone, for example, they are defining their role in providing habitat not just for early-successional species such as deer and elk, but also mid-successional species, such as woodpeckers that might require cavities. GIS gives us the wherewithal for examining ownership tradeoffs.

Policy Setting

With science now at the center of policy setting in natural-resource issues, GIS offers an excellent means of preparing information on management alternatives for elected and appointed officials whose time and perhaps attention span are short.

The Scientific Panel and Forest Ecosystem Management Assessment Team (FEMAT) exercises are again cases in point. Beyond identifying significant old growth, these scientific panels were charged with developing and evaluating alternatives for protecting late-successional forest ecosystems and associated species, with an eye toward potential solutions to the Pacific Northwest forest controversy. Both groups assessed, as best they could, ecological and economic cost-benefits. The Scientific Panel ultimately presented Congress with 34 land-management alternatives—a set of options and their tradeoffs, a range of possibilities and their consequences. FEMAT presented President Clinton with 10 alternatives. Yet, although (as previously mentioned) remote sensing helped both groups catalog old-growth acreage, GIS was not readily or universally available. The Forest Service had no single GIS database for Pacific Northwest forests; in fact, individual forests were at highly differing levels of sophistication with respect to their GIS systems. Interestingly, the Bureau of Land Management had more advanced GIS capabilities, which accelerated analysis for that federal ownership. Moreover, where GIS was available, the Scientific Panel discovered the hard way the importance of quality control. Though data for a significant number of alternatives had been digitized and mapped, map registry was poorly done, rendering the analysis worthless. The bottom line was that the many maps displayed to Congressional delegations and other groups had to be hand drawn, a tedious, labor-intensive process introducing, no doubt, many errors of transcription.

Working with information not uniformly entered on a GIS database was problematical not only for analysis but for display. It was difficult, when interacting with Congress, to manipulate the alternatives further—to "play" with them, to tweak this piece or that, to generate still more permutations in response to an on-the-spot query or supposition. A fully operational GIS should eliminate such presentation problems. And, indeed, the level of sophistication in GIS application was dramatically improved in the FEMAT exercise two years later (Thomas et al. 1993).

Research

Both remote sensing and GIS are critical to research: scientists cannot go into the field and conduct a series of replicated experiments at the landscape level, as they can at smaller scales. With GIS, researchers can attack the important questions involving larger spatial scales by synthesizing ecological data, developing concepts, and displaying the findings. Information about patches, edges, landscape matrix, connectivity, cumulative effects, and dispersed versus aggregated activities can be included.

Scientists and resource managers will find themselves using many spatially explicit models and software. For example, the SNAP model developed by John Sessions has the robustness to allow us to look at a number of landscape alternatives, incorporating variables like edge effects (see Chapter 5). Using the technology, we should be able to test different landscape configurations and then, on the basis of those modeling exercises, implement a GIS-generated alternative in the real world. Comparing prediction to reality (much like we do via monitoring) effectively becomes the experiment.

Consensual Decision-Making

Perhaps the single most important contribution of GIS will be to make stakeholders—various user groups—and the public part of the planning process, especially long-term planning like the 10-year cycles for the National Forests. Clearly, resource management in the future will have to be the result of collaborations among managers, scientists, and the interested and effective stakeholders. GIS provides a critical resource—spatially explicit information and the ability to readily manipulate that information—that is central to this collaboration. Imagine stakeholders and managers laying out a timber sale together: developing, displaying, and evaluating alternative ways of carrying out particular management activities on a landscape.

The key, of course, is that everyone must have access to and be working from the same databases toward the same set of objectives. The various models associated with GIS could predict not only short-term tradeoffs and cost-benefits but also consequences 50, 100, 200 years down the line. GIS, then, becomes a powerful tool not only for integrating, analyzing, and presenting information but also for gaining consensus before decision-making.

Looking Forward

GIS makes possible, for the first time, truly interactive collaborations among resource managers, policy setters, scientists, and stakeholders. The day of a professional developing and presenting an alternative to the public without incorporating current science, social perspectives, and economic interests is over. Professionals are going to be participants and facilitators, not experts offering a "magic bullet" or singular decision-makers.

There is, however, a significant problem that must be addressed: Who will oversee the GIS universe? That is, who will take responsibility for the newly generated maps, atlases, databases, and libraries of information? Who will maintain and update them, define standards, assure quality control? Who will instruct people in the use of these technologies? And who will see to it they are available to *everyone*? For only when everyone is playing with "the same deck of cards" will the full potential of these tools be realized to contribute as they should, not simply to the technical but to the social resolution of natural-resource issues.

This challenge—public availability of the tools and the databases and their continued maintenance—is central to realizing the full social (as well as technical) benefits of the incredible technologies now available to us. Let us not lose sight of this nontechnical challenge!

References

Johnson, K.N., J.F. Franklin, J.W. Thomas, and J. Gordon. 1991. Alternatives for management of late-successional forests of the Pacific Northwest. A report to the Agriculture Committee and the Merchant Marine Committee of the U.S. House of Representatives.

Thomas, J.W., et al. 1993. Forest ecosystem management: An ecological, economic, and social assessment. A report of the Forest Ecosystem Management Assessment Team. U.S. Government Printing Office 1993-793-071, Washington, D.C.

3

Challenges in Developing and Applying Remote Sensing to Ecosystem Management

Roger M. Hoffer

Resource managers and ecologists need a wide variety of information about the characteristics and condition of the forest ecosystems with which they are dealing. Remote sensing, in the form of aerial photography, has been used for more than half a century to obtain some types of information. The availability of multispectral scanner and even radar data obtained from satellite altitudes, along with a multitude of quantitative computer-aided analysis techniques for processing such data, has created a tremendous increase in the array of data, analysis procedures, and results that can be obtained using remote sensing capabilities. Recent and very significant developments in GIS technology, and the interrelationships between remote sensing and GIS, have created additional dimensions of complexity as well as opportunities to use various data sources and analysis techniques to provide the *information* needed by the resource managers. This chapter examines some of the critical issues that need to be addressed as we continue to develop and use remote sensing technology to meet some of the information needs of resource managers and ecologists.

Introduction

Remote sensing can be defined as the science and art involved with the gathering of data about the earth's surface or near surface environment, through the use of a variety of sensor systems that are usually borne by aircraft or spacecraft, and the processing of these data into information that can be used for understanding and/or managing our environment (both natural and cultural). As one examines this definition, it should be noted that the critical element involves the human-machine interactions involved in the conversion of *data* into useful *information*. In forestry, as in many other disciplines, there is clearly a need for accurate, reliable information of many types, some of which can be obtained using various remote sensing techniques. As stated so well in Chapter 1 of the Manual of Remote Sensing (Colwell 1983):

> "The earth is a finite planet with limited resources. As population continues to grow, placing ever increasing demands on this resource base, wise and prudent management of these resources is becoming increasingly important. Such management is best achieved if accurate resource inventories are made available to the resource manager at suitably frequent intervals. Although the identification, measurement, and inventory of the earth's resources is a considerable task, the technology of remote sensing does offer mankind the potential to produce a broadly consistent data base, at spatial, spectral, and temporal resolution, that is useful for resource managers. . . . When allied with cartography through the use of information systems, remote sensing techniques can rise above the level of a mere technology. This coupling can change our perceptions, our methods of data analysis, our models, and our paradigms. At its most fundamental level, remote sensing provides a means by which data can be produced and analyzed for an area and then incorporated in decision-making or problem-solving procedures." (Colwell 1983, p. 1)

Since the launch of Landsat 1 in 1972, a tremendous amount of activity has been focused on the development and refinement of methods for using multispectral scanner data obtained from satellite altitudes to generate the information needed by many disciplines. The results of these efforts have shown that, indeed, many types of information can be effectively obtained using such satellite data. Yet many questions remain concerning the actual state of the art of today's technology, and how to effectively couple the data obtained from remote sensor systems with the data obtained from GIS databases (such as topographic or soils data) in order to optimize the information content of the results obtained.

The purpose of this paper is to examine some of the issues that continue to confront us as we work toward refining and implementing the tremendous capabilities provided by remote sensing and GIS technologies. Four topic areas will be addressed: (1) requirements for effective use of remote sensing technology; (2) relative costs for acquiring remotely sensed data; (3) a look at our operational capabilities; and (4) the challenges and opportunities ahead.

Requirements for Effective Use of Remote Sensing Technology

For many years, remote sensing specialists and resource managers (as well as others) have often had difficulty in communicating effectively. This seems to start at the very basic level when the resource manager attempts to ascertain the capabilities of remote sensing technology in relation to the information needed about the resource with which he or she is dealing. The resource manager needs many different types of information. Can some of these information needs be met using remote sensing and/or GIS technologies? Which ones? At the same time, the remote sensing specialist may be hard pressed to define the realistic capabilities and limitations for obtaining a specific type of information with various remote sensing techniques, due to the tremendous variability of the ecosystems with which we are involved and the relative immaturity of modern remote sensing technology. I call this communication predicament the "Information Needs Definition Circle" (Figure 3.1). (We're starting to see some of the same communication difficulties develop in relation to the use of GIS.)

So how can we overcome the dilemma of the "Information Needs Definition Circle"? Clearly, more *effective communication is the key!* I would

Figure 3.1 The "Information Needs Definition Circle"

suggest that an appropriate beginning point would involve examining the user's basic requirements for spatially explicit earth resource information. One often finds the user community stating that they need (1) data and an analysis system (including the human analyst) that are capable of "doing the job" (i.e., obtain the information needed, in a suitable format, within the time required, and with an acceptable level of accuracy and reliability) and (2) data and analysis systems that are economically feasible.

The remote sensing specialist and the resource manager must then work together to determine, quite specifically, what type of information is needed (e.g., the types of earth surface cover that are involved and their characteristics; the level of detail needed; the location, size, and topographic or other characteristics of the area involved); the format of the information needed (e.g., maps? tables? what scale?); the time frame involved for both collecting and processing the data; the levels of accuracy and reliability required (which may be different for different cover types), as well as how the accuracy of the results will be determined; and finally, what costs will be involved for both obtaining and interpreting/processing the data. An effective mechanism to initiate a meaningful communication process involves the definition of an appropriate array of questions to pose to the resource manager (the user of the information), such as:

- What cover types or earth surface features are of concern? (certain crop species, forest cover types, water characteristics, etc.)
- What other cover types are found in association with those of primary concern? (possible sources of spectral confusion)
- What geographic area is involved? (how large, where located, its physical and climatic characteristics, etc.)
- What format is required for the information? (maps, tables, or both; scales of map)
- What time of year is of critical importance for obtaining the required data? (consider the phenological conditions of the various cover types at different times of the year, and normal cloud conditions at various times of year and at different times of the day)
- What accuracy and degree of reliability is absolutely *required* in relation to what would be *desirable*? (overall and for specific cover types)

- Is this type of information being obtained at present, and if so, what does it cost per unit area to obtain this information? (If not, what is it worth to obtain this information?)

From such a question and answer process would come a better understanding of the resource manager's actual information needs and restrictions (such as cost factors). This would then allow the remote sensing specialist to define the best data source and the most appropriate procedures to follow in meeting the user's information needs. The steps toward effective use of remote sensing technology include:

- Develop a clear definition of the problem and information needed.
- Evaluate potential for obtaining needed information from remotely sensed data.
- Define data acquisition requirements.
- Define data interpretation/analysis procedures and reference data required.
- Define output products.
- Define evaluation criteria.

The first step is a clear statement of the problem being addressed as well as the type and characteristics (e.g., scale, season, etc.) of the information needed, from the perspective of the resource manager or user. The second point needs to be emphasized. Sometimes, the type of information required by the user simply cannot be obtained using remote sensing techniques. When remote sensing is not the appropriate tool to be used, this needs to be clearly recognized and alternative sources or methods to obtain the information needed by the user can be sought. Assuming, however, that remote sensing *is* an appropriate technology to meet the user's information needs, then the remote sensing specialist must examine the various options available in terms of the types of sensors and acquisition variables available, analysis procedures, characteristics of various output products that can be obtained, and evaluation procedures to be followed.

Relative Costs for Acquiring Remote-Sensor Data

As previously indicated, resource managers require data collection and analysis capabilities that not only will "do the job" to obtain the needed information, but the data collection and analysis system must be economically viable. For most users, one of the primary reasons for using remote-sensor data and analysis techniques lies in the cost effectiveness of this technology. Type mapping using aerial photos is clearly a faster, more accurate, and more cost-effective method than trying to do the same thing from the ground. Smaller scale aerial photos (e.g., 1:40,000) would be more cost effective than larger scale photos (e.g., 1:10,000), since fewer photos would be required to cover the same area. However, in some cases, the smaller scale photo may not be appropriate because the desired information simply cannot be obtained from the smaller scale photo as it could be obtained from a larger scale photo. Because of the very large area covered by a single frame of satellite imagery, it is often considered a potentially very cost-effective source of data (particularly in the digital format, since this allows computer analysis techniques to be applied to the data). For example, a single frame of Landsat data covers an area of approximately 185 × 170 km (115 × 105 mi). Therefore, even though digital Landsat TM (Thematic Mapper) data costs $4,400 per frame, that amounts to only $0.14/km^2$ or $0.36/mi^2$. By way of comparison, 609 photos at a 1:40,000 scale or 9545 photos at a scale of 1:10,000 would be required to cover the same area as a single frame of Landsat data! As shown in Table 3.1, if one assumed a cost of $16 per photo, the total cost would be 35 times greater for the 1:10,000-scale photos or 2.2 times greater for the 1:40,000-scale photos than for the Landsat data of the same area.

Although there are considerable differences in cost per unit area among the different scales of aerial photography and the digital Landsat data, it should be emphasized that one source of data may be appropriately considered to be better than another only if the information needed can be obtained from that data source! If the needed information cannot be obtained from that particular source of data, it doesn't matter how economical or how expensive the data may be—it will not be useful data! In addition, we must not forget that the *data acquisition* costs account for only a portion (sometimes a very small portion) of the total costs to obtain the *information* needed. In the case of aerial photography, significant costs are involved in handling and storing hundreds or thousands of aerial photos, in addition to the costs of the photo interpretation and then perhaps conversion of those results into data layers in a GIS database, as well as the associated costs of the field work involved in the interpretation process. Computer-aided analysis of satellite data would also involve costs for the necessary field work, as well as the analyst's time, and various computer software and hardware costs.

TABLE 3.1 Comparative cost of aerial photography and Landsat digital data

Area covered (width × length in km)	Type of data	Number of photos required[1]	Total cost	Cost/sq. km
185 × 170	Landsat TM (digital-format)	1	$ 4,400	$0.14
Non-Stereo Photo Coverage				
185 × 170	1:40,000-scale photos	609	$ 7,308	$0.23
185 × 170	1:24,000-scale photos	1,680	$ 20,160	$0.64
185 × 170	1:10,000-scale photos	9,545	$114,540	$3.64
Stereo Photo Coverage				
185 × 170	1:40,000-scale photos	1,363	$ 16,356	$0.52
185 × 170	1:24,000-scale photos	3,744	$ 44,928	$1.43
185 × 170	1:10,000-scale photos	21,390	$256,680	$8.16

[1]In order to provide a more appropriate comparison to Landsat TM data, the number of photos required is first calculated for non-stereo coverage. Then, for information purposes only, the same comparison is made for stereo photo coverage. For the non-stereo coverage, 10-percent endlap between photos and 30-percent sidelap between flight lines were assumed; for stereo coverage, 60-percent endlap and 30-percent sidelap were assumed. In both cases, for the sake of simplicity, I made the assumption that a full frame of Landsat data is rectangular in shape, 185 km wide and 170 km long. Each frame of aerial photography was assumed to cost $16, which is the current cost of NAPP or other aerial photography that is in the USGS or USDA archives. The cost of digital Landsat data was used for this comparison (rather than the cost of a hard copy Landsat image) so that one could enlarge the data to a scale that would be somewhat comparable to the scale of the photos.

Thus, in summary, when discussing the cost effectiveness of remote sensing technology, or the relative costs of one type of data versus another, one must keep several key questions in mind, such as:

- What is the value of the information desired or required?
- What are the alternatives for acquiring the needed information?
- If remote sensing seems to provide a viable approach for obtaining the information needed, what are the costs for acquiring the *data* and what additional costs will be involved in converting the data into useful *information?*

A Look at Operational Capabilities with Remote Sensing

New technologies are often treated initially with great skepticism by the user community. Until the new technology has been proven useful and cost effective, people take a "show me" approach. This is, perhaps, as it should be, and an attitude of healthy skepticism is often beneficial. Unfortunately, when Landsat 1 was first launched in 1972, remote sensing from satellite was frequently "oversold," with the result that many people developed a negative attitude toward remote sensing in general and Landsat in particular. Remote sensing technology may still be suffering from some of those effects. In addition, many agencies and companies have been reticent to embrace Landsat technology because of questions concerning the continuity of the Landsat program and the continued availability of Landsat data, as well as questions concerning the accuracy and reliability of information obtained from this type of data.

Many of these concerns should be laid to rest. The recent federal legislation placing the Landsat program under the authority of NASA (National Aeronautics and Space Administration) and DoD (Department of Defense), in conjunction with the huge remote sensing industry that has developed, and the growing use of Landsat TM as well as SPOT data by various federal agencies virtually guarantees the continued availability of this type of data for the foreseeable future. The 15-meter panchromatic channel on Landsat 6 will make this source of data increasingly useful.

When considering the potential for obtaining various types of information from Landsat data, we must keep in mind that the current capabilities of the Landsat TM (Thematic Mapper) scanner far surpass the capabilities of the Landsat MSS (Multi-Spectral Scanner) system. TM data have 30-meter spatial resolution (vs. 80-meter for the MSS), better spectral and radiometric resolution, and data are obtained in the middle and thermal infrared portions of the spectrum. A summary of the Landsat TM, MSS, and SPOT characteristics is shown in Table 3.2. As can be seen, there are distinct differences in many of the data characteristics among the different sensor systems. In addition, the cost comparisons shown at the bottom of this table clearly indicate that there are significant cost differences between Landsat and SPOT data, both on a "per frame" and on a "per unit area" basis. It is good for the user, or potential user, to be aware of these differences.

It is also helpful to remember that neither Landsat TM nor SPOT data were in existence until about 12 years ago! During this relatively brief number of years, it has been shown that many information needs can be met very effectively using this type of satellite data, for example, mapping major land-cover types, including deciduous, and coniferous mixed forest types, updating maps to show recent land use changes, identifying and mapping agricultural crop species using multitemporal Landsat TM data, mapping forest clearcuts, and so on (Colwell 1983, Hoffer 1986). Detailed analysis work is also showing that more detailed information such as individual forest cover types or density or age classes can be obtained from Landsat TM data if appropriate analysis techniques are used (Green 1994). In many cases, such techniques involve the incorporation of GIS databases to assist in the analysis of the Landsat data. It has often been stated that remote sensing provides input to GIS databases—for example, cover-type maps, change-detection data, and a wide variety of other types of information from satellite data such as Landsat TM, as well as a tremendous amount of data normally obtained from aerial photos, such as topographic data, cartographic and hydrology data, transportation networks, the location of power lines, and so on. However, GIS databases are also being used more and more as input to aid in the effective analysis of remotely sensed data. Combining topographic data from a GIS file with Landsat TM data often enables much more accurate and reliable cover-type classifications to be obtained. (This particular example is not new—work with Landsat MSS data nearly 20 years ago first demonstrated the effectiveness of combining these types of data in the analysis process [Hoffer and staff 1975]).

In spite of many examples showing a capability to classify and map various forest cover types and ecosystem characteristics with reasonable levels of accuracy (Colwell 1983, Hoffer 1986), there remain many questions and concerns that need to be addressed in relation to the utility and

TABLE 3.2 Comparison of satellite sensor systems[1]

Satellite	Landsats 1,2,3	Landsats 4 & 5	SPOT 1,2,3	
Date(s) launched	July 23, 1972 July 21, 1975 March 5, 1978	July 16, 1982 March 1, 1984	February 21, 1986 January 21, 1990 September 25, 1993	
Altitude	920 km (570 mi)	705 km (438 mi)	832 km (517 mi)	
Cycle for complete coverage of the earth	18 days	16 days	26 days[2]	
Equatorial crossing time	8:50; 9:08; 9:31 a.m.	9:45 a.m.	10:30 a.m.	
Currently functional?	No	Yes	Yes	
Sensor	MSS[3] (Multispectral Scanner)	TM[4] (Thematic Mapper)	Multispectral HRV (High-resolution visible)	Panchromatic HRV
Ground resolution	80 m	30 m	20 m	10 m
Swath width	185 km	185 km	60 km	60 km
Wavelength region and bands				
Visible	0.5–0.6 μm 0.6–0.7 μm	0.45–0.52 μm 0.52–0.60 μm 0.63–0.69 μm	0.50–0.59 μm 0.61–0.68 μm	0.50–0.73 μm
Near IR (reflective)	0.7–0.8 μm 0.8–1.1 μm	0.76–0.9 μm	0.79–0.89 μm	

Middle IR (reflective)		1.55–1.75 μm		
		2.08–2.35 μm		
Thermal IR (emissive)		10.4–12.5 μm[5]		
1992 cost of data				
Total cost per frame[6]	$1,000	$4,400	$2,450	$2,450
Cost/sq. mile	$0.08	$0.36	$1.75	$1.75

[1]From Hoffer, R.M. 1988. "Remote Sensing from Space—Two Decades of Change." Invited Paper. *Resource Technology '88, An International Symposium on Advanced Technology in Natural Resource Management*. American Society for Photogrammetry and Remote Sensing, Falls Church, VA, pp. 4–17. (Updated with 1992 prices)

[2]The SPOT satellite can be pointed to obtain data from adjacent ground tracks, thereby providing a potential for obtaining data from a particular location several times per week.

[3]Landsats 1 & 2 also carred 3-band RBV (return beam vidicon) systems, but the data quality was not good so relatively little data was acquired. Landsat 3 carried a pair of panchromatic RBV cameras having 40-m ground resolution that acquired good quality black-and-white imagery for many regions of the world. Landsat 3 also had a thermal infrared channel as part of the MSS system, but this channel failed soon after the launch.

[4]Landsats 4 & 5 also carried a 4-band MSS system similar to that on Landsats 1–3.

[5]Ground resolution for the Thermal IR wavelength is 120 m.

[6]Digital data tapes, 1600 bpi, geometrically and radiometrically corrected, but not having cartographic control.

effectiveness of computer classification of satellite data. These concerns could be grouped into four areas:

- Repeatability
- Extendability
- Integration of data sources
- Effectiveness of analysts

Repeatability has to do with establishing a better understanding of our capability to obtain similar, reasonably predictable results from remotely sensed data while attempting to obtain the same type of information under a variety of circumstances. If one tries to define the operational capability to analyze remotely sensed data for a particular purpose, one often finds that there is considerable variability in the results being obtained by different analysts around the country or around the world. Often this is caused by different analysis methods (e.g., clustering into 50 vs. 200 spectral classes); temporal variability in the data (e.g., July vs. August or September data); software differences; differences in the evaluation methodologies used; differences due to the type of data (e.g., Landsat TM vs. SPOT, or one type of aerial photography vs. another, such as color vs. infrared, or one scale of aerial photography vs. another); or many other possible causes. There will probably always be variability among analysts, but if we are careful in documenting our methods and the characteristics of the data used, as more experience is gained collectively in the use of the various types of data and analysis techniques, we will be able to do a better, more predictable job of defining the capabilities and limitations for using remote sensing technology as applied to a variety of information needs.

Extendability gets into questions dealing with distance and time. How far (geographically) can one extend a set of training statistics without degrading classification accuracy? How effective is a particular analysis procedure in different ecosystems? What calibration techniques are best for minimizing data variability due to atmospheric conditions, topographic effects, temporal differences, and so on?

The integration of data sources involves questions such as how to best use various combinations of remotely sensed and GIS data, or different types of remotely sensed data (e.g., Landsat plus radar). There is tremendous potential in effectively integrating various types of GIS data with remotely sensed data, and in combining various types of remotely sensed data. However, we have much to learn in these areas.

Finally, there are questions involving the effectiveness of the analysts themselves. It must be recognized that the effective analysis of remotely sensed data is very much an art as well as a science. I believe that to be effective, an analyst must have:

- Knowledge of the discipline;
- Knowledge of the sensor systems and analysis techniques; and
- The capability to accurately and effectively interpret the data (based on knowledge, training, experience).

It has been said that 80 percent of the information obtained from a particular aerial photo or image will be a function of the interpreter's discipline—for example, given the same photo, a forester will obtain a different array of information than would be the case for a geologist or geographer or landscape architect. Thus, a thorough knowledge of one's discipline is critically important. Knowledge of the sensor systems and analysis techniques is obviously important, but experience provides an element or ingredient that enables an analyst to be truly effective. There simply is no substitute for experience! Thus, a really good analyst would possess an effective blend of all of the above attributes.

The Challenges and Opportunities Ahead

When one considers that Landsat 1 was launched just over 20 years ago (1972), and the Landsat TM type of data has only been in existence for approximately 12 years (as indicated in Table 3.2), it seems clear that tremendous progress has been made in a relatively short time in developing effective methods for processing and analyzing such data. However, the decade of the 1990s will see even more rapid advances, and a significant increase in the operational use of remotely sensed data. Much of this increased use will be due to the continued integration of remote sensing, GIS, and GPS (Global Positioning Systems) technologies. These three technologies form a powerful, interrelated combination, as indicated in Figure 3.2. As both the public and private sectors implement and use GIS, there will be a continued increase in the use of remotely sensed data to provide input to new GIS databases, to update existing databases, and for monitoring land-use/land-cover changes of various types. As previously indicated, not only does remote sensing provide input to GIS databases, but GIS data often can be very helpful in the analysis of remotely sensed data, enabling significant improvements in the classification accuracies achieved. At the same time, GPS capabilities will provide effective cartographic control to the GIS databases, and will enable field plots (per-

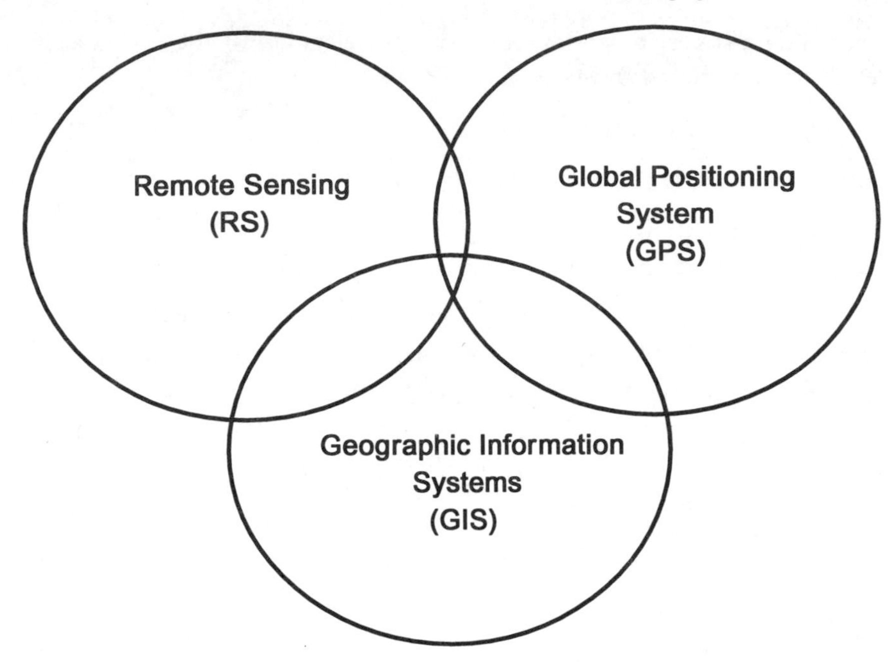

Figure 3.2 The Relationship between Remote Sensing, GIS, and GPS

haps used for training or test data in the analysis of remotely sensed data) to be located in the data efficiently and accurately.

One challenge that will continue to face us, but perhaps with increasing importance, involves the effective documentation of the characteristics of the data in a GIS database—both the digital map data and the associated attribute files. Defining and documenting the "pedigree" of the data in a GIS database is critically important. For example, was the vegetation type map in your GIS database developed from computer-aided analysis of Landsat TM data, or from photo interpretation of aerial photos? What scale and type of photos? When was the data obtained that was used to develop this data layer (year and season)? Many, many other questions could be posed as well. The point is that without adequately documenting the pedigree of a GIS data layer, other users who don't have firsthand knowledge of the origins of the data may inadvertently use it in some inappropriate manner, or assume that it is more accurate than is actually the case, with the results being the old "garbage in, garbage out" situation. Perhaps the most worrisome part of such situations is that the analyst or user might not know that the database being used is "garbage" and that the results obtained are therefore seriously flawed! To put it another way, if you don't know something, that's only a minor problem;

the big problem comes when you don't know that you don't know, because then some really poor decisions can result!

A few of the other challenges during the next few years will involve (1) the development of effective techniques to integrate and analyze data from multiple sources, such as AVHRR and Landsat TM or SPOT data, or Landsat TM plus satellite radar data; (2) converting research into operational applications in various disciplines; (3) developing effective expert systems to assist the analyst; and (4) *educating* as well as training the user community in the principals and theory of these technologies, so that they can use these powerful tools wisely, appropriately, and effectively. There are undoubtedly others that could be added to this list. Clearly, however, such activities represent many challenges as well as opportunities!

A brief summary of some of the major developments that I anticipate will occur during the next decade would include:

- Satellite optical sensor data having improved spatial, spectral, radiometric and temporal resolution, available at lower cost
- Operational multifrequency, multipolarization synthetic aperture radar data from satellite altitudes
- Continued developments in computer storage and processing capabilities
- Better understanding and use of combined data from optical, microwave, and other remote sensors
- Further integration of remote sensing, GIS, and GPS technologies
- Increased use of expert systems for data analysis

Such developments will provide many significant improvements in the quality and characteristics of the data and analytical capabilities available to the resource manager of the future. Indeed, I believe the combination of knowledgeable, well-educated *people* along with the technological tools and data available to them holds great promise for a fantastic future, thereby enabling resource managers to become much more effective stewards of this wonderful planet we call Earth!

References

Colwell, R.N. (ed.). 1983. Manual of remote sensing, 2nd ed. American Society for Photogrammetry and Remote Sensing, Falls Church, VA. 2440 pp.

Green, K. 1994. "The potential and limitations of remote sensing/GIS in providing ecological information." (this volume, Chapter 19).

Hoffer, R.M., and staff. 1975. Natural resource mapping in mountainous terrain by computer analysis of ERTS-1 satellite data. Agricultural Experiment Station Research Bulletin 919, and LARS Contract Report 061575, Purdue University, W. Lafayette, IN. 124 pp.

Hoffer, R.M. 1986. Digital analysis techniques for forestry applications. Remote Sensing Reviews. New York: Harwood Academic Press. pp. 61–110.

Hoffer, R.M. 1988. Remote sensing from space—Two decades of change. Invited Paper. Resource Technology '88, An international symposium on advanced technology in natural resource management. American Society for Photogrammetry and Remote Sensing, Falls Church, VA. pp. 4–17.

Part II

Case Studies in Major U.S. Forest Regions

Case 1

Maintenance of Late-Successional Ecosystem Values in the Pacific Northwest: Evolving Methodologies to Guide Habitat Planning and Policymaking

Introduction

James A. Rochelle

Late-successional forest ecosystems are the centerpiece of a controversy, the resolution of which is resulting in dramatic changes in the current and future management of federal forest lands in the Pacific Northwest. As proposed solutions to this controversy begin to emerge it is clear that unparalleled ecologic, economic, and sociologic changes are in store for northwest forests and their associated human communities.

Many issues contribute to this controversy. The federal forestlands in question have historically been a major part of the economic base for many northwest communities, the states of Washington and Oregon, and the region. Federal ownerships make up roughly 50 percent and 60 percent of the forestland in Washington and Oregon, respectively (Figure C1.1). As demands for wood products increase, the degree to which these lands contribute to meeting these demands will have a major influence on the overall timber economy. Direct impacts on forest-related employment levels, and the sociological implications of lifestyle changes in timber-dependent communities are major concerns.

At the same time, public expectations for the management of all forestlands are changing. Demographic change, and the growing economic importance of other segments of the economy, coupled with questions about ecosystem health in general and endangered species in particular, have generated high levels of public concern. While many factors contribute to documented declines in such familiar organisms as salmon stocks, the public attributes these declines, and the listing of increasing numbers of these stocks and other species as threatened or endangered, at least in part to forest management.

Legal mandates for maintenance of viable populations of individual species, including the Endangered Species Act, the National Forest Management Act, and other statutes, and legal challenges to national forest plans from various interest groups have placed major decisions about the future management of federal forests in the hands of the courts.

Media attention accompanying this controversy has resulted in heightened, but not necessarily technically well-founded, public concern about the management of both public and private forests. This concern manifests itself politically in the form of increasing levels of regulation of forest management practices.

Current Actions

Specific actions related directly to the maintenance of late-successional forests, or some of their key structural components, have taken place or are in process in the Pacific Northwest. These include set-asides of some 4.7 million acres of old-growth forest (Waddell et al. 1987) not counting additional old-growth area currently dedicated to northern spotted owl habitat. An additional 7.0 million acres of late-successional/old-growth

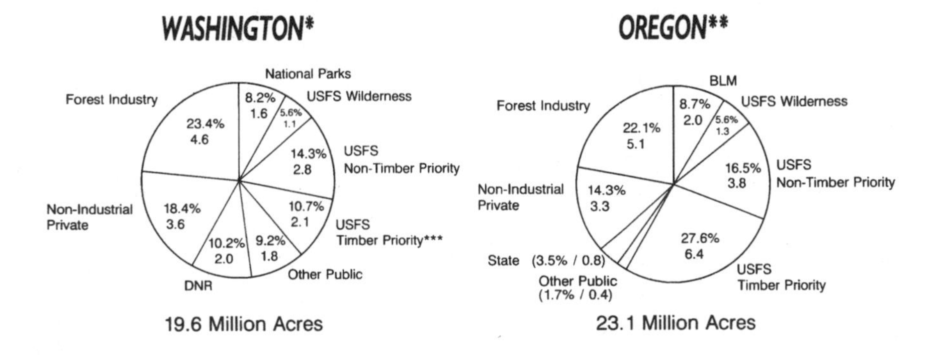

Figure C1.1 Pacific Northwest Forestland Ownership patterns, 1987 (Source: Waddell et al. 1989)

reserve may be permanently set aside as a result of President Clinton's April 1993 Pacific Northwest Forest Conference.

Regulatory provisions intended to provide for structural features of late-successional forests are in effect on state- and privately owned land in Washington and Oregon. These include riparian and wetland management zones, in which a portion of the existing stand is left to provide for recruitment of large trees for wildlife habitat and in-stream woody debris. Wildlife trees are also retained in conjunction with timber harvests in upland areas, with the intent of providing large live trees and snags through time. Voluntary set-asides are made in the state of Washington of areas that are difficult to operate, are of low productivity or low timber value, or are environmentally sensitive, many of which have current or future late-successional attributes. Alternative silvicultural treatments, in which portions of existing stands are retained to provide late-successional habitats, are being evaluated by some landowners (Rochelle and Hicks, in press).

Emerging Approaches

Ecosystem management, or using an ecological approach to maintain diverse, productive, and healthy ecosystems while sustaining the production of needed products, is currently viewed by many as the framework in which maintenance of late-successional values should occur. While several definitions of ecosystem management exist, at this time the specifics of this framework have not been developed. This approach will not be uniform across ownerships, in view of the differing objectives and mandates associated with each class of ownership. For example, in order for private lands to meet their mandate of profitable timber production, they will necessarily make different environmental contributions than public lands. There is general agreement among private landowners that the following are appropriate ecological contributions for industrial forestlands in western Washington:

- Maintenance of site productivity
- Protection of riparian areas and wetlands
- Provision of habitats for early and mid-successional species
- Assistance to public landowners in meeting habitat needs for species requiring extensive late-successional forests (Olympic Natural Resources Center, 1993)

An example of a landscape approach to resource management is found in the state of Washington's Watershed Analysis regulations. In this approach, watersheds are assessed both from the standpoint of resource sensitivities (e.g., fish habitat, water quality) and potential management-related hazards (e.g., landslides, surface erosion from roads). This information is then examined from the perspective of potential risk to resources from forest management activities and a set of prescriptions is developed consistent with minimizing those risks. Development of methodologies to address wildlife habitat management on a landscape scale is also underway in Washington.

The Role of Remote Sensing and Geographic Information Systems

The use of watershed, landscape, or ecosystem approaches has broad support as a means of realistically addressing resource use, protection, and integration objectives on a scale that is meaningful to the resources of interest. At the same time, addressing resource management issues like the maintenance of late-successional forest values on these large geographic scales requires enhanced technical capabilities and tools. Current and developing remote sensing technologies and geographic information systems (GIS) are critical to addressing the spatial and temporal interactions inherent to these large-scale management approaches in the Pacific Northwest.

To make management decisions and monitor their effectiveness, a reliable resource inventory is essential, as are tools that permit easy and effective access to the data. For example, assessments of biodiversity conditions require information on dominant forest cover, the variation existing within this overstory layer, as well as information on understory composition. And it is necessary to understand broad vegetative patterns, such as distribution of forest stand ages, as well as the occurrence and location of smaller-scale features such as streamside zones or snags. Thus GIS and remote sensing must provide information that is both accurate and precise, at multiple scales, for multiple resources.

A major strength of ecosystem or landscape-scale management lies in understanding the interactions between the various resources in the area of interest, in response to proposed management decisions. Relationships of interest might include indices of connectivity, dispersion, or fragmentation that provide measures of particular ecological conditions. Essential to this understanding is the ability to do spatial analysis, another capability that GIS can provide.

In the absence of complete or definitive data on the consequences of a particular resource management decision, adaptive management, in which the results of the decision are monitored and used to guide future decisions, is increasingly employed. While there are strong arguments in favor of adaptive management, it is costly both in terms of time and resources, and the range of alternatives that can be considered is limited. For some applications, GIS can provide a means of predicting the outcomes of alternative courses of action, from both a spatial and temporal perspective, in a timely and cost-effective manner, but does not preclude the need for monitoring on-the-ground results to guide future management adjustments.

An important, nonanalytical use of GIS and remote sensing products is in communication. While the outputs are only as good as the data from which they were developed, and assumptions and limitations must be pointed out, GIS provides a means of displaying complex information in ways that are readily understandable.

The chapters that follow provide examples of applications of remote sensing and GIS to planning at the watershed level, as tools in assessing historical and current status of late-successional forests in the Pacific Northwest, and in achieving the multiscale spatial and temporal perspective necessary for understanding and maintaining these forests.

References

Olympic Natural Resources Center. 1994. The role of industrial forestlands in the management of western Washington's forest ecosystems, A discussion paper. J. Forestry 92(5):18–19.

Rochelle, J.A., and L.L. Hicks. The role of private forestlands in the management of biological diversity. In: Szaro, Robert C. (ed.). Biodiversity in managed landscapes. New York: Oxford University Press, 1993, in press.

Waddell, K.L., D.W. Oswald, and D.S. Powell. 1989. Forest statistics of the United States, 1987. USDA Forest Service. Resource Bulletin PNW-R8-168.

4

Ecological Perspective: The Nature of Mature and Old-Growth Forest Ecosystems

Thomas A. Spies

Introduction

Over the last 20 years intense scientific and public interest has developed around the fate of existing mature and old-growth forest ecosystems in uplands and riparian zones on public lands in the Pacific Northwest. Prior to this time, mature and old-growth forests were typically viewed as sources of large timber. Ecological studies during the past two decades have revealed that these late-successional ecosystems provide many ecological values (Franklin et al. 1981). More recently, the importance of multiscale landscape perspectives has been recognized in the conservation and management of forest ecosystems of all ages (Franklin and Forman 1987, Hansen et al. 1991, Swanson and Franklin 1992). Our ability to maintain and manage these ecosystems depends on the development of new scientific information concerning ecosystem structure and function, new concepts and tools of ecosystem management (such as landscape ecology, remote sensing, and GIS), and experience gained in applying these concepts and tools in management.

The objectives of this chapter are to (1) briefly review what is known about the ecological characteristics and conservation of mature and old-growth forest ecosystems in uplands and riparian zones of the Pacific Northwest, west of the crest of the Cascades; (2) identify major gaps in our knowledge of these ecosystems; and (3) discuss the role of remote sensing and GIS in meeting research and management needs.

Historical Overview

Prior to settlement of the region by people of European descent in the mid-1800s, the region was dominated by coniferous forests that originated following fires of various severities and frequencies. Estimates of natural rotations of high-severity fires in the region range from about 150 years in drier areas of the region (Morrison and Swanson 1990) to more than 450 years on moist sites (Hemstrom and Franklin 1982). In moist, coastal areas of the Olympic Peninsula, natural fire rotations may have been more than 750 years (Agee and Edmonds 1992). Given this range of fire frequencies, it is difficult to estimate what proportion of the landscape would typically have been covered by mature and old-growth forests (defined here as forests dominated by trees more than 80 years old) prior to 1800. If we assume a regional average natural fire rotation of 250 years, then about 70 percent of the region on average may have been covered by forests older than 80 years, assuming a negative exponential age-class distribution (Van Wagner 1978). Average current ages for old-growth trees in forests of Washington and Oregon are 350 to 450 years (Spies and Franklin 1991).

In the late 1800s and early 1900s extensive clearcut logging began at low elevations, primarily on privately owned lands. Logging of old-growth and mature forests on public lands that occupy the mid to upper elevations did not begin until after the mid-1940s. As a result of past logging, less than 20 percent of the current old-growth forest exists on private lands in the region and most of the remaining mature and old-growth forest occurs on Forest Service, BLM, and National Park Service lands (Bolsinger and Waddell in press). Estimates of the percentage of the landscape currently in mature and old-growth forests on all lands in the region vary depending on inventory definitions and techniques. Recently adopted U.S. Forest Service definitions of old-growth (Old-growth Definition Task Group 1986) are based on the live and dead structure and composition of old forests. Some of the elements of these definitions, such as numbers and sizes of fallen trees, were not measured in plot-based inventories and are not possible to measure with remotely sensed imagery. Consequently, definitions based on age or characteristics of the tree canopies are often used to estimate acreage of this forest condition (Haynes 1986, Spies and Franklin 1988, Bolsinger and Waddell in press, Booth 1991). Published estimates place the amount of existing mature and old-growth forest in the region on all forest lands at 13 to 25 percent of the original amount, assuming that about 60 to 70 percent of the region was covered by mature and old-growth forests before logging (Spies and Franklin 1988, Booth 1991, Bolsinger and Waddell in press). Whatever definition or inventory method is used, estimates of the current acres of this ecosystem type indicate that considerably less mature and old-growth forest exists now than at the time of settlement and that the recent trend, if continued, would result in replacement of almost all of these old forests with managed plantations that would contain few of the characteristics of the mature and old-growth forest ecosystems they replaced (Hansen et al. 1991).

Processes and Stages of Forest Succession and Stand Development

Forest succession and forest stand development are two related ecological processes that result in change in the biotic and some of the physical characteristics of an ecosystem. "Forest succession" refers more to the gradual changes in species composition that occur on a site over time, and the term "forest stand development" has been used to describe the changes in tree population and stand structure that occur over time. In the Pacific Northwest these processes require 400 to 1000 or more years to occur, primarily because of the great longevity of the dominant tree species (Franklin and Waring 1980). Much of the interest in the later stages of these ecological processes has focused on old-growth stages; however, recent conservation efforts have used the term "late successional" to refer to a broader ecological condition that includes mid-aged natural forests (typically 80 to about 200 years) and is less tied to the specific definitions (Johnson et al. 1991) that have been developed for old-growth in the region. In this chapter I will focus primarily on the ecology and management of mid-age and old-age forests, which I term mature and old-growth (see below).

The characteristics of mature and old-growth forest ecosystems are directly dependent on the structure and dynamics of the dominant primary producers in the ecosystem—the trees. Several authors have identified stages of forest development following stand-replacement disturbance (Bormann and Likens 1979, Oliver 1981), which include establishment, thinning, understory reinitiation, transition, and steady state. The term "old growth" has typically been applied to the later two stages (Oliver and Larson 1990). While the general model of development stages has proved useful in a wide variety of forests (Peet and Christensen 1987), it lacks sufficient resolution to characterize the protracted nature of stand development and succession in coniferous forests of the Pacific Northwest (Spies and Franklin in press). Consequently, I break the later two stages into three: "maturation," "transition," and "shifting-gap."

The maturation stage is characterized by a slowing of the rate of height growth and crown expansion of the overstory trees. The overstory may contain both shade-tolerant and shade-intolerant species, which were established at about the same time following the initial disturbance. Heavy limbs begin to form and gaps between crowns become larger and more stable or expand from insect and pathogen mortality. Large dead and fallen trees begin to accumulate and the understory may be characterized by seedlings and saplings of shade-tolerant tree species, which may also occur as overstory trees. In Douglas-fir stands on the western slopes of the Cascade Range, this stage typically begins between 80 and 140 years, depending on site conditions and stand history.

During the transition stage, the original cohort of overstory trees approaches its maximum height and crown diameter, and diameter growth is slow. Tree crowns open and become irregular in shape, contain heavy limbs, and broken, dead, and decaying portions of tree crowns become common. Old trees are relatively resistant to fire of low to moderate intensity, depending on species—crown bases are high above the understory and bark is relatively thick. During this stage understory trees form multiple canopy layers, and coarse woody debris accumulates to relatively high levels. Low to moderate intensity disturbances from insects, disease, wind, and fire create patchy openings, accumulations of dead and down trees, and frequently promote establishment or advancement of understory trees that eventually fill the holes in the canopy. In westside Douglas-fir stands this stage typically begins between 150 and 250 years and may last for an additional 300 to 600 years depending on site conditions and species.

The shifting-gap stage forms when the last of the original cohort of overstory old-growth trees dies and all trees in the canopy have become established, following smaller gap disturbances of various types. This stage corresponds to the classic concept of climax or true old growth (Oliver and Larson 1990), where regeneration consists primarily of shade-tolerant trees. It can be broken into an early and a late stage (Spies and Franklin in press) based on the presence or absence of large dead trees of the original cohort. In western Oregon and Washington this stage is relatively uncommon because the overstory trees are relatively long-lived and stand-replacing fires typically occur before the initial cohort of trees dies. However, it is more common in the northern part of the region, where western hemlock, a shorter-lived tree than Douglas-fir, is dominant and fires are less frequent.

These idealized stages assume that forest development follows disturbances that kill all or most of the overstory trees, but in fact a large proportion of the "stand-replacing" disturbances in the Pacific Northwest appear to leave patches and scattered individuals of live overstory trees that become remnants in the young stand that is established following the disturbance (Morrison and Swanson 1990). Thus, many natural stands that currently exist in the Pacific Northwest cannot be neatly fitted into the idealized stages described above. The stand development stages still may have utility, however, because many complex, multiaged stands are composites of smaller patches that approximate these stages in process and structure.

Distinctive Structural, Compositional, and Functional Features of Mature and Old-Growth Ecosystems

The greatest changes in ecosystem structure, composition, and function during stand development and succession occur early in the process, when tree canopies grow together. Vegetation structure, microclimate, productivity, plant and animal composition differ dramatically between open, early-successional forests and closed, later-successional forests. After canopy closure, changes occur more gradually, eventually leading to the development of mature and old-growth ecosystems. The features that distinguish mature and old-growth forests from earlier developmental stages vary depending on which stages are compared. While we know of numerous differences between old-growth forests and early-successional forests (e.g., Schowalter 1988), especially recent clearcuts, the differences between mature and old-growth and younger closed-canopy forests are more subtle and less well-known. Nevertheless, a number of structural, compositional, and functional differences occur and have been described (Franklin et al. 1981, Ruggiero et al. 1991, Harmon et al. 1990) (Table 4.1). In a natural forest development process, many of the structural and compositional elements associated with mature and old-growth forests can be found in low abundance or immature growth forms in younger forests.

While we probably know more about the ecological characteristics of old-growth Douglas-fir forests than any other old-growth forest in the world, there are still major gaps in our understanding of old-growth and mature ecosystems. Major information gaps include (1) composition and function of canopy and below-ground invertebrates, and nonvascular plants; (2) differences in structure, composition, and function between natural forests and forest plantations; (3) disturbance and developmental history of old-growth forests across the landscape; and (4) differences, similarities, and interactions between riparian and upland forest ecosystems.

TABLE 4.1 Distinctive structural, compositional, and functional features of mature and old-growth Douglas-fir forest ecosystems relative to younger closed-canopy conifer forests

Structure

Old, relatively large, live trees characterized by:

- Broken, twisted crowns and heavy and dead limbs
- Thick bark
- Slow growth rates
- Presence of disease or insect activity
- Presence of scars from disturbances

Large variation in sizes of trees

Large standing dead trees in various stages of decay

Large fallen tree boles in various stages of decay

Large canopy gaps

High tree canopy leaf area

Patchy tree regeneration, and patches of shrubs and herbs, beneath canopy or in canopy gaps

Composition[1]

Vascular Plants: 21 trees, shrubs and herbs

Nonvascular Plants: 2 lichens, 4 fungi

Vertebrates: 30 birds, 14 mammals, 11 amphibians

Ecosystem Processes and Functions

Tree regeneration

- Primarily shade tolerants in canopy gaps

Sources of canopy tree mortality

- Primarily diseases and breakage
- Wind
- Low intensity fire some ecosystems

Productivity

- Relatively stable biomass accumulations

Inputs of nitrogen

- Relatively high inputs via nitrogen-fixing epiphytic lichens

Atmospheric interactions

- Relatively high canopy interception of snow and fog moisture
- Low albedo
- High levels of carbon storage
- Buffered microclimate

[1]Numbers of species by life form occurring more abundantly in old growth and mature or old growth alone than in other closed canopy developmental stages (From Ruggiero et al. 1991)

Riparian and Aquatic Ecosystems

The forested areas of the Pacific Northwest are typically characterized by mountainous terrain and deeply dissected drainage networks in which narrow riparian ecosystems develop. These ecosystems occupy a relatively small portion of the total landscape, but contribute greatly to overall ecological diversity. For example, several species of vertebrates are riparian specialists, plant diversity is higher in forested streamsides than on forested uplands nearby, and the streams themselves are habitat to valuable anadromous fish and other aquatic organisms (Gregory et al. 1991).

Vegetation conditions near the streams influence water temperatures, through shading, and contribute fine litter to stream food webs and large deadwood that creates valuable habitat complexity and structure within the stream. Forest development along streamsides is less well understood than in uplands and is controlled by a potentially more diverse set of factors than in uplands. Floods and landslides are characteristic disturbances that frequently affect riparian areas, and fire intensity appears to be more variable in riparian areas than in uplands. Many of the deciduous tree and shrub species, which are relatively less common than coniferous species in the region, reach their greatest dominance in riparian areas. This occurs in part because riparian zones stay relatively moist during the dry season, favoring the more moisture-demanding deciduous species, and because recent disturbances from fires, landslides, roads, and logging have created early-successional conditions that are typically dominated by deciduous vegetation.

Although the same general processes of stand development and succession occur in riparian forests and upland forests, riparian developmental pathways may be more diverse than in uplands. Old-growth coniferous forests often reach their greatest biomass and herbaceous species richness on productive streamside terraces (Franklin and Waring 1980). Some pathways of succession may be more common in riparian zones than in uplands. For example, in the Coast Range of Oregon it appears that alder forest succession on many sites leads to dominance by shrubs such as *Rubus spectabilis* rather than by coniferous trees (Carelton 1988). Aquatic ecosystems surrounded by deciduous shrubs have different species diversity, productivity, and structural complexity than those surrounded by tall conifers. Maintenance of riparian ecosystems in managed forest landscape typically consists of protecting buffers of width ranging from about 20 to 100 meters, depending on the size of the stream (Gregory and Ashenas 1990). The long-term stability and ecological function of these buffers is not known and is a management concern. If forest buffers are subject to higher rates of disturbance than intact forest areas, succession could lead to an increase in deciduous shrub species in some areas and a decline in the quantity and quality of late-successional riparian vegetation.

Landscape and Regional Ecology

In recent years considerable interest has developed in the ecological characteristics of landscapes or large land areas composed of heterogeneous mosaics of smaller ecosystems, watersheds, vegetation patches, and land-use types. Spatial patterns of habitat conditions such as patch size and arrangement appear to affect habitat quality for spotted owls (Carey 1992) and forest microclimate (Chen et al. 1993). The pattern of disturbance and forest cutting also can affect the timing and magnitude of stream flow (Swanson and Franklin 1992) and is hypothesized to affect the occurrence of fire, windthrow, and insect and forest diseases in a landscape. The practice of dispersed cutting on National Forests has led to an increase in edge habitat and a decline in interior forest habitat (Franklin and Forman 1987, Spies et al. in press) (Figure 4.1). At the watershed scale, roads and early successional clearcuts have increased erosion potential and decreased quality of stream ecosystems associated with late-successional forests (Swanson et al. 1982). Although we have learned much about landscape-level effects on ecosystem quality in recent years, our scientific knowledge and management experience at this scale is still extremely limited. For example, we have little information about the long-term effects of landscape pattern of seral stages, disturbances, and old-growth forest fragmentation on animal and plant populations, disturbance regimes, and ecosystem productivity.

Regional-scale ecological patterns and processes are also important to understanding and managing old-forest ecosystems. The Pacific Northwest is an ecologically and socially diverse region. Environmental gradients of geology, precipitation, growing season, and disturbance regime are steep and complex. Physiographic diversity is high (Franklin and Dyrness 1973). Land ownership varies from large blocks of federal land to small parcels of private lands; strong social differences exist between rural communities, where people tend to value forests for their extractive commodities, and large metropolitan communities, where people tend to value forests for their recreational and scenic values. The maintenance of biological diversity requires that the diverse Pacific Northwest region be stratified into ecological regions for purposes of planning and manage-

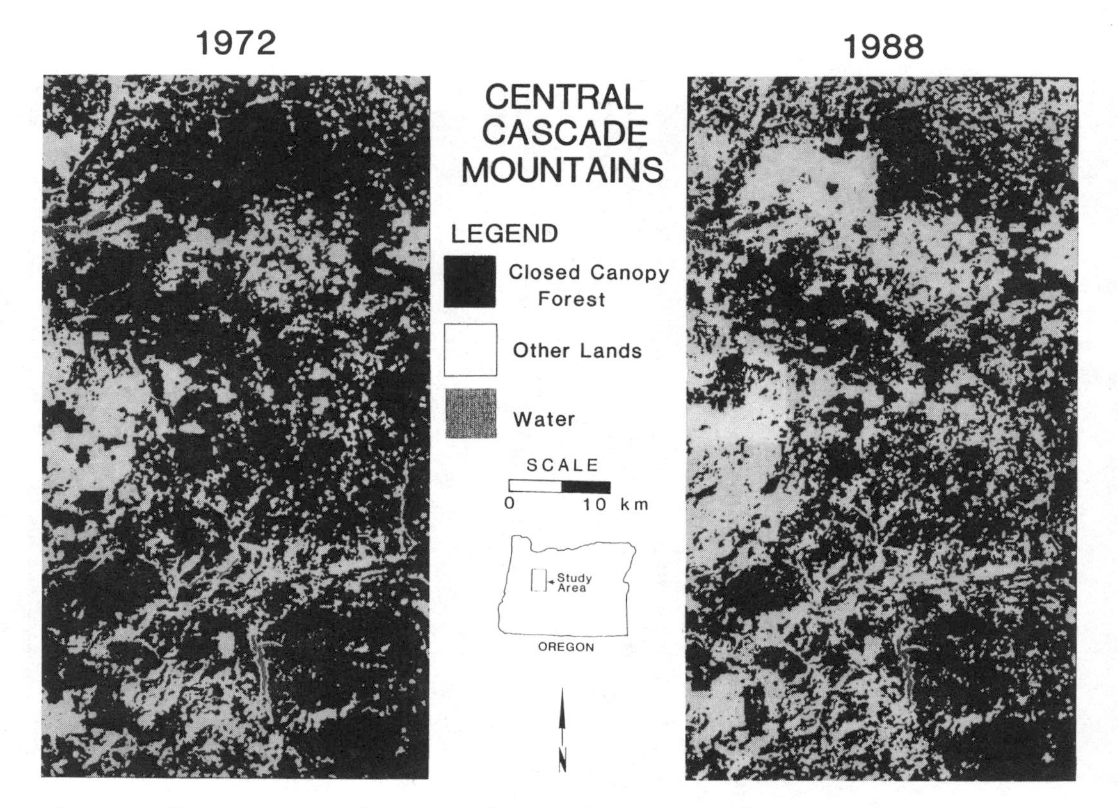

Figure 4.1 Distribution of closed-canopy conifer forests (primarily naturally regenerated conifer forests >40 years old) in a multi-ownership landscape in western Oregon in 1972 and 1988. (From Spies et al. in press.)

ment. Development of conservation strategies for spotted owls, anadromous fish, and old-growth ecosystems have been based on subregional ecological units (USDI 1992), and reserve design at the regional scale is partly based on genetic and population demographic considerations as well as land ownership patterns.

Incorporating landscape- and regional-scale perspectives into the maintenance of late-successional ecosystems presents considerable challenges to management and planning in the region. Since the vast majority of the mature and old-growth forest remaining in the region occurs on federal land, this ownership will have to bear the primary economic burden of maintaining this ecosystem for the region as a whole. However, it is now recognized that land-use practices on private lands can have effects on the ecosystems of federal lands, especially where the two ownerships are interspersed in alternating mile-square sections. In such landscapes it may not be possible to maintain late-successional upland and riparian ecosystem quality on federal lands alone without some modification of land-use practices on the private lands. The social and economic aspects of multiownership planning and management for biological diversity are extremely complex and are primarily in the discussion stages in the region (Anonymous 1993).

Role of Remote Sensing and GIS in Research and Management in the Pacific Northwest

Many of the scientific and management issues in the long-lived coniferous forests of the Pacific Northwest require information and understanding at large spatial scales and over long time periods. Tools such as remote sensing and GIS can help meet the needs of scientists and managers. Many ecologically important characteristics of the upper canopy of mature and old forests are quite distinctive from younger forests (Spies and Franklin 1991) and can be sensed with fair to good reliability with satellite remote sensing (Cohen and Spies 1992). Landscape patterns of older forests can also be readily quantified with remote sensing and GIS (Ripple et al. 1991, Spies et al. in press). However, our experience in applying these relatively new technologies in research and management is still very limited, and we are a long way from realizing their full benefits and limitations.

While some of the structural features of mature and old-growth forest ecosystems are sensible with current remote-sensing technologies, most are not. Old-growth ecosystems are highly layered, so that upper layers obscure lower layers. While some correlation between layers exists (Spies and Franklin 1991), variation is probably too high to use characteristics of the upper canopy to reliably predict characteristics of lower layers. Many attributes, such as standing dead trees or scattered old-growth trees, are too small to be seen with coarse-resolution sensors. Other attributes such as species composition may not be spectrally distinct (Cohen and Spies 1992). Riparian zones and buffers, which are frequently less than 100 meters wide, typically occur in deep, topographically shaded valleys and are often a heterogeneous mosaic of deciduous and coniferous vegetation that can be difficult to characterize with current satellite sensors.

Although remote sensing and GIS offer much promise to the study and management of forest ecosystems, they cannot be the only tools of ecosystem research and management. Future advances in the use of these technologies will require merging of remote sensing with ground-based sampling and monitoring systems to characterize fine-scale and hidden features of these forest ecosystems. Remote sensing is typically viewed as use of Landsat satellite imagery; however, advances in landscape analysis will require development of analysis techniques for other satellite imagery such as radar, and other airborne digital imagery such as infrared video and photographs, which have higher spatial resolution. A fuller application of GIS technologies to ecological problems in the region requires research and development in the areas of spatial ecological analysis, spatial simulation, interfacing of GIS with remote sensing, spatial statistics, and spatial database management.

Remote-sensing and GIS research, development, and application are particularly needed in the following areas:

- Quantifying abundance, distribution, and spatial pattern of forest structure and composition in closed-canopy forests with high leaf areas.
- Characterizing watershed conditions such as vegetation cover, unstable slopes, road locations, and riparian zones at multiple spatial scales and across ownerships.
- Identifying appropriate spatial resolution and number of attribute classes for land-cover maps.
- Monitoring and displaying ecological change resulting from succession, disturbance, and climate change with integrated remote-sensing and GIS tools.
- Modeling and simulating ecological change in GIS resulting from disturbance, succession, and movement of organisms;

water and habitat features such as large woody debris; and landslides over decades and centuries.

- Understanding patterns and sources of spectral and spatial variation in land cover at stand, landscape, and regional scales.

Conclusions

Mature and old-growth forest ecosystems are distinctive and important components of the biological diversity of the Pacific Northwest. These ecosystems are characterized by complex structures and ecological relationships closely linked to the growth and death of large, long-lived conifers. The structures and dynamics of these ecosystems span a wide range of spatial and temporal scales. Research on the ecology and management of mature and old-growth forest ecosystems is less than 20 years old, and despite recent advances in our knowledge of these ecosystems, we have much to learn. The maintenance of these ecosystems in the region requires advances in our understanding of forest development and disturbance and their ecological effects in uplands and riparian ecosystems, as well as whole landscapes and regions. Remote sensing and GIS are just beginning to find application in our efforts to understand and maintain these ecosystems in the long term.

References

Anonymous. 1993. Forest ecosystem management: an ecological, economic, and social assessment. Report of the Forest Ecosystem Management Assessment Team. U.S. Government Printing Office 1193-793-071.

Agee, J.K., and R.L. Edmonds. 1992. Forest protection guidelines for the northern spotted owl. In: Final draft recovery plan for the northern spotted owl. Washington, D.C., U.S. Department of the Interior. Appendix E.

Bolsinger, C., and K.L. Waddell. In press. Area of old-growth forests in California, Oregon, and Washington. Resour. Bull. PNW-RB-XXX. Portland, OR. USDA Forest Service, PNW Research Station.

Booth, D.E. 1991. Estimating prelogging old-growth in the Pacific Northwest. J. Forestry 89:25–29.

Bormann, F.H., and G.E. Likens. 1979. Pattern and process in a forested ecosystem. Springer-Verlag, New York. 253 pp.

Carleton, G.C. 1988. The structure and dynamics of red alder communities in the central Coast Range of western Oregon. M.S. thesis. Oregon State University, Corvallis, OR.

Carey, A.B., S.P. Horton, and B.L. Biswell. 1992. Northern spotted owls: influence of prey base and landscape character. Ecological Monographs 62:223–250.

Chen, J., J.F. Franklin, and T.A. Spies. 1993. Contrasting microclimates among clearcut, edge, and interior of old-growth Douglas-fir forests. Agricultural and Forest Meteorology 63:219–237.

Cohen, W.B., and T.A. Spies. 1992. Estimating structural attributes of Douglas-fir/western hemlock forest stands from Landsat and SPOT imagery. Remote Sens. Environ. 41:1–17.

Franklin, J.F., and R.T. Forman. 1987. Creating landscape patterns by forest cutting: ecological consequences and principles. Landscape Ecology 1:5–18.

Franklin, J.F., and C.T. Dyrness. 1973. Natural vegetation of Oregon and Washington. Oregon State University Press, Corvallis, OR.

Franklin, J.F., K. Cromack, Jr., W. Denison, A. McKee, C. Maser, J. Sedell, F. Swanson, and G. Juday. 1981. Ecological characteristics of old-growth Douglas-fir forests. General Technical Report PNW-118, USDA Forest Service, Pacific Northwest Forest and Range Experiment Station, Portland, OR. 48 pp.

Franklin, J.F., and R.H. Waring. 1980. Distinctive features of the northwestern coniferous forest: development, structure, and function. In: Waring, R.H. (ed.). Forests: fresh perspectives from ecosystem analysis. Proceedings of the 40th Annual Biological Colloquium, Apr. 27–28 1979. Corvallis: Oregon State University Press. pp. 59–86.

Gregory, S.V., F.J. Swanson, W.A. McKee, and K.W. Cummins. 1991. An ecosystem perspective of riparian zones. Bioscience 41(8):540–555.

Gregory, S.V., and L. Ashkenas. 1990. Riparian management guide, Willamette National Forest. U.S. Department of Agriculture, Forest Service. Willamette National Forest, Eugene, OR.

Hansen, A.J., T.A. Spies, F.J. Swanson, and J.L. Ohmann. 1991. Conserving biodiversity in managed forests. Bioscience 41:382–392.

Harmon, M.E., W.K. Ferrell, and J.F. Franklin. 1990. Effects on carbon storage of conversion of old-growth forests to young forests. Science 247:699–702.

Haynes, R.W. 1986. Inventory and value of old-growth in the Douglas-fir region. USDA Forest Service, Pacific Northwest Research Station Note PNW-437, Portland, OR. 19 pp.

Hemstrom, M.A., and J.F. Franklin. 1982. Fire and other disturbances of the forest in Mount Rainier National Park. Quaternary Research 18:32–51.

Johnson, K.N., J.F. Franklin, J.W. Thomas, and J. Gordon. 1991. Alternatives for management of late-successional forests of the Pacific Northwest. Report to the Agriculture Committee and the Merchant Marine and Fisheries Committee of the U.S. House of Representatives. 59 pp.

Morrison, P.H., and F.J. Swanson. 1990. Fire history in two forest ecosystems of the central western Cascade Range, Oregon. General Technical Report PNW-GTR-254. USDA Forest Service, Pacific Northwest Research Station, Portland, OR.

Old-Growth Definition Task Group. 1986. Interim definitions for old-growth Douglas-fir and mixed-conifer forests in the Pacific Northwest and California. Research Note PNW-447. USDA Forest Service, Pacific Northwest Research Station, Portland, OR. 7 pp.

Oliver, C.D. 1981. Forest development in North America following major disturbances. Forest Ecology and Management 3:153–168.

Oliver, C.D., and B.C. Larson. 1990. Forest stand dynamics. New York: McGraw-Hill. 467 pp.

Peet, R.K., and N.L. Christensen. 1987. Competition and tree death. Bioscience 37:586–595.

Ripple, W.J., G.A. Bradshaw, and T.A. Spies. 1991. Measuring forest landscape patterns in the Cascade Range of Oregon, USA. Biological Conservation 57:73–88.

Ruggiero, L.F., L.C. Jones, and Keith B. Aubry. 1991. Plant and animal habitat associations in Douglas-fir forests of the Pacific Northwest: An overview. In: L.F. Ruggiero, K.B. Aubry, A.B. Carey, and M.H. Huff (tech. coords.), Wildlife and vegetation of unmanaged Douglas-fir forests. General Technical Report PNW-GTR-285 USDA Forest Service, Portland, OR.

Schowalter, T.D. 1988. Canopy arthropod community structure and herbivory in old-growth and regenerating forests in western Oregon. Can. J. For. Res. 19:318–322.

Spies, T.A., and J.F. Franklin. 1988. Old-growth and forest dynamics in the Douglas-fir region of western Oregon and Washington. Natural Areas Journal 8(3):190–201.

Spies, T.A., and J.F. Franklin. 1991. The structure of natural young, mature, and old-growth forests in Washington and Oregon. In: L.F. Ruggiero, K.B. Aubry, A.B. Carey, and M.H. Huff (tech. coords), Wildlife and vegetation of unmanaged Douglas-fir forests. General Technical Report PNW-GTR-285, USDA Forest Service, Portland, OR. pp. 91–121.

Spies, T.A., and J.F. Franklin. In press. The diversity and maintenance of old-growth forests. In: R.C. Szaro (ed.), Biodiversity in managed landscapes: theory and practice. Cary, NC: Oxford University Press.

Spies, T.A., W.J. Ripple, and G.A. Bradshaw. In press. Dynamics and pattern of a managed coniferous forest landscape in Oregon. Ecological Applications.

Swanson, F.J., and J.F. Franklin. 1992. New forestry principles from ecosystem analysis of Pacific Northwest forests. Ecological Applications 2(3):262–274.

Swanson, F.J., S.V. Gregory, J.R. Sedell, and A.G. Campbell. 1982. Land-water interactions: the riparian zone. In: R.L. Edmonds (ed.), Analysis of coniferous forest ecosystems in the western United States. U.S./I.B.P. synthesis Series 14. Hutchinson Ross, Stroudsburg, PA. pp. 267–291.

U.S. Department of Interior. 1992. Recovery plan for the northern spotted owl. Portland, OR. 662 pp.

Van Wagner, C.E. 1978. Age-class distribution and the forest fire cycle. Can. J. For. Res. 8:220–227.

5

Resource Management Perspective: Geographic Information Systems and Decision Models in Forest Management Planning

John Sessions, Sarah Crim, and K. Norman Johnson

Introduction

Forest planning operates on the interface between management actions and ecosystems. It requires understanding the existing ecological state of the forest and projecting, controlling, and manipulating vegetation over time and space to attain management objectives. Geographic information systems (GIS) can provide a high-quality spatial database to describe the state of the forest, but must be combined with spatial and temporal decision models to create alternatives for forest planning. These decision models need the capability to represent and control a variety of spatial and temporal considerations including location and size of disturbances and seral stages, location of wildlife corridors, and relations between aquatic and upland ecosystems.

Recognition of the importance of spatial relationships in soil, water, and wildlife resources has grown rapidly during the past decade. Accompanying the growing awareness has been the need to inventory and map resources. Rapid development of GIS software, increasing access to high-quality graphics hardware, and decreasing processing costs have enabled great progress in spatial data and mapping capability.

Efficient data storage and display facilities are a necessary but not sufficient condition for solving today's forest planning problems. Forest planners must also examine the effects of collections of management treatments on forest ecosystems as well as the ability of these management actions to produce outputs. Based on this information they can then derive ranges of alternatives that are spatially and temporally feasible and that attain, to the extent possible, desired mixes and levels of outputs. Data storage and display, when used alone, will not be adequate to find feasible solutions to today's forest planning problems.

Problem Description

To illustrate the type of spatial and temporal relationships planners generally consider, we describe a typical planning problem faced by many public-land managers in the Pacific Northwest. Pursuant to the National Forest Management Act of 1976 (NFMA), the Forest Service has developed land and resource management plans for each of the National Forests. These strategic plans establish broad direction for management of National Forest lands. The land base for suitable timber production and a ceiling on sustainable harvest levels are established as part of national forest plans. While forest plans have a life expectancy of 10 years, 150- to 200-year planning horizons are employed in developing forest plans to ensure harvest levels are sustainable and to review future effects of management actions taken today.

In the Pacific Northwest, these strategic plans contain limited spatial definition due to their broad scope. Many spatial details have been left for resolution during implementation of the plans, including the location, timing, type, and extent of harvesting. During plan implementation, resource planners must therefore determine whether or not it is possible to develop a harvest plan for a subarea of the forest that achieves the desired output levels established in the overall forest plan while providing adequate protection for wildlife, fisheries, soils, and other ecological concerns.

Harvest plans for subareas of the forest usually focus on the short term and are referred to as tactical plans. The size of the tactical problem varies both in area and potential harvest units. However, it is not unreasonable to consider forest areas as large as 10,000 hectares consisting of 500 to 1000 potential harvest units to be scheduled for treatment over one to four time periods. In connection with harvest decisions, different parts of the transportation system must be scheduled for construction, reconstruction, restoration, or left unbuilt. A typical transportation system consists of 1000 to 3000 potential road segments.

Since there is usually more than one plan to harvest a particular unit, silvicultural treatment and logging system selection must be considered simultaneously with the development of the transportation system. If it is possible to find more than one alternative that meets all of the ecological and social objectives, the less costly of the alternatives must also be identified.

Spatial and Temporal Relationships

A specific requirement of these tactical plans for each national forest is that it contain sufficient spatial and temporal detail to define where to locate the harvest units with respect to one another, when the harvest units will be treated relative to one another, where the units are located with respect to the transportation system, when they will be harvested relative to road construction and reconstruction, and when and where the areas will be treated relative to key ecological variables. These considerations are needed for at least three reasons: (1) harvest cannot take place on two adjacent parcels until vegetation on the previously harvested parcel has reached some specified condition; (2) accurate transportation costing requires explicit spatial and temporal definition of the transportation plan; and (3) assessing impacts on ecosystem health requires establishing when and where harvesting and road construction will occur in relation to key locations such as stream reaches, wildlife forage and cover areas, and wildlife corridors. To understand what these requirements may entail, consider the following examples of resource planning problems.

Location and Size of Disturbances

Many wildlife, watershed, and soil resource relationships are affected by the location and size of disturbance regimes. When a unit of mature sawtimber is harvested, an opening will be created as the unit's seral stage changes from late-successional to an early-successional grass or forb condition. If the distance to surrounding cover is not excessive, this new grass/forb area may provide forage for certain types of wildlife. If the harvest unit is located in the middle of a block of interior old-growth habitat, its movement from a late-successional seral stage to a grass/forb state may result in the fragmentation of habitat for other types of wildlife.

Many social objectives also revolve around size and location of disturbance regimes. The extent of an opening, for example, might need to be controlled to meet visual management objectives. Using important viewing points, the forest may be divided into zones such as foreground, middle ground, and background. In the foreground, smaller openings may be permitted and the new vegetation may need to grow taller before adjacent areas can be harvested than for areas harvested in the visual background category.

When these concepts are applied to even-age management regimes, the length of time or number of seral stages the harvest unit must move through before it emerges from an opening is sometimes called the "green-up period." The concept, however, is more general. An opening can mean any contiguous group of units sharing some common attributes and seral stages. For example, it could mean not thinning too large an area at one time to prevent windthrow. In this case, the seral stages defining an opening would be those that involved a thinned stand for several periods until the stand closed up again. It is often necessary to control many different types and sizes of openings and green-up periods simultaneously across the landscape.

The openings must be controlled by scheduling units individually or in groups in such a way that the maximum permitted size of opening is not exceeded, time period after time period. A snap shot in time is not sufficient.

Manual methods and close manual interaction with GIS have been used to verify that vegetative treatments in the first time period and possibly the second time period were spatially feasible. Map coloring has been used to demonstrate that it is possible to spatially locate units to harvest with respect to simple adjacency rules. In general, these approaches have had only limited success with respect to overall resource management objectives. Approaches are needed that can simultaneously consider a variety of spatially defined objectives over multiple time periods.

Wildlife Corridors

Wildlife corridors are connected sets of acceptable seral stages that provide a "pathway" between two or more points in the landscape. They are sometimes required to prevent isolation, provide migration routes, or limit maximum distance to hiding cover. To accommodate different wildlife or social uses of the forest, a number of different corridors may be needed simultaneously across the landscape. For example, a specific fur-bearer may require corridors containing one set of seral-stage conditions connecting its critical habitat, while recreationists require connections with a different set of seral stages. Since the forest is dynamic, these pathways may need to shift over time.

Seral Stage Distributions

A number of resource issues involve limiting the maximum number of hectares of some seral stage or group of seral stages in different areas of the forest. An example is habitat maintenance for the northern spotted owl in the Pacific Northwest. One set of guidelines requires that for each quarter-township there must be approximately 50 percent of the land with trees at least 25 centimeters in diameter and at least 40 percent crown closure at any given point in time. Another example is that to protect watershed health, no more than 25 percent of a drainage basin can be in certain seral stages during any time period. Both of these examples have two elements in common. First, they apply to a specific geographical area. Second, they require that no more than a maximum number of hectares can exist in one seral stage or group of seral stages during any time period. To ensure these distributional conditions are met, management treatments must be linked across both space and time.

Upland Management—Aquatic Ecosystems

The health of aquatic ecosystems is tied to management occurring on the upslope areas. Often the stream system must be considered so that water temperature and quality linkages between management treatments, transportation system development, and fisheries impacts can be measured and controlled. Locating harvest units upslope from stream reaches rich in the number of pools per kilometer, for example, creates quite different impacts than locating the units near stream reaches devoid of complex aquatic structures. With the emphasis on fisheries health in the Pacific Northwest, the location of harvesting relative to stream systems needs to be modeled so that impacts on aquatic ecosystems can be evaluated.

Analysis

There are two key pieces of technology forest planners need at their disposal to develop management alternatives that incorporate the spatial and temporal relationships outlined in the previous section: (1) GIS and (2) spatial- and temporal-based decision models. To illustrate how resource planners might organize their analysis to generate management alternatives that incorporate these spatial and temporal relationships, we present an example using the Scheduling and Network Analysis Program, SNAP (Sessions and Sessions 1992), a decision model being used in a number of locations in the western United States.

The hypothetical project area is Snap Creek (Figure 5.1a). It is composed of 175 polygons, each representing a unique seral stage and each having a choice of three possible types of initial silvicultural treatments. Each of the polygons or land parcels are also unique combinations of attributes such as slope class (high, medium, low), visual class (foreground, background), and risk class (high- and low-risk soils). Other necessary information includes the potential road network of 214 links that might be constructed, reconstructed, or left unbuilt (Figure 5.1b) and the stream network (Figure 5.1c) in Snap Creek identified by 100 individual stream reaches and their attributes.

Information used to construct these types of planning problems can be extracted from a geographical information system. The management unit and stream attributes, for example, can be retrieved from different layers in a GIS. The Department of Natural Resources (DNR) in the state of Washington has constructed linkages between ARCINFO and SNAP to facilitate data transfer between the two systems.

Returning to our example, planners want to know, among other things, if it is possible to harvest 4500 Mbf (thousand board feet) of timber from Snap Creek during the next four time periods. Their initial set of ecological requirements for the area are: (1) a late-successional seral stage wildlife corridor must be maintained on the area through time; (2) 30 percent of Snap Creek must be in a mature or late-successional seral stage at any given point in time; (3) openings created in the foreground and background areas cannot exceed 10 acres and 40 acres, respectively; (4) a maximum of 25 percent of the area can be in a treated or disturbed state in any time period; and (5) regeneration harvesting cannot occur on high-risk soils. The beginning seral-stage pattern for Snap Creek is shown in Figure 5.2a, the critical points for the wildlife corridor that must be maintained are shown in Figure 5.2b, and the relationship between seral stages and created openings is illustrated in Figure 5.2c.

With these objectives and requirements in hand, the model is run to determine if harvesting can take place, where it can occur, when it will occur, the type of silvicultural treatments selected, and the set of associated logging systems and road activities needed. Five feasible harvest patterns are generated, each producing close to 4500 Mbf in each of the first three time periods. In the fourth time period it is not possible to produce 4500 Mbf due to cumulative spatial constraints, even though a substantial volume of mature timber remains. Taking a look at one particular pattern, we can see the harvest units and transportation system selected (Figure 5.3a), the silvicultural treatments employed (Figure 5.3b), and the location of the wildlife corridor (Figure 5.3c).

At this point, resource planners can take the solution and export it back out to GIS for further analysis. We have found, however, that before planners export solutions to a GIS environment, they often want to perform a critical amount of this post-analysis directly in the decision-model software: they want to be able to recycle solutions quickly without

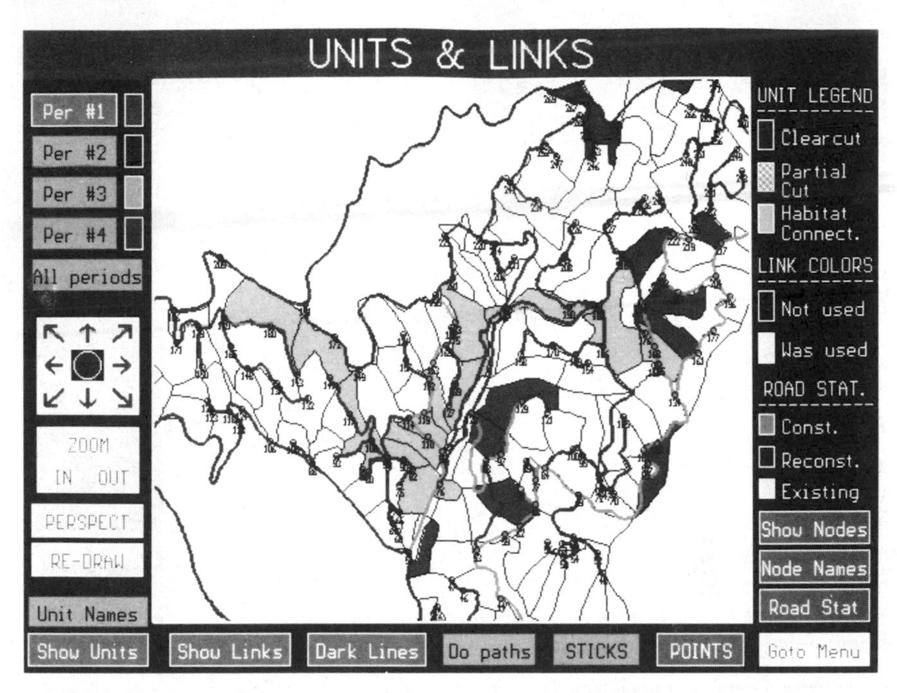

Figure 5.3a Solution showing selected harvest units and road links (segments) for period 1. Shaded units are harvest units.

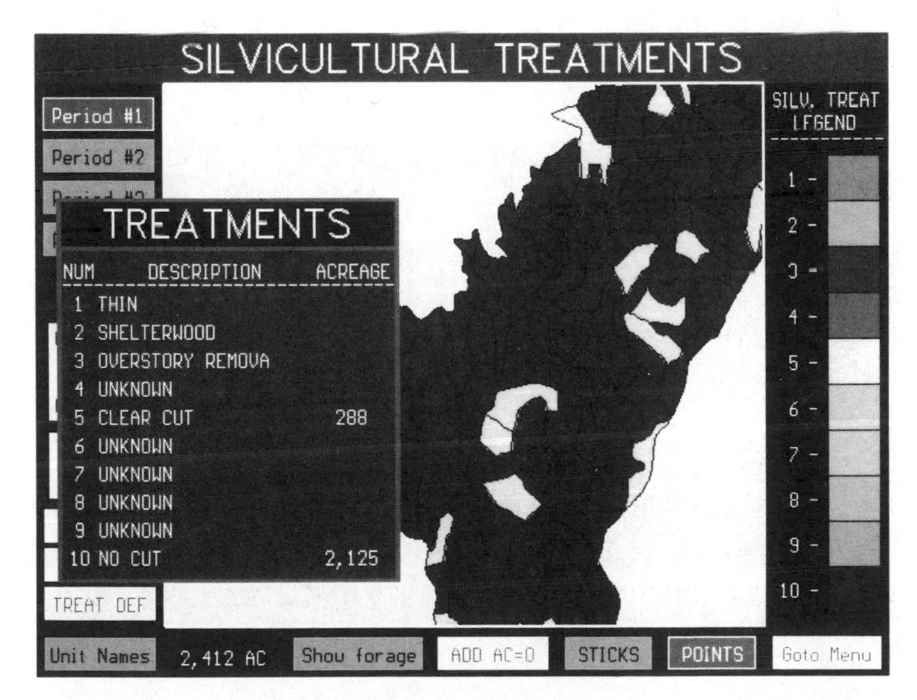

Figure 5.3b Solution showing selected silvicultural treaments. Pop-up box summarizes acres treated by silvicultural treatment in period 1.

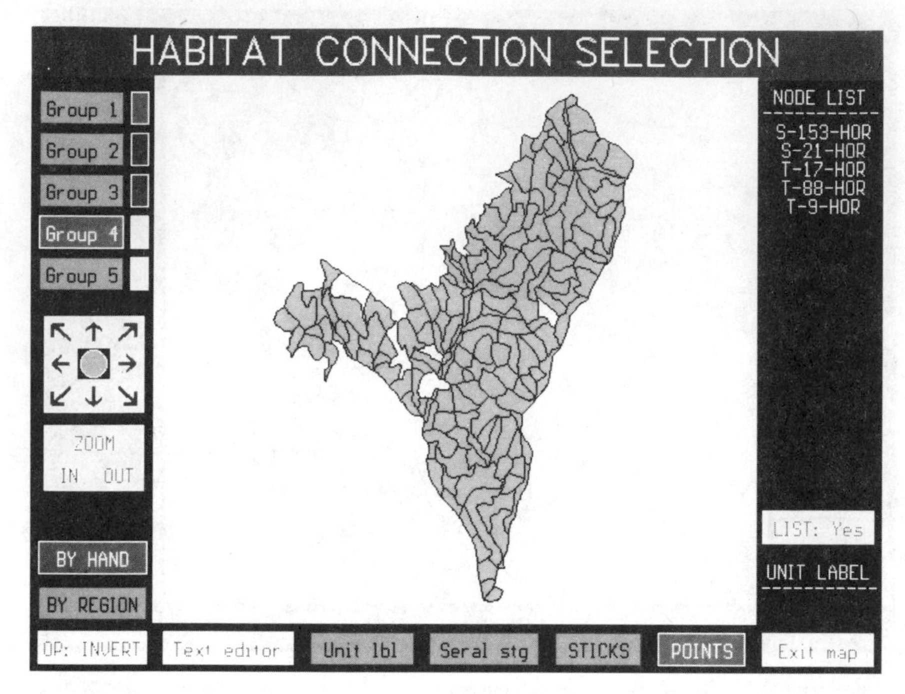

Figure 5.2b Snap Creek Critical Points for Habitat Connection. A feasible solution must connect the critical points by a pathway of acceptable vegetative characteristics (seral stages).

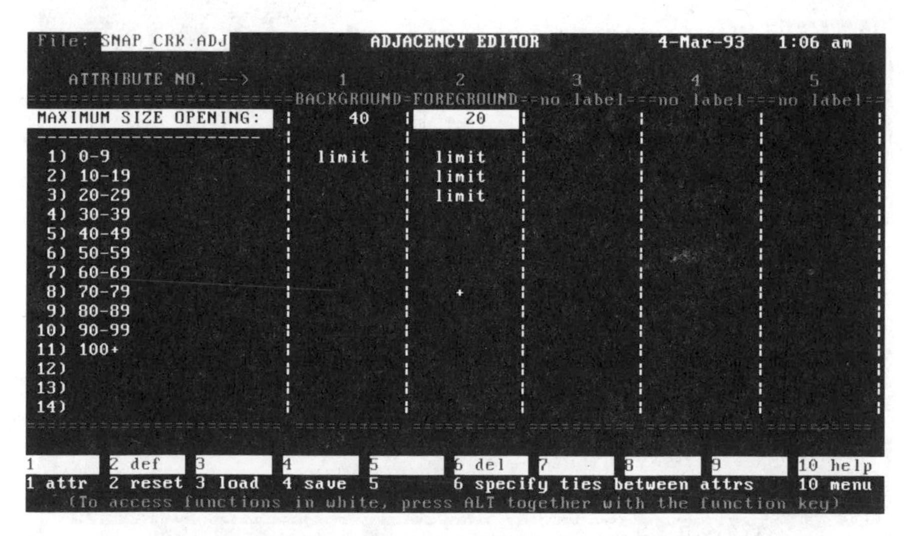

Figure 5.2c Snap Creek Relationships of Seral Stages to Created Openings. The maximum size of opening is specified by polygon attribute and the "green-up" period is indicated by seral stage "limit."

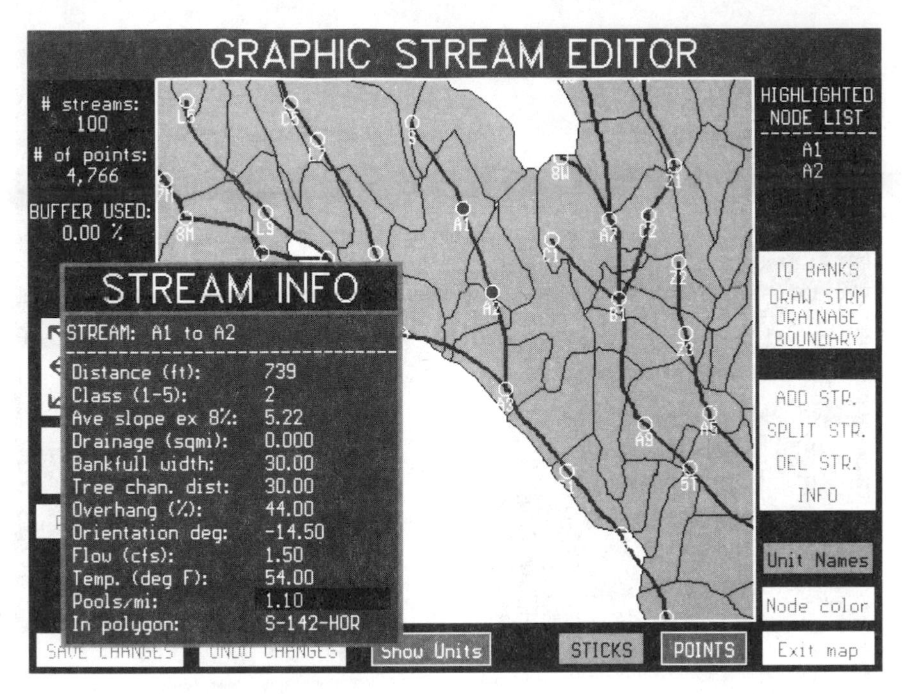

Figure 5.1c Snap Creek Project Area Stream System generated by the Scheduling and Network Analysis Program (SNAP II). Pop-up box summaries stream information for stream segment A1–A2.

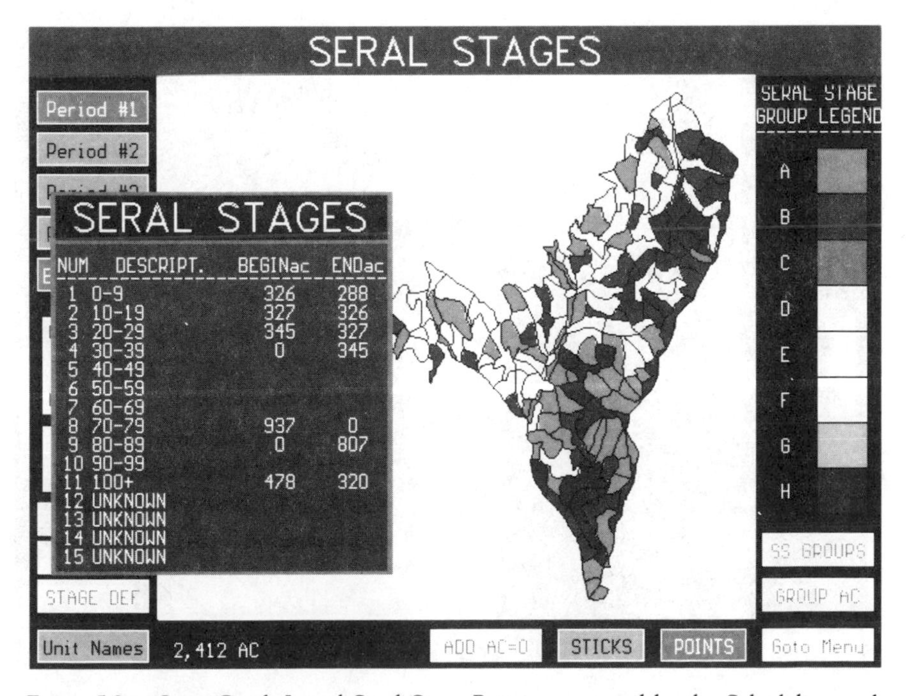

Figure 5.2a Snap Creek Initial Seral Stage Pattern generated by the Scheduling and Network Analysis Program (SNAP II). Pop-up box summarizes acreage by Seral Stage at the beginning and end of planning period 1.

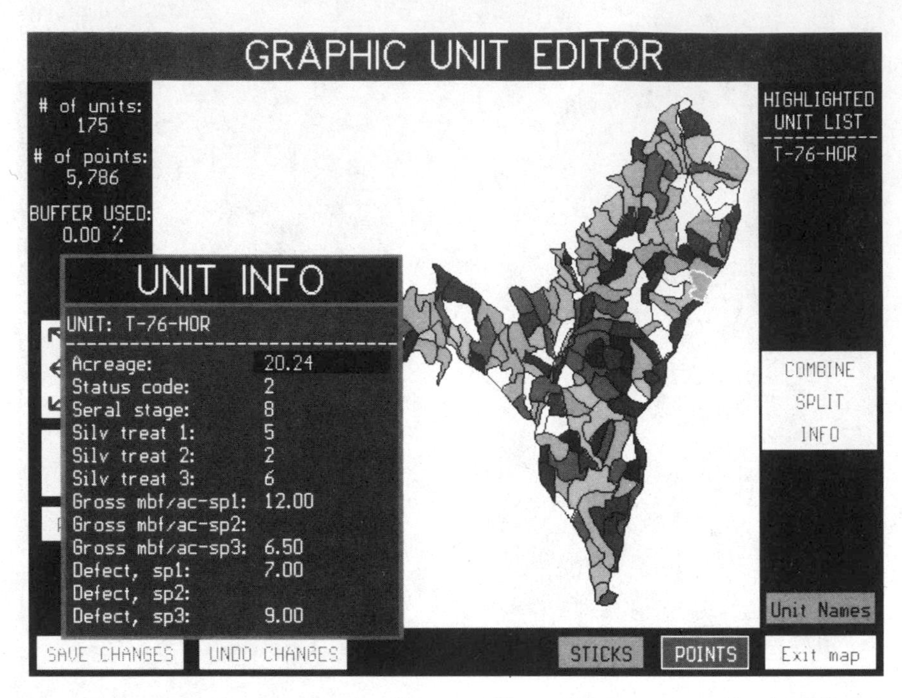

Figure 5.1a Snap Creek Project Area Potential Harvest Units generated by the Scheduling and Network Analysis Program (SNAP II). Pop-up box summarizes polygon information for polygon T-76-HOR.

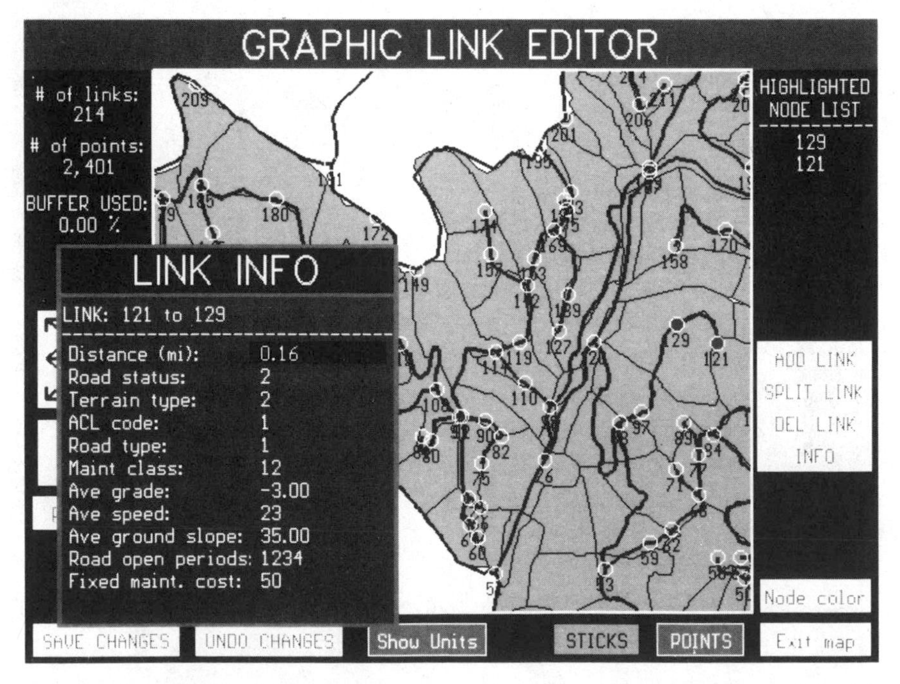

Figure 5.1b Snap Creek Project Area Potential Road Network generated by the Scheduling and Network Analysis Program (SNAP II). Pop-up box summarizes road information for road link (segment) 121–129.

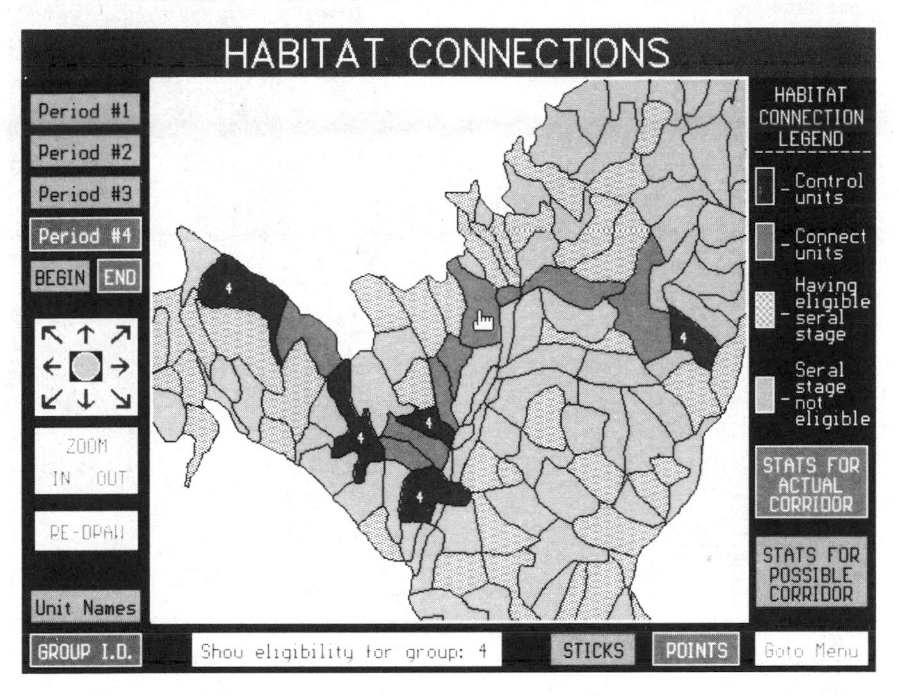

Figure 5.3c Solution showing the selected habitat connection during time period 4. Black units are critical polygons that must be connected by polygons of acceptable vegetation. Polygons with wavy pattern have been selected by SNAP to connect the critical polygons during period 4.

switching analysis environments. Examples of post-analysis to aid wildlife biologists, landscape architects, and fisheries biologists are briefly illustrated below.

To aid wildlife biologists, the area and effectiveness of acceptable big-game cover at various distances from forage can be calculated during post-analysis review in SNAP for any solution and time period using techniques similar to those described by Wisdom et al. (1986). Similarly, the area of effectiveness of forage at various distances from acceptable cover can be reported. Figures 5.4a and 5.4b display the habitat effectiveness and road closure information for one particular harvest pattern in Snap Creek.

To aid in assessing visual objectives, the planning area can also be viewed in perspective from user-defined points inside or outside the planning area after each time period for any solution identified by SNAP.

Additional post-analysis evaluations can be done on aquatic resources. Figure 5.4c displays a stream temperature model that evaluates the effects

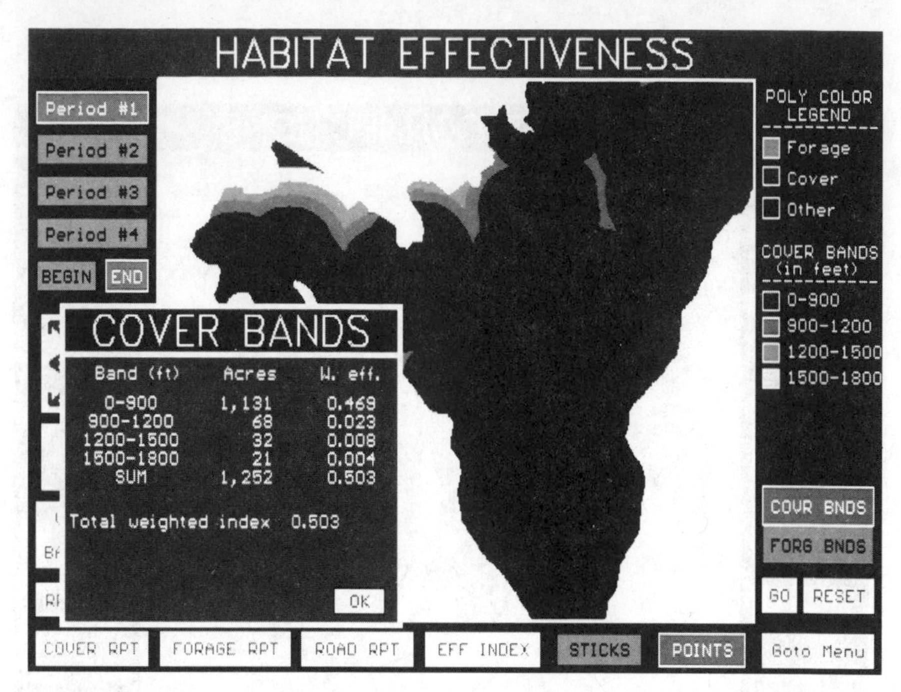

Figure 5.4a Post Analysis showing analysis for habitat cover effectiveness at the end of period 1. Pop-up box shows the area of various categories of cover within 1800 feet of the perimeter of forage.

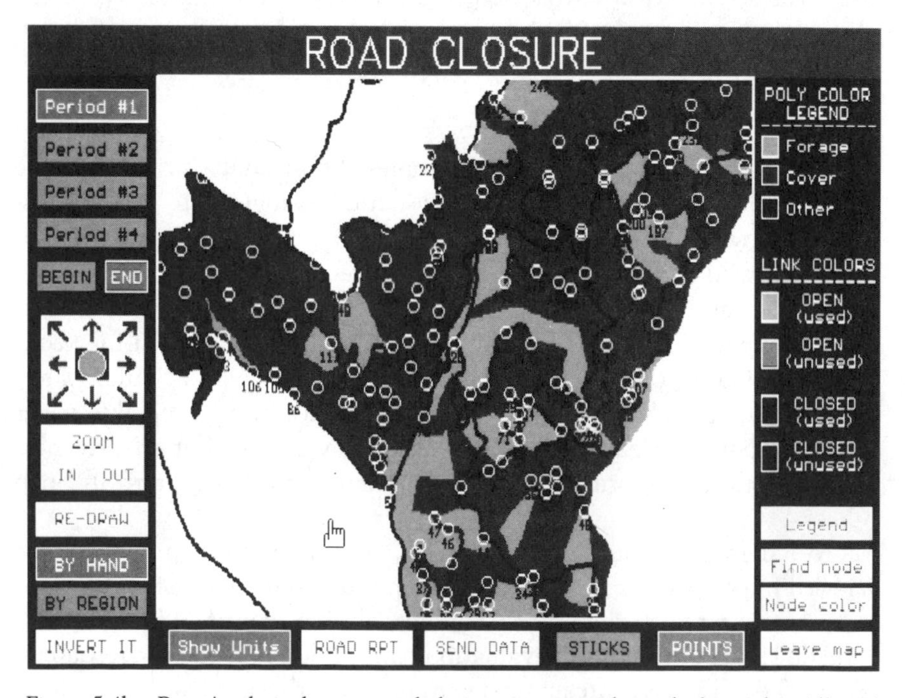

Figure 5.4b Post Analysis showing road closure strategy at the end of period 1. All road segments shown are designated to be closed between harvest entries.

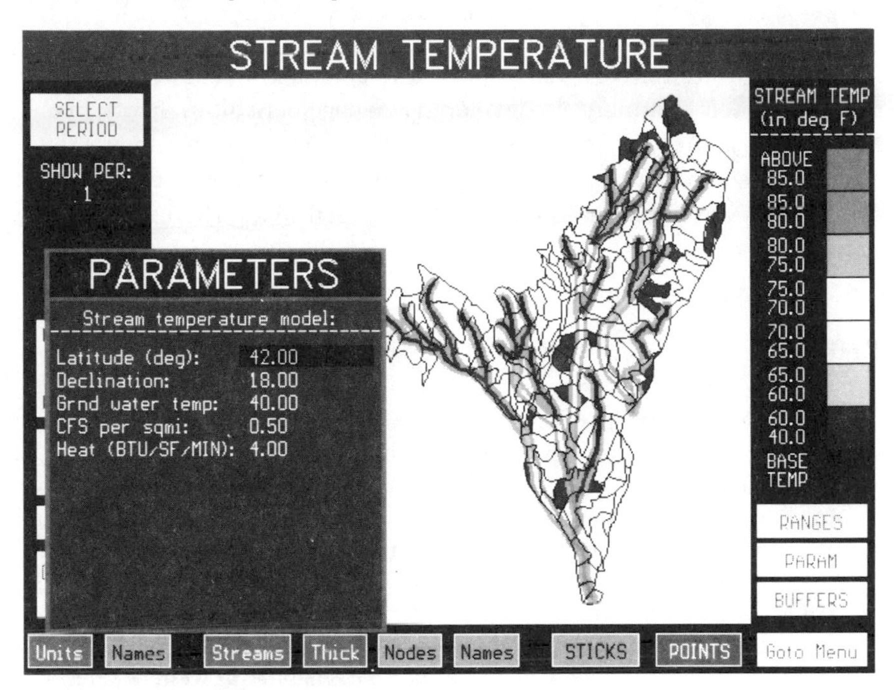

Figure 5.4c Post Analysis showing stream temperature evaluation at the end of period 1. Results of stream temperature calculation are color coded according to the bar code at right. Pop-up box summarizes project location, declination of sun at day of low stream flow, ground-water temperature, runoff coefficient, and heat rate.

of harvesting on stream temperature loadings, with the stream temperature model based on techniques developed by Park (1994). In addition to looking at temperature loadings, fishery biologists can evaluate the proximity of harvest activity to individual stream reaches. If the biologists feel, for example, particular harvest units are too close to a stream segment rich in aquatic structure, they can make these land parcels off limits to harvesting and quickly generate new solutions that take these additional requirements into account.

Conclusion

Forest planners need both GIS and decision models at their disposal if they are going to meet the complex challenge ahead of them. GIS serves an important role as a spatial database and mapping tool. For a snapshot in time, it can be used to evaluate the juxtaposition of mapped activities and resources and thereby aid in spatial feasibility determinations. Decision models, however, are needed to generate spatially and temporally explicit management alternatives that capture the complex, dynamic nature of forests and their treatment. In developing management alternatives, planners need to identify mixes and schedules of management treatments that attain a broad range of social, economic, and ecological objectives. Decision models serve an important function in identifying these mixes and schedules.

References

Park, Chris. 1994. SHADOW, Stream Temperature Management Program. USDA Forest Service, Region Six, Portland, OR.

Sessions, J., and J.B. Sessions. 1992. Tactical forest planning using SNAP 2.03. In: Proceedings, Computer Supported Planning of Roads and Harvesting Workshop, International Union of Forestry Research Organizations, August 26–28, Munich, Germany, pp. 112–120. Available through Department of Forest Engineering, Oregon State University, Corvallis.

Wisdom, M.J., L. Bright, C. Caret, W. Hines, R. Pedersen, D. Smithey, J. Thomas, and G. Witmer. 1986. A model to evaluate elk habitat in western Oregon. USDA Forest Service, Region Six, Portland, OR. 36 pp.

6

GIS Applications Perspective: Development and Analysis of a Chronosequence of Late-Successional Forest Ecosystem Data Layers

Peter H. Morrison

Coniferous forests covered most of the presettlement landscape in the Pacific Northwest. Although a natural mosaic of successional stages existed, historical studies indicate that late-successional forests dominated much of the area. Forest management practices of this century have created a landscape much different from the presettlement landscape. The rapid rate of loss of late-successional forests has contributed to the population decline of many species.

Reliable information on the historical and current status of late-successional forest ecosystems is essential to forest planning and policy development in the region. In this chapter, several techniques are described for the development and analysis of such information within a geographic information system (GIS) environment. Historical information on forest condition was developed using old field-plot data, archived aerial photography, and historical forest-cover maps. Early MSS (multispectral scanner) satellite imagery provided a source of information on forest condition beginning in 1972. Current information on the status of late-successional forest ecosystems was developed using aerial photography, field data, and various types of satellite imagery.

Results from three studies indicate that landscape condition varies over the region and is dependent on environmental factors and management practices. A study of the Olympic National Forest (WA) indicated a 60-percent decline in late-successional forests during a 48-year period, from 1940 to 1988. Fragmentation of late-successional forests through dispersed clearcuts was well established by 1962. By 1988, logging had resulted in a total loss of large (>10,000 hectares) old-growth forest patches. A similar study of Vancouver Island (Canada) indicated a 49-

percent decline of late-successional forests between 1954 and 1990. Forest fragmentation occurred but had less impact than in the prior study due to a practice of progressive clearcutting. An on-going study of the Okanogan National Forest indicates initial higher fragmentation due to environmental conditions, and widespread alteration of late-successional forests due to fire suppression and logging.

Introduction

Forest ecosystem management can benefit greatly from GIS-based analyses and data derived from remote sensing of forest resources. Reliable and current information on ecosystem characteristics at a landscape level provides the foundation for intelligent planning, management, and conservation. Likewise, historical data at stand and landscape levels can provide a much needed perspective on the condition of the premanagement landscape. The development of a chronosequence of landscape-level views developed in a GIS environment leads to insight into the functioning of the premanagement ecosystem, habitat requirements for native fauna, and the current state of ecosystem health (Morrison 1992a).

The maintenance of late-successional ecosystem values in the Pacific Northwest is a forest ecosystem management goal that has received considerable attention lately. Forest planning and policy decisions in this area have long been made in an information-poor environment. Although progress has been made in the last few years in developing landscape-level information on some ecosystem characteristics, many ecosystem attributes are still not well inventoried on a landscape level. Information on fish and wildlife population density, mortality, and fecundity can be developed on a landscape level with aid from GIS analysis and modeling capability. Likewise data on the spatial distribution of native flora, with a priority on rare plants, can be developed from field data, remote sensing, and ecological modeling in a GIS environment. Better data are needed on forest characteristics such as tree age and size class distributions of forest stands and amounts of standing dead and down trees. These data can be developed in a GIS environment with the aid of remote sensing and ecological analysis and modeling. GIS is the optimal environment for analysis of landscape patterns and processes including the size and configuration of forest patches, spatial density of forest edge environments, survival and dispersal of fauna dependent on interior forest habitat, and the spread of disturbance phenomena.

Although an emphasis on development of current information on forest ecosystem characteristics is a priority, the temporal aspect of forest ecosystem analysis and management must not be neglected. Too often, in our current culture, we tend to become preoccupied with the current dilemma, focusing our data gathering and analysis efforts within the time frame spanning the immediate past to the near future. When the maintenance of late-successional ecosystem values is at stake in a landscape that has been highly altered by human activities, this short temporal perspective is inadequate. Development of a chronosequence of landscape-level forest ecosystem data layers can provide valuable insight into the original nature of the ecosystem and the cumulative effect of human activities on ecosystem integrity and health.

A Historical Overview of Forest Conditions in the Pacific Northwest

There has been no comprehensive study of the overall extent and condition of presettlement forests in the Pacific Northwest, but there is ample evidence that mature coniferous forests blanketed most of the western portion of the region for a long time (Brubaker 1990, Franklin and Dyrness 1973, Norse 1991). Surviving trees and decaying stumps from the presettlement late-successional forests can still be found throughout the region in second- and third-growth managed forests, partially cleared farmland, and even in some city parks.

Around the turn of the century the U.S. Geological Survey conducted a set of extensive surveys of the newly designated federal forest reserves (Dodwell and Nixon 1902, Langille and others 1903, Lieberg 1899, Plummer 1900, 1902). The resulting maps and reports document extensive mature and older forests covering the Cascades and Coast Ranges, with small areas of younger stands, recent burns, and limited early logging. It is humbling to realize that this extensive work was accomplished without the aid of any remote sensing or GIS technology.

Although the coniferous forests that originally covered much of the Pacific Northwest were composed of a natural mosaic of successional stages resulting from a variety of disturbance mechanisms, late-successional forests dominated much of the area. These late-successional forests are often more than 200 years old and may contain mixtures of several age classes resulting from natural successional processes and a variety of natural disturbance mechanisms, including variable-intensity fire and windstorms.

A historical perspective on Pacific Northwest forests is incomplete without some understanding of the natural disturbance regimes operating in various parts of the region. Along the coast of Washington and British Columbia natural fire is rare, and climax forests often develop relatively undisturbed for many centuries (Henderson and others 1990, Huff and Agee 1980). In the coastal region, wind is the most common disturbance phenomenon. In the western Cascade Mountains natural stand-replacement fire is relatively infrequent (natural fire rotation of 150 to 800 years), allowing old forests to develop and persist on the landscape (Hemstrom 1979, Hemstrom and Franklin 1982). The natural fire regime in much of the central and southern parts of the Cascades and Coast Ranges is characterized by a fire regime of variable-intensity fire with understory burns occurring at greater frequency (Morrison and Swanson 1990, Teensma 1987, Yamaguchi 1986). A canopy of residual older trees often survived the low- to moderate-intensity fires of this subregion, and other old-growth characteristics often persisted through successive disturbances. The ponderosa pine and Douglas-fir forests of eastern Washington and Oregon experienced frequent low-intensity fire prior to fire-suppression efforts (Hall 1980, Keen 1937). Large fire-resistant trees persisted through successive disturbances in these forests as well. Although the above studies have been field-based, remote sensing and GIS can aid in assessing disturbance patterns and modeling disturbance processes.

Major human alteration of this forest landscape began in the late 1800s with early logging, clearing for agriculture, and the development of urban centers in the region. Much of the original forest on private lands was logged during the first half of this century and the remainder has been logged in the last 40 years. Logging on state and federal lands began in earnest in the late 1940s. It is now estimated that the only about 10 to 15 percent of the presettlement late-successional forests remain (Morrison and others 1991, Norse 1990).

The use of satellite imagery to develop a historical chronosequence of forest condition is limited to a reconstruction of the last 20 years (the temporal window of the Landsat satellites). Forest condition maps produced from MSS satellite imagery acquired in the early 1970s usually represent a mid-point in the cutting cycle on federal land and close to an end point on private land in the Pacific Northwest. Old aerial photography can be used to reconstruct forest condition in the late 1930s or 1940s in some areas. Old forest-type mapping, survey, and timber plot data can also aid in partial reconstruction of the presettlement forest.

Development of a Historical Chronosequence on the Olympic Peninsula with GIS and Remote Sensing Technology

The first detailed historical reconstruction of a Pacific Northwest landscape was completed for the Olympic Peninsula in Washington State (Morrison 1990). In this study older forests were mapped over a chronosequence of nearly 50 years on the Olympic National Forest in the central part of the peninsula. A more cursory analysis based on MSS satellite imagery was conducted for the rest of the peninsula. Historic aerial photographs, timber inventory plot data, field ecology plot data, satellite imagery, and a variety of digital data layers were used to study the changing condition of this landscape. This work was accomplished using GIS and remote sensing techniques. It was one of the first forest landscape analyses to integrate raster and vector GIS development and analysis (Morrison 1992b).

The historical chronosequence covered five time intervals: early 1940s, 1962, 1980, 1982, and 1988. A sequence of aerial photography, field plot data, and satellite imagery was used to determine the condition of the forests during these periods. GIS data layers were developed representing forest stand conditions for each time interval. In addition GIS layers were developed for administrative boundaries (including National Park, Wilderness, and National Forest), forest roads, elevation, slope, aspect, environmental zone, and forest series (derived from an ecological model from Henderson and others [1990] implemented in GIS).

Old forests blanketed most of the Olympic National Forest in the early 1940s. By 1988 only 40 percent of the older forests remained after nearly five decades of logging. Only about 38,000 hectares (24 percent) of the most biologically productive and diverse old-growth forest in the western hemlock (*Tsuga heterophylla*) and pacific silver fir (*Abies amabilis*) zones remained. GIS analysis revealed that only about 12 percent of the remaining old-growth forest was protected in Wilderness or Research Natural Areas. The Olympic National Park contained about 91,000 hectares of old-growth forest.

GIS analyses indicated that logging roads extended 4500 kilometers across the Olympic National Forest landscape. Further investigation and GIS-based analyses indicated that the ecological impact of these roads might be profound and persistent. On the Olympic National Forest salvage logging, snag removal, and firewood cutting altered the standing dead and down tree component of forests adjacent to roads. Roads (and the vehicles traveling on them) provide a migration corridor and dispersal mechanism for nonnative weedy species. Wildlife disruption, vehicle collisions with wildlife, hunting, and poaching along roads are major factors affecting wildlife populations. Roads may also be a barrier to dispersal of some species. GIS analyses indicated that many of the roads on the Olympic National Forest were built on steep slopes, where road-cut slopes are often precipitous. In these cases, roads may even be an impediment to the movement of large mammals. Roads contribute to fragmentation of the forest landscape. Thirty-five percent of the National Forest was in a 114-meter edge zone adjacent to roads.

Further examination of the combined edge effects of roads, clearcuts, and young plantations together with an analysis of historic and current patch size distributions of forest stands indicated that the remaining forest is now highly fragmented. In 1988 more than 54 percent of the older forest was within 114 meters of a recent clearcut, logging road, or young plantation. Only 18.5 percent of the original old-growth forest present in 1940 was left uncut or unaltered by edge effects.

An analysis of the patch size distribution of old-growth forest stands indicated an even more dramatic change. In the early 1940s more than 80 percent of the old-growth forests existed in patches greater than 10,000 hectares (Figure 6.1). By 1988 the patch size distribution had been dramatically altered (Figure 6.2) and no old-growth patches greater than 10,000 hectares remained. Only one large patch of about 4000 hectares remained in 1988 (Figure 6.3).

Following the Olympic Peninsula study, we began a series of GIS-based studies to analyze the condition of late-successional forests on 11 additional National Forests in Washington, Oregon, and northern California (Morrison and others 1990, 1991). Our focus was to develop current, reliable, and uniform information on the condition of the forests on the west side of the Cascade Range covering an area of about 6 million hectares. These studies also focused on the distribution of remaining late-successional forests with respect to existing protected areas and proposed habitat conservation areas for the threatened northern spotted owl (*Strix occidentalis caurina*) (Thomas and others 1990). These studies also used GIS and remote sensing approaches. Forest condition mapping of four forests was based on interpretation of aerial photography, updated in a GIS with current MSS satellite imagery. On eight forests, mapping was based on analysis of MSS satellite imagery using topographic normalization (Civco 1990, Eby 1987, 1990) and a stratified, unsupervised classification approach.

This expanded study indicated that about 1.5 million hectares of late-successional forest remain on the 12 national forests examined by the

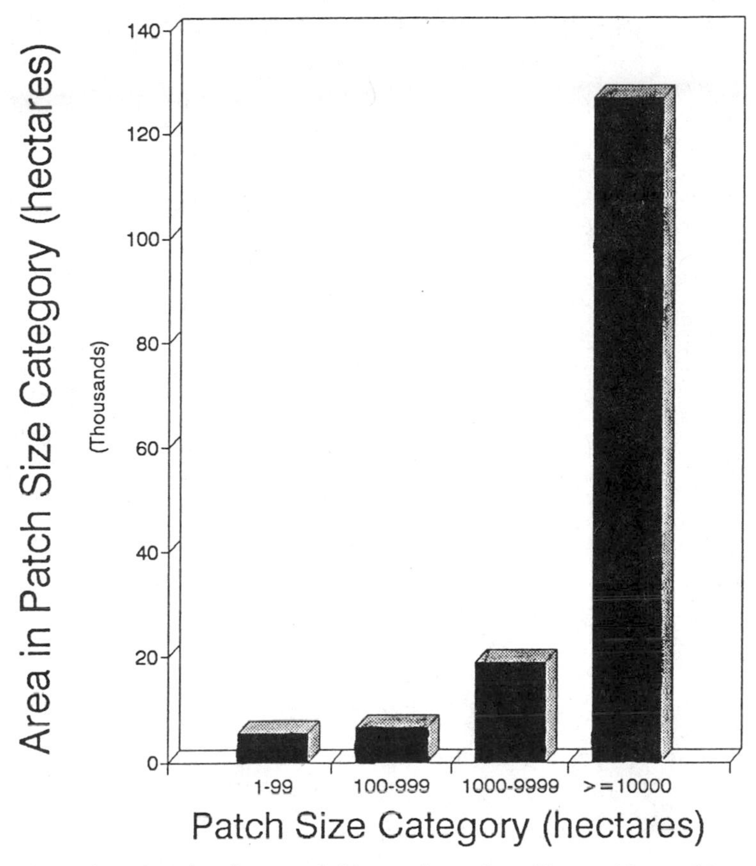

Figure 6.1 Size distribution of old-growth patches, Olympic National Forest, 1940

study. Less than 25 percent of that amount is in congressionally designated wilderness or other areas protected from timber cutting. About 280,000 hectares of biologically rich late-successional forest is found at low elevations, and only about 10 percent is protected. The GIS analysis indicates that the proposed northern spotted owl habitat conservation areas would protect additional late-successional forest but that more than 46 percent of the remaining forest would still remain unprotected.

The comprehensive computer-based maps, analyses, and information developed in these studies have been used by a wide audience to aid in planning, policy, and legal decisions regarding the ancient forest/logging dilemma in the Pacific Northwest.

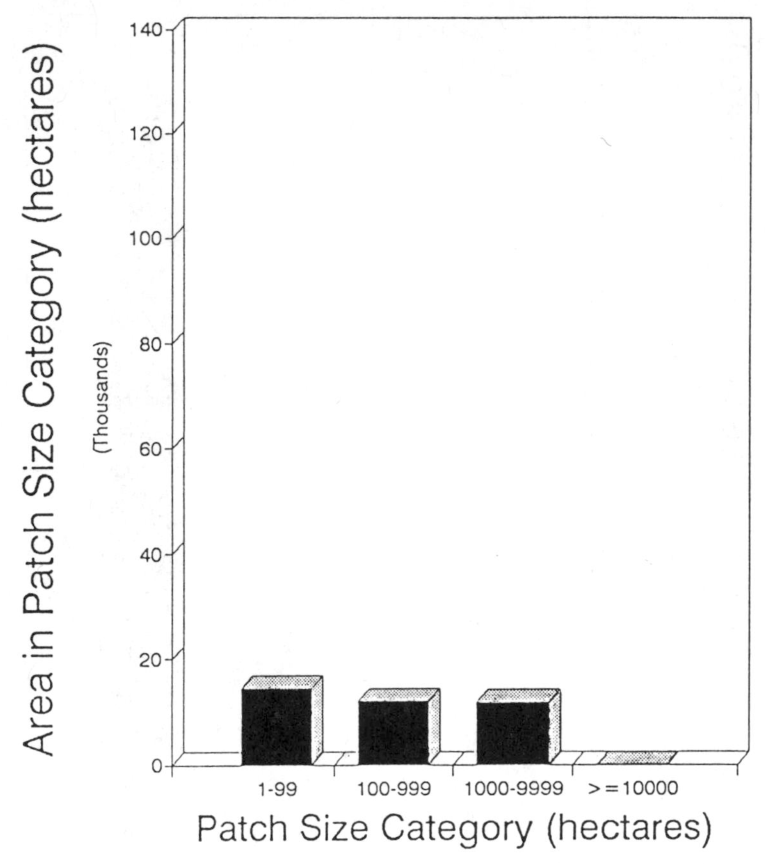

Figure 6.2 Size distribution of old-growth patches, Olympic National Forest, 1988

The Role of GIS in Evaluating the Current and Historical Distribution of Late-Successional Forests on Vancouver Island, British Columbia

The Sierra Club of Western Canada, The Wilderness Society, and British Columbia Ministry of Forests and Ministry of Environment initiated a cooperative GIS-based study of the historical and current condition of the forests on Vancouver Island. In early 1991, no current information existed about the condition of forests on a 3.2-million hectare island. Multispectral scanner satellite imagery from 1990 and 1972–74 was used in conjunction with recent aerial photography, biogeoclimatic subzone

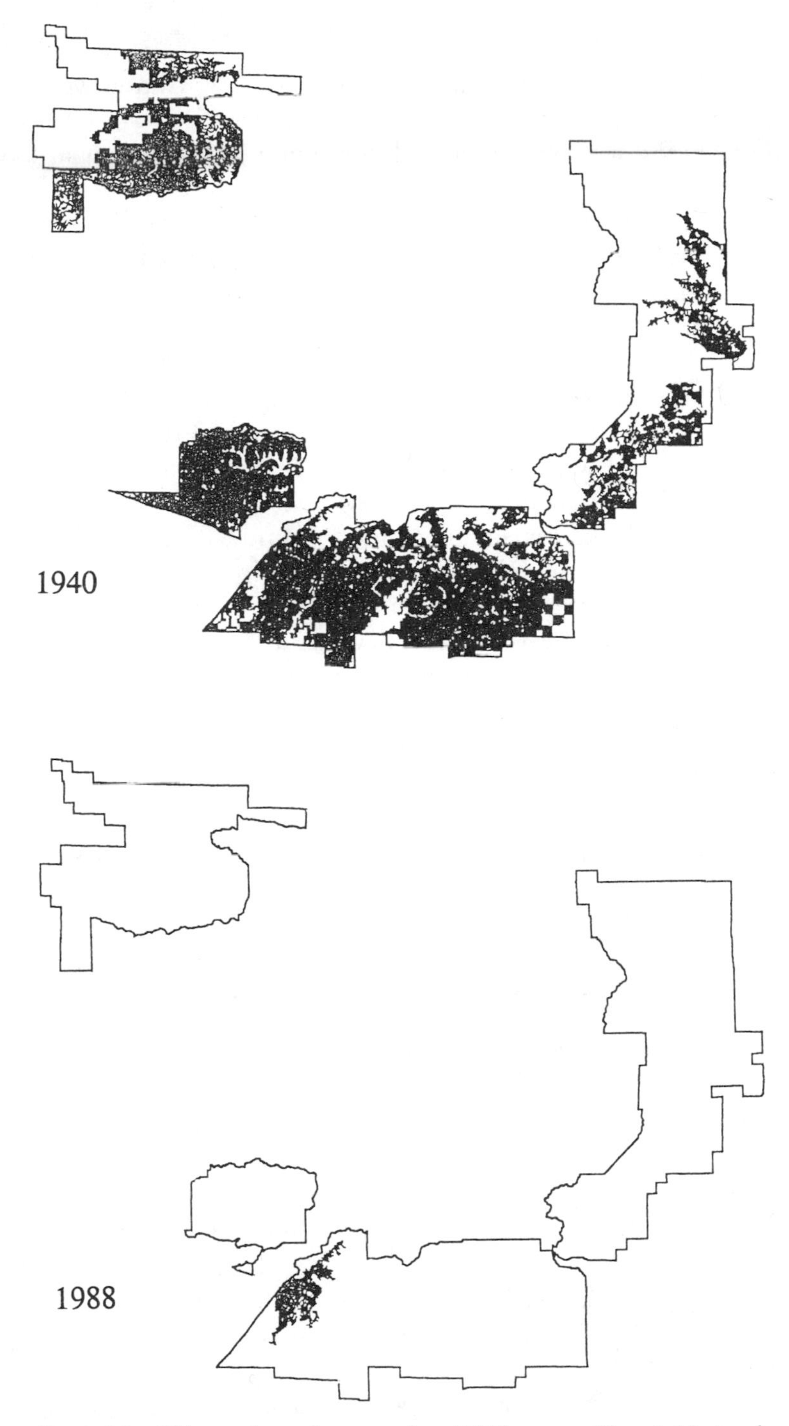

Figure 6.3 Old-growth patches more than 4000 hectares, Olympic National Forest, 1940 and 1988

maps, and forest cover-type maps developed in 1954. Current vegetation maps delineating urban areas, clearcut areas, second-growth forests, unproductive bog forests, nonforested areas, and late-successional forests were developed.

The study indicated a 49-percent loss of late-successional forests due to human activity during the period from 1954 to 1990 (Sierra Club of Western Canada and The Wilderness Society 1991). Much of the eastern and southern portions of the island were logged prior to 1954. Logging has taken the form of progressive clearcutting, often covering entire watersheds. While this practice has often resulted in severe degradation of aquatic ecosystems, the landscape is less fragmented than similar forest landscapes in the western United States. Many large blocks of late-successional forest remain today. In a few cases large watershed, or multiple watershed blocks, remain in pristine condition. These large blocks of forest are some of the few places left on the planet where late-successional forests span the entire distance from the ocean to alpine regions.

Recent satellite imagery reveals a new trend toward dispersed clearcutting and accelerated road-building on Vancouver Island. This trend will lead to the rapid fragmentation of the remaining large blocks of late-successional forest. The project demonstrated the power of remote sensing and GIS in developing data essential to conservation, planning, and policy development. The GIS databases are now being used extensively by both the conservation community and the British Columbia provincial government.

Spatial Texture Analysis in Remote Sensing

Development of Techniques for Late-Successional Forest Mapping on the Okanogan National Forest in Eastern Washington

The Okanogan National Forest asked for assistance in conducting a late-successional forest inventory and in developing a GIS database and map layers of late-successional forest attributes. A cooperative project was initiated with the goal of developing remote sensing and GIS methods that take maximum advantage of the spectral and spatial domains in thematic mapper and SPOT HRV panchromatic satellite imagery in combination with a GIS expert system approach involving ecological modeling.

Forests east of the Cascade Range have greater ecological complexity than forests on the west side due to environmental extremes of precipitation and temperature. Consequently, there is great variation in the structure and composition of late-successional forests in the subregion. Late-successional forests in the Okanogan National Forest range from very open-canopy ponderosa pine forests growing on shallow soils to closed-canopy mixed conifer forests on more mesic and productive sites.

More than 150 field plots were taken over the course of two field seasons. The plot locations were entered into a GIS and the extensive plot data were entered into a linked database. A computer program was developed to analyze and classify the textural variation in the SPOT panchromatic satellite image based on a statistical approach to texture analysis developed by Wang and He (1990). The output of the texture analysis has been combined with spectral information from a TM satellite image, a managed-zone GIS layer, and an ecological zone layer developed by the author in a previous study (Almack and others 1992).

The final analysis of this work is still in progress, but results indicate that the combination of spatial and spectral analysis has resulted in development of accurate information on late-successional forest attributes that would not have been possible using either approach alone. Preliminary results also indicate that the forest is now highly fragmented by at least five decades of forest-cutting and road-building. Forest fragmentation was also relatively high in the premanagement forest due to interspersion of nonforested areas occurring on steep south-facing slopes, at the shrub-steppe/forest ecotone, in rocky areas, and at high elevations.

Conclusion

Wise management decisions require good information about the current condition of forested ecosystems. The management and protection of rare and endangered species and ecological management of late-successional forests require adequate information about the current and original distribution of native vegetation, the distributions of critical species, and species–habitat relationships. Remote sensing, geographic information systems (GIS), and ecological modeling on a landscape scale have much to offer in providing reliable information on which to base critical resource allocation and management decisions. Without this information, decision-makers lack key facts and cannot be expected to provide sound solutions for the management and protection of these ecosystems.

The studies described in this chapter provide an example of projects that have benefited from the use of remote sensing and GIS technology. They also provide an example of the usefulness of developing a historical perspective to guide policy decisions regarding critical resources and endangered species. Analysis of landscape pattern was made possible through use of GIS technology. This analysis revealed that the degree and pattern of forest fragmentation differed markedly in the three study areas. Management history and response to environmental gradients contributed to the variation in forest landscape pattern observed in the Pacific Northwest.

References

Almack, J.A., W.L. Gaines, P.H. Morrison, J.R. Eby, and others. 1992. North Cascades Grizzly Bear Ecosystem Evaluation—Final report : A report to the Interagency Grizzly Bear Committee. Washington Dept. of Wildlife and USDA Forest Service. 145 pp.

Brubaker, L.B. 1990. Climate change and the origin of Douglas-fir/western hemlock forests in the Puget Sound Lowlands. In: Proceedings from Old-Growth Douglas-Fir Forests: Wildlife Communities and Habitat Relationships, a Symposium. March 1989, Portland, OR.

Civco, D. L. 1989. Topographic normalization of Landsat Thematic Mapper digital imagery. Photogram. Eng. and Remote Sens. 55: 1303–1309.

Dodwell, A., and T.F. Nixon. 1902. Forest conditions of the Olympic Forest Reserve, Washington. USDI Geological Survey Prof. Paper No. 7., Series H, Forestry 4. Washington D.C.

Eby, J.R., and M.C. Snyder. 1990. The status of old growth in western Washington: a Landsat perspective. Washington Dept. of Wildlife, Wildlife Management Division, Olympia. 34 pp.

Eby, J.R. 1987. The use of sun incidence angle and infrared reflectance levels in mapping old-growth coniferous forests. ASPRS Technical Papers, 1987 ASPRS-ACSM Fall Convention. Reno. pp. 36–44. Am. Soc. for Photogram. and Remote Sensing, Bethesda, MD.

Franklin, J.F., and C.T. Dyrness. 1973. Natural vegetation of Oregon and Washington. USDA. For. Serv. Gen. Tech. Rpt. PNW-8. USDA Forest Service. PNW Research Station, Portland, OR. 417 pp.

Hemstrom, M.A. 1979. A recent disturbance history of forest ecosystems at Mount Rainier National Park. Ph.D. dissertation. Oregon State Univ., Corvallis. 66 pp.

Hemstrom, M.A., and J.F. Franklin. 1982. Fire and other disturbances of the forests in Mount Rainier National Park. Quaternary Res. 18:32–51. (Contains stand age maps and fire history maps; field plots also available.)

Henderson, J. A. 1990. Forested plant associations of the Olympic National Forest. USDA Forest Service, Pacific Northwest Region, (R6-ECOL-TP-001-88). 502 pp.

Huff, M.H., and J.K. Agee. 1980. Characteristics of large lightning fires in the

Olympic Mountains, Washington. Fire and Forest Meteorology Conf. Proc. 6:117–123.

Langille, H.D., F.G. Plummer, A. Dodwell, T.F. Nixon, and J.B. Lieberg. 1903. Forest conditions in the Cascade Range Forest Reserve, Oregon. USDI. Geological Survey Prof. Paper No. 9, Series H, Forestry 6. Washington D.C. 298 pp.

Lieberg, J.B. 1898. Cascade Range and Ashland forest reserves and adjacent regions. USDI Geological Survey, 21st Annual Report, 1899–1900: Part 5. Washington D.C.

Morrison, P.H. 1992a. Biodiversity and monitoring is goal of Wilderness Society ecosystem mapping with GIS/remote sensing data. Earth Observation Magazine 1:28–34.

Morrison, P.H. 1992b. Mapping endangered ecosystems and species using raster and vector GIS integration. Paper for the Twelfth Annual Environmental System Research Institute Users Conference, Palm Springs, June 1992 (in press).

Morrison, P.H. 1988. Old-growth forests in the Pacific Northwest: A status report. The Wilderness Society, Washington D.C. 67 pp.

Morrison, P.H. 1990. Ancient forests on the Olympic National Forest: analysis from a historical and landscape perspective. The Wilderness Society, Washington D.C.

Morrison, P.H., D. Kloepfer, D.A. Leversee, C.M. Milner, and D.L. Ferber. 1990. Ancient forests on the Mt. Baker-Snoqualmie National Forest: analysis of forest conditions. The Wilderness Society, Washington D.C. 19 pp.

Morrison, P.H., D. Kloepfer, D.A. Leversee, C.M. Socha, and D.L. Ferber. 1991. Ancient forests in the Pacific Northwest: analysis and maps of twelve national forests. The Wilderness Society, Washington D.C. 24 pp.

Morrison, P.H., and F.J. Swanson. 1990. Fire history and pattern in a Cascade Range landscape. USDA Forest Service General Technical Report PNW-254, Portland, OR., Pacific Northwest Research Station, 77 pp.

Norse, E.A. 1990. Ancient forests of the Pacific Northwest. Covelo, CA: Island Press. 327 pp.

Plummer, F.G. 1900. Mount Rainier Forest Reserve, Washington. In: USDI Geological Survey, 21st Annual Report, 1899–1900: Part 5. Forest reserves.Washington D.C. pp. 81–143.

Plummer, F.G. 1902. Forest conditions in the Cascade Range, Washington, between the Washington and Mount Rainier forest reserves. USDI Geological Survey, Washington D.C.

Sierra Club of Western Canada and The Wilderness Society. 1991. Ancient rainforests at risk: an interim report by the Vancouver Island Mapping Project. Sierra Club of Western Canada, Victoria, B.C.

Teensma, P.D. 1987. Fire history and fire regimes of the central western Cascades of Oregon. Ph.D. dissertation. Eugene, OR.: University of Oregon.

Wang, L., and D.C. He. 1990. A new statistical approach for texture analysis. Photogram. Eng. and Remote Sens. 56:61–66.

Yamaguchi, D.K. 1986. The development of old-growth Douglas-fir forests northeast of Mount St. Helens, Washington, following an A.D. 1480 eruption. Ph.D. dissertation. Univ. of Washington, Seattle. 100 pp.

7

GIS Applications Perspective: Current Research on Remote Sensing of Forest Structure

Warren B. Cohen

Introduction

The Pacific Northwest (PNW) region of the United States has received much attention in the past several years concerning the amount and distribution of late-successional forest conditions. This concern has prompted a number of studies to define, characterize, and map forest cover and structural attributes in the region. The mapping efforts have had at their core the use of satellite and other remotely sensed data (e.g., The Wilderness Society 1991, Congalton et al. 1993) and have proceeded with little or no scientific research component. Rather, the emphasis has been on map creation, using primarily established image analysis and GIS techniques. Although such techniques are useful and can provide reasonable and often acceptable results, research can help bring about a better understanding of the relationships between image characteristics and forest attributes. In turn, improved methods and models, and therefore more accurate map representations, can result. Equally important, research can help us better recognize the limitations of current remotely sensed data and of algorithms designed to extract the needed forest information. The latter is extremely important because, increasingly, maps derived from remotely sensed data are being commissioned by policymakers and others who have little knowledge of the technology.

Cohen and Spies (1990) are addressing the need for remote sensing research in the PNW region, and this chapter is a summary of two specific and focused research projects to that end. For further details on this research, see Cohen et al. (1990) and Cohen and Spies (1992).

Our first project involved the use of a geostatistical technique known as the semivariogram (Cohen et al. 1990). Semivariograms enabled exploration of the image *spatial domain*, which, depending on image spatial resolution, may contain substantial amounts of information with respect to forest structure. This is because forest structure is largely a spatial phenomenon. For the second research effort, the spatial domain was further explored by evaluating relationships between image texture and structural attributes (Cohen and Spies 1992). Most of the research in remote sensing of forested ecosystems has focused on the image *spectral domain;* that is, the analysis of multispectral images using statistics-based decision rules for determining the identity, with respect to forest stand characteristics of interest, of each pixel in the images. This holds significant promise; therefore, we incorporated analyses of image spectral properties as a part of this second study.

The primary emphasis in these research projects, thus far, has been on Douglas-fir (*Pseudotsuga menziesii*) forests of the western hemlock (*Tsuga heterophylla)* zone described by Franklin and Dyrness (1988), which dominates the PNW region. As a result of past fires, steeply dissected terrain, and intensive forest-management practices, the western hemlock zone has a complex pattern of stand structural conditions. Initially, the focus was on stands having a closed canopy (those in which at least 85 percent of horizontal space is occupied by trees), a condition that commonly occurs within the first 25 years or so after a major disturbance. In closed-canopy conifer forests of the region, structural conditions can be characterized on a continuum from simple in young, even-aged stands to complex in old-growth stands. Structural differences among these successional stages are described by Spies and Franklin (1988). Generally, simply structured stands have a single canopy layer of similar-sized small trees with few canopy gaps, high tree density, and low basal area. Complex structured stands commonly have multiple canopy layers with numerous gaps and a variety of tree sizes, and relatively low tree density and high basal area in the upper canopy layers.

Image Spatial Domain

Implicit in an analysis of image spatial properties is a recognition that image pixels exist within a neighborhood of other pixels, and that the spatial variability of the spectral properties of images contains information about the spatial characteristics of the ground scene. As forest structure is largely a set of spatial characteristics, we hypothesized that the spatial domain of images should contain valuable information about forest structure. Two major questions were: (1) "What structural attributes can be reliably estimated by spatial algorithms?" and (2) "What is the effect of image spatial resolution on attribute estimation?"

Image Semivariograms

A semivariogram is a graphical representation of the spatial variability in a given set of data. The semivariogram, or $\gamma(h)$, is calculated as

$$\gamma(h) = \frac{1}{2(n-h)} \sum_{i=1}^{n-h} [Z(x_i) - Z(x_i + h)]^2 \qquad [1]$$

where h is the lag (or distance) over which γ (semivariance) is measured, n is the number of observations used in the estimate of $\gamma(h)$, and Z is the value of the variable of interest at spatial position x_i (Webster 1985, Journel and Huijbregts 1978). The quantity $Z(x_i + h)$ is the variable value at distance h from x_i. Thus for spectral data, $\gamma(h)$ estimates the variability of radiance Z as a function of spatial separation.

Typically, the shape of a semivariogram is such that γ increases with h until it reaches a maximum, or *sill* (Figure 7.1). The lag at which the sill is reached is called the *range*. The range and the sill are the two most important parameters of the semivariogram used to describe the data. The range can be used as a measure of spatial dependency, or homogeneity, whereas the sill reflects the amount of spatial variability. For a more in-depth discussion of semivariograms and related theory see Matheron (1971), Clark (1979), Journel (1989), Oliver et al. (1989a and 1989b), and Webster and Oliver (1990). For examples of semivariogram usage in remote sensing other than that presented here see Yoder et al. (1987), Atkinson and Danson (1988), Curran (1988), Jupp et al. (1988a,b), Woodcock et al. (1988a,b), Curran and Dungan (1989), Ramstein and Raffy (1989), Wald (1989), Webster et al. (1989), de Miranda and MacDonald (1990), Rubin (1990), Atkinson et al. (1990), and Townshend et al. (1992).

The primary objective for our use of semivariograms was to evaluate their potential utility for distinguishing among forest stands having different canopy structures (Cohen et al. 1990). Digitized aerial true-color videography, at a nominal 1-meter pixel size, was used. As we also wanted to examine the potential utility of semivariograms for use with SPOT HRV panchromatic and Landsat TM data, analyses were repeated after the video data were spatially degraded to 10 and 30 meters.

Images of five forest stands were selected for study. The only quantified forest stand attribute was tree crown size. The stands were selected primarily on a qualitative basis so that they had different canopy structures, labeled as young, mature, old-growth, young-mix, and mature-mix. The

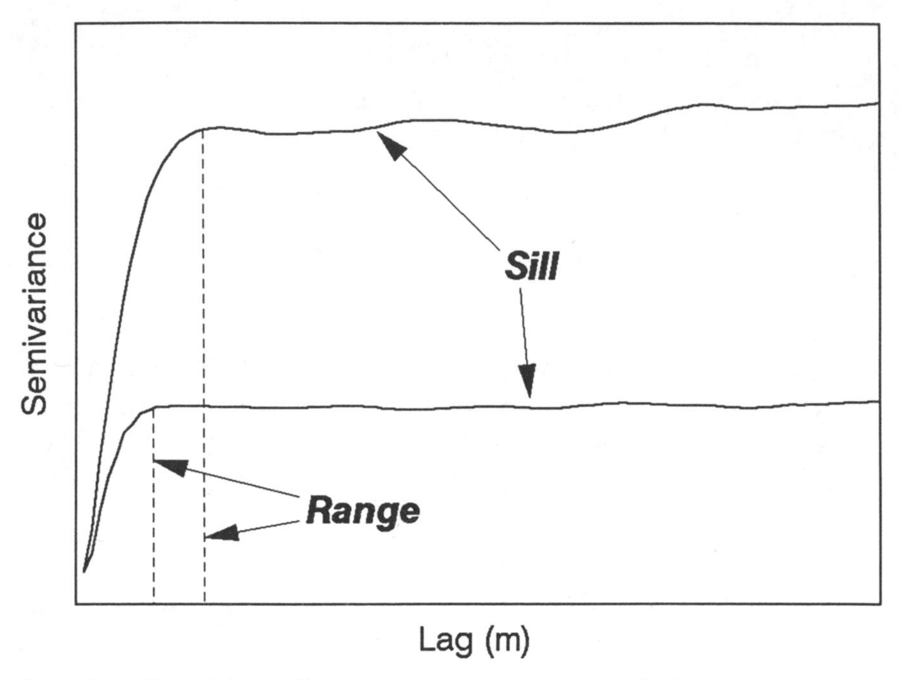

Figure 7.1 Typical shape of semivariograms with ranges and sills shown

young and mature stands had relatively simple canopy structures and the old-growth and mixed stands had complex structures. Semivariograms were calculated for each stand using the digital numbers (DN) in the red band of the images. The analysis was undertaken in two separate ways: (1) using a transect method that evaluates single transects of pixel DN, and (2) a matrix method that uses DN of the full two-dimensional pixel matrices in the images.

We concluded from this study that semivariograms of image data should be a useful means of evaluating canopy structure in the Douglas-fir forests of the PNW region. The matrix semivariogram should provide fairly accurate estimates of stand structure parameters, but will not readily permit the evaluation of patterns in stand structure. Because transect semivariograms exhibit periodicity, they may permit the detection of patterns in forest stands. However, as transects represent only a sample of data values, transect semivariograms will not depict stand structure parameters as accurately as matrix semivariograms.

It was clear from this study that the utility of semivariograms for evaluating within-stand conifer canopy structure from remotely sensed images is greatly influenced by image spatial resolution. At a spatial resolution of 1 meter, radiant energy sensed from all tree crowns, excluding those of saplings, will be expressed in the DN of several pixels. Because of this, the range of the 1-meter matrix semivariogram is a valuable measure of tree crown size in a conifer forest stand. At a 10-meter pixel size (e.g., SPOT HRV panchromatic data) the range of the matrix semivariogram is less useful, yielding only a coarse estimate of crown size. As few trees have crowns that are 30 meters in diameter or greater, the range of the semivariograms using Landsat TM data should not be useful for estimating tree crown sizes.

Sills of the 1-meter matrix semivariograms depict the presence of canopy layering and gaps in forest stands. Because the sill responds to both percent canopy cover and canopy layering, however, their use may not always facilitate distinguishing mixed-stand structures from old-growth structures. Likewise, these categories of forest structure are not always clearly distinguishable on the ground. Because the sills are reduced with increasing units of regularization, like the range, the sills are less informative as image pixel size increases. For the 30-meter data the sills were very similar in magnitude, again indicating a limitation for the potential usefulness of semivariograms of Landsat TM data for stand structure analysis.

To apply the semivariogram technique to image data for within-stand structure analysis, separate semivariograms must be calculated for each stand. This requires either the use of image segmentation algorithms (e.g., Woodcock and Harward 1992) or digitization of photointerpreted polygons. Image segmentation is a critical yet largely underdeveloped area of remote sensing research. Until a number of reliable image segmentation algorithms are developed, the use of semivariograms may be impractical on a full HRV scene that may contain hundreds, or perhaps thousands, of individual stands. If this procedure were to be executed, however, the sills could be used as an index of stand structural complexity. An alternative and maybe more practical approach would be to create a new data layer from the existing image data (e.g., HRV or merged HRV and TM) that largely distinguishes each stand's structural complexity from that of the other stands. This data layer could be used in the multispectral image processing context. Some measure of local variance for an image containing numerous stands would be helpful. Several measures of local variance, or texture, are possible (e.g., Woodcock and Strahler, 1987; Rubin, 1990; Wang and He, 1990), and adapting or developing one that performs well was a focus of our second research project.

Image Texture

A wide variety of texture algorithms have been used to process digital imagery (Irons and Petersen 1981, Gool et al. 1985, and Mather 1987). From these we chose the standard deviation and absolute difference algorithms described by Rubin (1990). Using these algorithms, texture images were created from an unenhanced HRV 10-meter image, and from 30-meter TM data (Cohen and Spies 1992). Initial evaluation indicated that the absolute difference algorithm provided a greater dynamic range of texture information and therefore had potentially more discriminating power.

From the absolute difference images, the boundaries of more than 40 forest stands representing a variety of conifer closed-canopy conditions were digitized. Then, those pixels corresponding to the ground location of each stand were extracted from the images. Ground data were collected by Spies and Franklin (1991) and used to calculate a number of attribute values for those stands relating to mean tree size, tree size variability, tree density and basal area, two newly developed canopy structural indices, and stand age. One new structure index was the Canopy Height Diversity index (CHD), which is based on theoretical concepts that describe the relative volume of "ecological space" occupied by trees in a stand. Development of the other index, the Structural Complexity Index (SCI), was inspired by the fact that most of the stand attributes evaluated are highly correlated in a given forest stand. Thus, the SCI is a means with which to capture in a single stand attribute the variability found in several stand attributes.

Results of this effort indicated that many of the stand attributes evaluated were strongly correlated with the mean texture number of a forest stand, as calculated from HRV data (Table 7.1). Texture numbers of TM data were not well correlated with stand attributes. An important finding was that for the mean tree size attributes and tree density relationships with texture were strong when only the dominant and codominant trees were considered. This is because trees in the understory layers are highly variable in size and number and, coincidentally, are not visible to the sensor. The tree size variability attributes, stand age, the CHD, and the SCI also were well correlated with HRV texture.

These results supported our hypothesis that spatial algorithms using 10-meter data are useful, and that 30-meter data are not, for analysis of conifer canopy structure of forests in the region. We found further support for this from Woodcock and Strahler (1987), who state that the value of

TABLE 7.1 Value of correlation coefficients for the relationships of stand structural attributes and variables derived from the satellite data (texture of SPOT 10-m HRV imagery, and the TM Tasseled Cap brightness, greenness, and wetness axes)

Stand attribute	HRV texture	Brightness	Greenness	Wetness
DBH (mn, all)	0.55	0.44	0.44	–0.60
DBH (sd, all)	0.88	0.53	0.55	–0.87
DBH (mn, upper)	0.88	0.43	0.47	–0.87
CD (mn, all)	0.67	0.33	0.35	–0.69
CD (sd, all)	0.72	0.27	0.25	–0.67
CD (mn, upper)	0.88	0.42	0.27	–0.88
HGT (mn, all)	0.45	0.37	0.37	–0.51
HGT (sd, all)	0.88	0.30	0.42	–0.81
HGT (mn, upper)	0.86	0.35	0.43	–0.85
DNY (all)	–0.62	0.49	0.51	–0.69
DNY (upper)	–0.84	0.39	0.51	0.87
BA (all)	0.73	0.53	0.52	–0.69
BA (upper)	0.73	0.47	0.53	–0.71
AGE	0.87	0.55	0.63	–0.90
SCI	0.88	0.55	0.65	–0.86
CHD	0.75	0.53	0.54	–0.69

Source: Adapted from Cohen and Spies 1992.

DBH is tree diameter at breast height, CD is crown diameter, HGT is total tree height, DNY is tree density, BA is basal area, AGE is stand age, SCI is the Structural Complexity Index, CHD is the Canopy Height Diversity Index, mn is mean, sd is standard deviation, all is all trees in the forest stand, and upper is only trees in the dominant and codominant canopy positions.

texture measures in image processing depends on whether image resolution cells are smaller than the elements of interest in the scene. The basic element of interest in our study is the individual tree or, as viewed from above, the tree crown. In the young, simply structured Douglas-fir stands evaluated by us the dominant tree crown is approximately 5 meters in width, whereas in mature (moderately complex) and old-growth (complex) stands it is roughly 10 and 15–20 meters, respectively (Cohen et al. 1990, Spies et al. 1990). Even though young closed-canopy stands have tree crowns that are subpixel size, the fact that more complex stands have numerous tree crowns that are at least as large as the pixels helps to discriminate simply structured stands from more complex stands. Likewise, because stands with moderately complex structures have tree crowns roughly equivalent in size to the image pixels and complex structured stands have crowns larger than one pixel, these two conditions also are distinguishable using textural measures of 10-meter data. With the 30-meter data at least two or three large trees generally appear in many pixels, severely diminishing the ability of TM texture to discriminate.

Image Spectral Domain

Remotely sensed images consist of pixels that contain data, in any number of bands, on the electromagnetic spectral properties of a ground scene. Analyses of image data that evaluate pixels irrespective of their neighbors operate solely in the spectral domain. Excluding radiometric and geometric preprocessing considerations, the common choices for such analyses are to work either with the unprocessed band data or to first create one or more vegetation indices (Perry and Lautenschlager 1984, Cohen 1991a). The normalized difference vegetation index (NDVI) is by far the most commonly used. However, the NDVI does not take full advantage of the TM data, as it uses only two spectral bands. Nonetheless, a strong tendency exists by users of remotely sensed data to compute only the NDVI and then use it in further analyses. Why this has happened is not clear from the literature, but the reasons seem to be rooted in the fact that NDVI is simple to compute and exhibits a strong relationship with a number of vegetation characteristics (e.g., Tucker 1979, Ajai et al. 1983).

Other indices that received a significant amount of attention in the early days of digital image analysis still continue to experience some, albeit relatively little, application. The *Tasseled Cap brightness* and *greenness* indices are perhaps the most important of these (Kauth and Thomas 1976). The Tasseled Cap was adapted to TM data by Crist and Cicone (1984) and an additional index, or axis, was defined. That axis has been called *wetness* (Crist et al. 1986). The TM Tasseled Cap indices were designed to take optimal advantage of the original six TM bands, and together these first three axes account for as much as 85 percent or more of the spectral information of a vegetated TM scene.

Brightness is a weighted sum of the six reflectance bands of the TM imagery, and greenness is a contrast between the near-infrared band (TM4) and the three visible bands (TM1, TM2, and TM3). Use of the terms brightness and greenness is well accepted within the remote sensing community, and there is a substantial body of literature suggesting that the names of these spectral features of TM data are consistent with the information they represent. Wetness is a contrast of the mid-infrared bands (TM5 and TM7) with the other four bands, and has been shown to correlate with the amount of moisture in a scene (Crist et al. 1986, Musick and Pelletier 1988, Cohen 1991a). Wetness has received little attention in an image processing framework.

We turned to the Tasseled Cap set of spectral indices for evaluations of the TM spectral domain. This was done in concert with our analysis of the HRV texture data; thus, the same general methodology as described earlier applies here as well. Results indicated that of the three Tasseled Cap indices, only wetness was strongly correlated with the stand attributes (Table 7.1, shown earlier). This was because brightness and greenness responded more to topographic variation than to stand condition, whereas wetness did not. The attributes most strongly correlated to wetness were generally the same as those for HRV texture. In fact, the relative strengths of the attribute relationships with HRV texture and TM wetness were generally consistent. This was somewhat surprising, as two apparently fundamentally different phenomena drive these image algorithms.

As described earlier, our understanding of the relationships between HRV texture and stand structure was fairly sound; but we did not understand what structural phenomena were influencing wetness. Thus, we consulted the literature and collected ground radiometer data of individual scene components (Cohen and Spies 1992). The radiometer we used was a Barnes Modular Multispectral Radiometer (MMR) equipped with filters that replicate the six TM reflectance bands. In-situ reflectance measurements were obtained from canopies of the most prevalent tree species, for foliage of herbaceous and hardwood plant species commonly found in the western hemlock zone, and for deadwood, tree bark, soil, and epiphytic canopy lichens. Spectra of the same components were also collected in completely shaded conditions.

The scant literature on the subject revealed that both theory and empirical evidence are in support of each other. In essence, wetness responds to both the amount of water and shadows in the ground scene, with these relationships being positive. After giving this topic additional thought we discovered an apparent incongruency with respect to how we would expect the wetness index to respond to different types of canopy structures. In forest canopies the amount of "water-filled" foliage viewed by the sensor is closely linked to the degree of canopy shadowing. In general, the older and more complex the stand, the greater the proportion of shadow present. Because the amount of shadow is greater in more complex stands, we can safely assume that the amount of foliage per unit area visible to the TM sensor is less in these stands than in closed-canopy stands with more simple structures. This is in fact true. Therefore, with increasing structural complexity, wetness should increase due to greater proportions of canopy shading; however, wetness should decrease due to less foliage being viewed by the sensor. It might appear that the effect of less water is dominant, since wetness clearly decreases with increasing structural complexity. In reality however, wetness values for young, simply structured sunlit Douglas-fir and western hemlock foliage and for shaded components are very similar (Cohen and Spies 1992), depriving this whole argument of practical significance.

In Cohen and Spies (1992) we hypothesized that the controlling factor may be in part the amount of age-related canopy die-back, tree death, and increase in epiphytic lichen. This is based on the fact that with increasing age, individual tree crowns in the upper layers begin to die, thereby losing foliage and exposing bark and deadwood. At the same time, canopy lichens begin to grow in increasingly greater amounts. The tops of many of the trees break, as others become snags. Cohen (1991b) demonstrated that woody and other nongreen plant materials have wetness values considerably lower than does green foliage. The same result can be inferred from Guyot et al. (1989) and Elvidge (1990). Our radiometer data from this study indicate that canopy lichen, bark, and deadwood have significantly lower wetness values than sunlit Douglas-fir and western hemlock foliage and all shaded components. Our hypothesis is further supported by evidence that the inner half-radius of an old Douglas-fir tree crown can be covered by as much as 75 percent canopy lichen, as viewed from above (Bruce McCune, pers. comm.).

If our hypothesis for the response of wetness to vegetation senescence holds true, the question must be asked, "Is use of the term wetness appropriate?" Viewing a wetness image of a landscape containing distinct water bodies reveals that water is not the wettest feature. This alone is cause for dropping use of the term wetness and searching for an alternative, more appropriate name for this important spectral feature of TM data—one that is robust enough to describe a general vegetation condition, or process, across numerous ecosystem types. In Cohen and Spies (1992) we made the case for a term such as maturity index. Recently, we have noticed that stands not particularly aged in years but which exhibit structural characteristics similar to old-growth forest stands have wetness values like those of old stands. Such stands tend to grow on rocky, steep slopes that are drier, and therefore prematurely age the forests growing on them.

Topographic Effects

Effects of topography on image data are well documented, and several means exist with which to try to ameliorate their effects (e.g., Holben and Justice 1980, Justice et al. 1981, Ahern et al. 1987, Civco 1989, Colby 1991). Most methods are designed to correct the image data by a cosine function to force an incident energy level that would exist if the sun's position were normal to the surface (Smith et al. 1980). Although this methodology may be sufficient for some relatively simple vegetation structures like a field of corn or wheat, it is generally inappropriate for forest canopies. This is because the interaction of solar angle, topographic position, and vegetation structure is completely ignored, and it can be this interaction that is most important. Alternative means for minimizing effects of topography are needed. One method is to capitalize on this interactive relationship by its explicit incorporation into a classification scheme (Eby 1987), or to use something like a geometric-optical reflectance model (Li and Strahler 1985). Another method is to use algorithms that create new indices or images that do not exhibit strong topographic effects. Texture algorithms produce images that are generally insensitive to topography because local variability is evaluated rather than local means. Wetness images are not very sensitive to topography for at least two probable reasons.

First, the Tasseled Cap transformation is similar to a principal component analysis (PCA). As such, brightness, the first axis, accounts for most of the spectral variability in the data. Likewise, the greenness, or second axis accounts for the next largest proportion of spectral variability. Together brightness and greenness account for about 75 percent of the spectral variation in a TM image. As a topographic effect is so prominent in digital imagery, it tends to dominate the first two PC, or Tasseled Cap, axes. The third axis is therefore "free" to reflect a different source of spectral variability. We believe this source is largely related to maturity of vegetation.

The second probable reason we observed wetness to be insensitive to topography is likely that maturity in forested systems commonly causes changes visible in the upper canopy layers. As such, unless the topography is extremely steep so that the upper canopy layers are in almost total shadow, the tree crowns remain visible. Where there are topographic shadows wetness values should be high regardless of vegetation cover.

Current Research

We now are faced with the opportunity to map forest attributes over large geographic areas in the PNW region, and to quantify change in the forest over the more than 20-year Landsat data record. This brings with it some additional challenges.

For one, it would be best if we could have maximum flexibility for defining strata in the imagery. That is, we prefer to use an ecoregion (Omerlink and Gallant 1988) or some other concept to identify strata for image processing purposes; rather than having the strata be defined by image boundaries. For this, a method of radiometric rectification is needed. Radiometric rectification is used here to refer to some sort of normalization applied to image data to make multiple data sets more radiometrically compatible. Several options are available, but we are first testing a method developed by Hall et al. (1991a). This method "matches" digital numbers of each band of a subject image to those of a reference image. First, dark and bright radiometric control sets are selected from the reference and subject images in brightness-greenness space. The control sets are selected interactively by viewing images and highlighting potential control-set pixels in the images. Then, raw band-to-band linear transformations are developed from relationships between control sets of the subject and reference images, and the resultant linear transformations are applied to rectify each band of the subject image.

When we attempted to apply the Hall et al. (1991a) methodology, we were immediately confronted with an unexpected challenge. Our brightness-greenness histograms had a considerably different shape than those illustrated in Hall et al. (1991a). Whereas their histograms had a distinctly triangular shape, our histograms had several "tails" making up the brightness axis or "leg" of an otherwise triangular shape. The dark end of the brightness axis was close to what we expected, but the bright end of this axis contained several bifurcations. This raised the issue of where to select control sets.

After experimenting with control-set selection we found that the bifurcation that most closely resembled the Hall et al. (1991a) brightness axis did not provide good results for our images. Rather we found that one or more bifurcations below the brightness axis (i.e., into negative greenness) gave better results. Because the bifurcation that worked best was variable, we could not consistently get the desired result simply by picking what appeared to be the best brightness control set. Thus, we found it necessary to select three to five candidate bright control sets and evaluate which worked best. Furthermore, we found that the control set that worked best was band-dependent. In the final analysis, we decided to use the control set that worked best for a given band to rectify that band. This was determined from a combination of visual assessment and comparison of digital numbers from test sites of ground areas that overlapped in the subject and reference images.

The correction is imperfect, but does provide significantly improved radiometric matching among images. This should permit us to make spatial mosaics of images and to do temporal analyses on spatially coincident images, without having to overly concern ourselves about differing radiometric properties.

Another challenge is associated with change detection, which when done in a spatially explicit manner involves spatial overlay of two or more images from different dates (Hall et al. 1991b, Sader and Winne 1992). Three types of change detection algorithms are commonly used: (1) difference, (2) ratio, and (3) PCA. The difference and ratio algorithms require simply subtracting one image from another, or dividing one image by the other, respectively. These two algorithms are generally limited to comparison of two images, and should give similar results except for how the resultant image is scaled. The PCA algorithm can be applied to any number of images, with each PC axis generally representing change between two distinct time periods. Other algorithms have been applied but, to date, only in isolated situations.

We have just begun to experiment with change detection, and thus we have nothing to report here. We are aware, however, that spatial misregistration will cause erroneous results around "sharp" edges such as clearcut and forest boundaries. Initial experimentation reveals that we can filter these narrow boundaries out of the change image.

References

Ahern, F.J., R.J. Brown, J. Cihlar, R. Gauthier, J. Murphy, R.A. Neville, and P.M. Teillet. 1987. Radiometric correction of visible and infrared remote sensing data at the Canada Centre for Remote Sensing. Int. J. Remote Sens. 8:1349–1376.

Ajai, D.S., G.S. Kamat, A.K. Chaturvedi, A.K. Singh, and S.K. Sinha. 1983. Spectral assessment of leaf area index, chlorophyll content, and biomass of chickpea. Photogram. Eng. Remote Sens. 49:1721–1727.

Atkinson, P.M., and F.M. Danson. 1988. Spatial resolution for remote sensing of forest plantations. In: Proceedings, IGARSS '88 Symposium, Edinburgh, Scotland, 13–16 September 1988, ESA Publ. Division, pp. 221–223.

Atkinson, P.M., P.J. Curran, and R. Webster. 1990. Sampling remotely sensed imagery for storage, retrieval and reconstruction. Prof. Geographer 42:345–353.

Civco, D.L. 1989. Topographic normalization of Landsat Thematic Mapper digital imagery. Photogram. Eng. Remote Sens. 55:1303–1309.

Clark, I. 1979. Practical geostatistics. Elsevier, London.

Cohen, W.B. 1991a. Response of vegetation indices to changes in three measures of leaf water stress. Photogram. Eng. Remote Sens. 57:195–202.

Cohen, W.B. 1991b. Chaparral vegetation reflectance and its potential utility for assessment of fire hazard. Photogram. Eng. Remote Sens. 57:203–207.

Cohen, W.B., and T.A. Spies. 1990. Remote sensing of canopy structure in the Pacific Northwest. Northwest Environ. J. 6:415–418.

Cohen, W.B., T.A. Spies, and G.A. Bradshaw. 1990. Semivariograms of digital imagery for analysis of conifer canopy structure. Remote Sens. Environ. 34:167–178.

Cohen, W.B., and T.A. Spies. 1992. Estimating structural attributes of Douglas-fir/western hemlock forest stands from Landsat and SPOT imagery. Remote Sens. Environ. 41:1–17.

Colby, J.D. 1991. Topographic normalization in rugged terrain. Photogramm. Eng. Remote Sens. 57:531–537.

Congalton, R.G., K. Green, and J. Teply. 1993. Mapping old-growth forests on National Forest and Park Lands in the Pacific Northwest from remotely sensed data. Photogram. Eng. Remote Sens. 59:529–535.

Crist, E.P., and R.C. Cicone. 1984. A physically based transformation of Thematic Mapper data—the TM Tasseled Cap. IEEE Trans. Geosci. & Remote Sens. 22:256–263.

Crist, E.P., R. Laurin, and R.C. Cicone. 1986. Vegetation and soils information contained in transformed Thematic Mapper data. In: Proceedings, IGARSS '86 Symposium, Zürich, Switzerland, 8–11 September 1986, ESA Publ. Division, SP-254:1465–1470.

Curran, P.J. 1988. The semivariogram in remote sensing. Remote Sens. Environ. 24:493–507.

Curran, P.J., and J.L. Dungan. 1989. Estimation of signal-to-noise: a new procedure applied to AVIRIS data. IEEE Trans. Geosci. Remote Sens. 27:620–628.

de Miranda, F.P., and J.A. MacDonald. 1990. A variogram study of SIR-B data in the Guyana Shield, Brazil. In: Proceedings, Image Processing '89, 23 May 1989, Sparks, Nevada, pp. 66–77, ASPRS, Falls Church, VA.

Eby, J.R. 1987. The use of sun incidence angle and infrared reflectance levels in mapping old-growth coniferous forests. In: Technical Papers, 1987 ASPRS-ASCM Fall Convention, Reno, Nevada, pp. 36–44, ASPRS, Falls Church, VA.

Elvidge, C.D. 1990. Visible and near-infrared reflectance characteristics of dry plant materials. Int. J. Remote Sens. 11:1775–1796.

Franklin, J.F., and C.T. Dyrness. 1988. Natural vegetation of Oregon and Washington. Oregon State University Press, Corvallis, OR, 452 pp.

Gool, L.V., P. Dewaele, and A. Oosterlinck. 1985. Texture analysis anno 1983. Comput. Vision, Graph., and Image Process. 29:336–357.

Guyot, G., D. Guyon, and J. Riom. 1989. Factors affecting the spectral response of forest canopies: a review. Geocarto Int. 4(3):3–18.

Hall, F.G., D.E. Strebel, J.E. Nickeson, and S.J. Goetz. 1991a. Radiometric rectification: toward a common radiometric response among multidate, multisensor images. Remote Sens. Environ. 35:11–27.

Hall, F.G., D. Botkin, D.E. Strebel, K. Woods, and S.J. Goetz. 1991b. Large-scale patterns of forest succession as determined by remote sensing. Ecology 72:628–640.

Holben, B., and C. Justice. 1980. The topographic effect on the spectral response of nadir-pointing sensors. Photogramm. Eng. Remote Sens. 46:1911.

Irons, J.R., and G.W. Petersen. 1981. Textural transforms of remotely sensed data. Remote Sens. Environ. 11:359–370.

Journel, A.G. 1989. Fundamentals of geostatistics in five lessons. Short Course in Geology: Volume 8, American Geophysical Union, Washington.

Journel, A.G., and Ch.J. Huijbregts. 1978. Mining geostatistics. Academic Press, New York.

Jupp, D.L.B., A.H. Strahler, and C.E. Woodcock. 1988a. Autocorrelation and regularization in digital images: I. Basic theory. IEEE Trans. Geosci. Remote Sens. 26:463–473.

Jupp, D.L.B., A.H. Strahler, and C.E. Woodcock. 1988b. Autocorrelation and regularization in digital images: II. Simple image models. IEEE Trans. Geosci. Remote Sens. 27:247–258.

Justice, C.O., S.W. Wharton, and B.N. Holben. 1981. Application of digital terrain data to quantify and reduce the topographic effect on Landsat Data. Int. J. Remote Sens. 2:213–230.

Kauth, R.J., and G.S. Thomas. 1976. The tasseled cap—a graphic description of the spectral-temporal development of agricultural crops as seen by Landsat. In: Proc. Symp. on Machine Processing of Remotely Sensed Data, 4b:41–51, 6 June–2 July, 1976, LARS Purdue Univ., West Lafayette, IN.

Li, X., and A.H. Strahler. 1985. Geometric-optical modeling of a conifer forest canopy. IEEE Trans. Geosci. Remote Sens. 23:705–721.

Mather, P.M. 1987. Computer processing of remotely-sensed images. John Wiley, New York.

Matheron, G. 1971. The theory of regionalized variables and its application. Les Cahiers du Centre de Morphologie Mathématique de Fontainbleau, No. 5, Published by École Nationale Supérieure des Mines de Paris.

Musick, H.B., and R.E. Pelletier. 1988. Response to soil moisture of spectral indices derived from bidirectional reflectance in Thematic Mapper wavebands. Remote Sens. Environ.

Oliver, M., R. Webster, and J. Gerrard. 1989a. Geostatistics in physical geography. Part I: Theory. Trans. Inst. Br. Geogr. N.S. 14:259–269.

Oliver, M., R. Webster, and J. Gerrard. 1989b. Geostatistics in physical geography. Part I: Applications. Trans. Inst. Br. Geogr. N.S. 14:270–286.

Omerlink, J.M., and A.L. Gallant. 1986. Ecoregions of the Pacific Northwest. USEPA, ERL, Corvallis, OR. 39 pp.

Perry, C.R., and L.F. Lautenschlager. 1984. Functional equivalence of spectral vegetation indices. Remote Sens. Environ. 14:169–182.

Ramstein, G., and M. Raffy. 1989. Analysis of the structure of radiometric remotely sensed images. Int. J. Remote Sens. 10:1049–1073.

Rubin, T. 1990. Analysis of radar texture with variograms and other simplified descriptors. In: Proceedings, Image Processing '89, Sparks, NV, 23 May 1989, ASPRS, Falls Church, VA, pp. 185–195.

Sader, S.A., and J.C. Winne. 1992. RGB-NDVI color composites for visualizing forest change dynamics. Int. J. Remote Sens. 13:3055–3067.

Smith, J.A., T.L. Lin, and K.J. Ranson. 1980. The Lambertian assumption and Landsat data. Photogramm. Eng. Remote Sens. 46:1183–1189.

Spies, T.A., and J.F. Franklin. 1988. Old growth and forest dynamics in the Douglas-fir region of western Oregon and Washington. Natural Areas J. 8:190–201.

Spies, T.A., J.F. Franklin, and M. Klopsch.1990. Canopy gaps in Douglas-fir forests of the Cascade Mountains. Can. J. For. Res. 20:649–658.

Spies, T.A., and J.F. Franklin. 1991. The structure of natural young, mature, and old-growth forests in Washington and Oregon. In: Wildlife and vegetation of unmanaged Douglas-fir forests. L.F. Ruggiero et al. (eds.). USDA Forest Service Gen. Tech. Rep. PNW-GTR-285, Portland, OR, pp. 90–109.

Townshend, J.R.G., C.O. Justice, C. Gurney, and J. McManus. 1992. The impact of misregistration on change detection. IEEE Trans. Geosci. Remote Sens. 30:1054–1060.

Tucker, C.J. 1979. Red and photographic infrared linear combinations for monitoring vegetation. Remote Sens. Environ. 8:127–150.

Wald, L. 1989. Some examples of the use of structure functions in the analysis of satellite images of the ocean. Photogram. Eng. Remote Sens. 55:1487–1490.

Wang, L., and D.C. He. 1990. A new statistical approach for texture analysis. Photogram. Eng. Remote Sens. 56:61–66.

Webster, R. 1985. Quantitative spatial analysis of soil in the field. Advances in Soil Science 3:1–70.

Webster, R., P.J. Curran, and J.W. Munden. 1989. Spatial correlation in reflected radiation from the ground and its implications for sampling and mapping by ground-based radiometry. Remote Sens. Environ. 29:67–78.

Webster, R., and M. Oliver, M. 1990. Statistical methods in land and resource survey, Oxford University Press, Oxford.

Wilderness Society. 1991. Ancient Forests in the Pacific Northwest. The Wilderness Society, Washington, D.C.

Woodcock, C.E., and A.H. Strahler. 1987. The factor of scale in remote sensing. Remote Sens. Environ. 21:311–332.

Woodcock, C.E, A.H. Strahler, and D.L.B. Jupp. 1988a. The use of variograms in remote sensing: I. Scene models and simulated images. Remote Sens. Environ. 25:323–348.

Woodcock, C.E, A.H. Strahler, and D.L.B. Jupp. 1988b. The use of variograms in remote sensing: II. Real digital images. Remote Sens. Environ. 25:349–379.

Woodcock, C., and V.J. Harward. 1992. Nested-hierarchichal scene models and image segmentation. Int. J. Remote Sensing 13:3167–3187.

Yoder, J.A., C.R. McClain, J.O. Blanton, and L. Oey. 1987. Spatial scales in CZCS-chlorophyll imagery of the southeastern U.S. continental shelf. Limnol. Oceanogr. 32:929–941.

Case 2

Analysis and Mapping of Late-Successional Forests in the American Southwest

Introduction

Margaret M. Moore

The biotic communities of the American Southwest are diverse. The late-successional stages of these communities have unique structural, functional, and compositional characteristics that may be essential for the survival of certain plant and animal species. Yet species habitat preferences, landscape patterns, patch dynamics, minimum viable populations, and forest successional pathways of these southwestern forest ecosystems are not well understood. In the Southwest, the late-successional forest types receiving substantial attention as habitat for threatened, endangered, or sensitive species (particularly birds) include the oak and piñon pine/juniper woodlands and the ponderosa pine, mixed conifer (including aspen), and spruce-fir forests. All of these types provide important habitat for many migratory neotropical birds, and several are important for raptors such as northern goshawk and Mexican spotted owl. While the amount and distribution of late-successional habitat necessary to maintain viable local populations of threatened or endangered species has not been determined, the controversy is increasing.

As has occurred in other regions, the late-successional and old-growth portions of the forested landscapes have sharply declined in the Southwest. This decline has been primarily due to timber harvesting, but there have been other alterations to these ecosystems. One hundred and twenty-five years of heavy livestock grazing and fire suppression associated with Euro-American settlement have brought about substantial changes in tree densities, vegetation patchiness, crown closure, shrub and herbaceous biomass, forest fuels, and nutrient cycling. A critical issue for researchers and managers to consider is whether the forest ecosystems of today are "natural" and whether they are sustainable. (The word "natural" does not exclude humans as part of the system, and although the definition varies, the time prior to Euro-American settlement is intended as the benchmark for this discussion [Kilgore 1985]). The ecological conditions that shaped individual species over thousands of years have been changed in many ecosystems over the last one hundred years through management actions. The range of conditions (critical structural, functional, and compositional conditions) that some species require for success may have changed or may no longer exist. Will plant or animal species be added to the threatened, endangered, or sensitive species lists tomorrow because we are managing for today's conditions without regard for the past or future?

While most of our natural resource inventory questions are spatial in nature (What is it? How much exists? Where is it located? What is the arrangement?), the inventories themselves often are not designed to address ecological questions across the relevant spatio-temporal scales of the ecosystem in question. In addition, they tend to be counts (e.g., "tree tallies" or "spotted owl tallies"), not true ecosystem inventories. Tree tallies often begin at the "stand level" and are entered into some database at that level. However, key questions, such as what is the spatial arrangement often begin at the substand level. In the uplands of the ponderosa pine type of the American Southwest, for example, there is additional horizontal complexity because of the patchy nature of old-growth (presettlement) pine groups, young-growth (post-settlement) pine and bunchgrass openings. For old-growth ponderosa pine management or wildlife habitat analysis of certain species, it is critical to understand the past and current vegetation patterns at this detailed level. Yet these patterns may be too small to detect with large-area inventory methods or too costly to collect with more intensive methods. However, the bottom line is that for ecosystem management at the landscape level we need to recognize and emphasize the relationships between spatial patterns and ecological processes, over time, and at several hierarchical scales.

The continual plea is for inventory and monitoring techniques that suffice for ecosystem management at the landscape scale, which usually involves remote sensing and GIS technologies. However, the problems are more fundamental than just providing new tools to managers. The problems will not go away or become any simpler just because there are image-processing and GIS capabilities on every U.S. Forest Service Ranger district (although this might help).

These technologies have important potential in landscape-level management and planning. Ecosystem classification, determination of key ecological parameters, or modeling techniques using remote sensing and GIS techniques are important tools in the entire inventory, monitoring, and decision-making process. Aerial photography, one form of remote sensing, has long been used by resource managers to complement and reduce field work. Why should satellite remote sensing and GIS techniques be used any differently? Several old-growth inventories have used remote sensing and GIS technologies as screening mechanisms by which unnecessary ground visits are avoided; others have modeled "probable old growth," since much of what defines old-growth forest ecosystems lies below the canopy. Modeling old growth often includes a combination of crown closure, cover type, size classes, and topography (the basic structural characteristics mentioned above as inventory definitions). We should also investigate (and many research projects are doing this) other variables that are not typical inventory measurements, such as leaf area index, normalized difference vegetation index (NDVI), texture, branchwood-to-leaf ratio, shadows, and canopy nitrogen or phosphorus content. GIS models and decision support systems (DSS) should be used to explore the alternatives, such as which management action will result in the least forest fragmentation over time.

This case study will focus on some of the basic ecological and planning needs for landscape-level management and planning in the American Southwest. The potential role of satellite imagery and GIS in analyzing and solving landscape-level management issues will be explored. One specific project that used Landsat TM data to determine the distribution and abundance of probable old-growth areas in the Jemez Mountains on the Santa Fe National Forest of New Mexico will be featured. This project's classification accuracies as well as limitations and problems, and its usefulness for ecosystem management and planning, will be discussed.

Reference

Kilgore, B. 1985. What is "natural" in wilderness fire management? USDA Forest Service Gen. Tech. Rpt. INT 182.

8

Ecological Perspective: Linking Ecology, GIS, and Remote Sensing to Ecosystem Management

Craig D. Allen

Introduction

Awareness of significant human impacts on the ecology of Earth's landscapes is not new (Thomas 1956). Over the past decade (Forman and Godron 1986, Urban et al. 1987) applications of geographic information systems (GIS) and remote sensing technologies have supported a rapid rise in landscape-scale research. The heightened recognition within the research community of the ecological linkages between local sites and larger spatial scales has spawned increasing calls for more holistic management of landscapes (Noss 1983, Harris 1984, Risser 1985, Norse et al. 1986, Agee and Johnson 1988, Franklin 1989, Brooks and Grant 1992, Endangered Species Update—Special Issue 1993, Crow 1994, Grumbine 1994). As a result agencies such as the U.S. Forest Service, U.S. Fish and Wildlife Service, and National Park Service are now converging on "ecosystem management" as a new paradigm to sustainably manage wildlands and maintain biodiversity. However, as this transition occurs, several impediments to implementation of this new paradigm persist, including (1) significant uncertainty among many land managers about the definition and goals of ecosystem management, (2) inadequate ecological information on the past and present processes and structural conditions of target ecosystems, (3) insufficient experience on the part of land managers with the rapidly diversifying array of GIS and remote sensing tools to effectively use them to support ecology-based land management, and (4) a paucity of intimate, long-term relationships between people (including land managers) and the particular landscape communities to which they belong. This chapter provides an ecological perspective on these issues as applied to ecosystem management in a southwestern U.S. landscape.

An Ecological Perspective on the Conceptual Basis of Ecosystem Management

Ecosystem management is typically presented as an effort toward "sustainable" land management, often defined as the persistence or restoration of diverse, healthy, and productive ecosystems (Salwasser 1991, Robertson 1992, USDI Fish and Wildlife Service 1994, Christensen 1994). While ecosystems may potentially be manipulated into a variety of ecological states, they are not infinitely malleable. So what range of conditions should land managers target in order to achieve sustainability? A reference template to evaluate the ecological status of a site may be derived from the envelope of historic and prehistoric site conditions, as revealed by ecological history and paleoecological research on ecosystem dynamics at multiple spatial scales. This historic range of variability in ecological processes and structures (often called presettlement, "natural," native, or ecological reference conditions) presumably can be used to guide sustainable management action. For certain ecosystems it can be documented that this range of ecological conditions persisted through significant periods of time. Land managers, working through the public participation process, can then develop a "desired future condition" from within the intersection of that ecologically defined template of sustainable conditions and societal aspirations (Figure 8.1).

Many practical applications of wildland ecosystem management seek to manipulate localized, site-scale, ecosystem patches (e.g., a 50-hectare forest stand dominated by relatively uniform vegetation) within the mosaic of patches found in a larger landscape context (e.g., a 5000-hectare watershed, or a 500,000-hectare national forest) to achieve desired future conditions. This typical ecosystem management scenario requires ecological information and operational management attention at two relatively distinct spatial scales: (1) the local site-scale patch, and (2) the landscape mosaic of patches (Figure 8.2).

At both patch and landscape scales, managers require understanding of current and historic ecological processes and structures, as well as their interactions through time, in order to define the envelope of historically sustainable conditions. Managers need information on site-specific ecological processes such as microclimate, water and mass movement, soil development, nutrient cycles, food webs, biotic movements, and disturbance regimes (the frequency, intensity, areal extent, and seasonality of disturbances like fire, windthrow, and insect outbreaks). The ubiquity of disturbances (Sousa 1984) and their significance as formative processes in most terrestrial ecosystems (Pickett and White 1985) lends particular

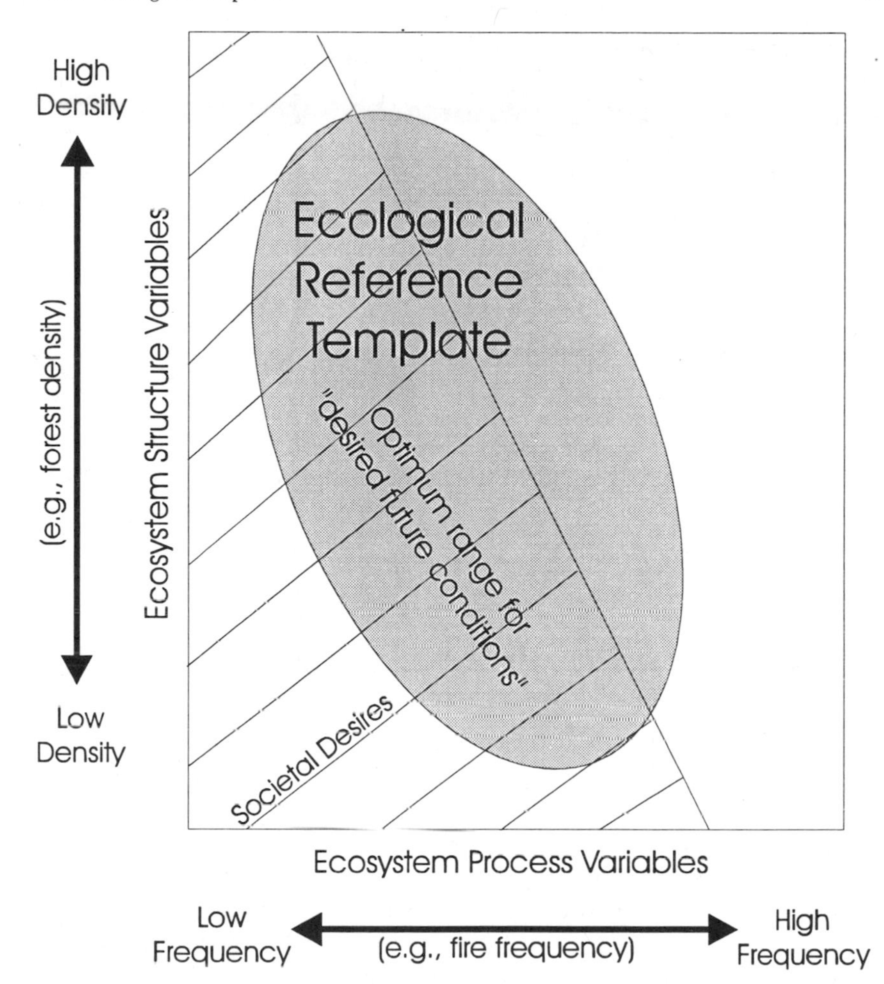

Figure 8.1 Conceptual diagram of an ecological reference template as a limited range of states in ecosystem process and structural variables. The specified ecological reference template is based on historic site conditions and defines an envelope that presumably encloses sustainable ecosytem states. This reference template may be applied in ecosystem management to develop "desired future conditions" by defining ecosystem states within the intersection between the ecological template and societal desires.

importance to collecting information on disturbances so that management may attempt to mimic the role of native disturbance regimes (e.g., Agee 1993). Information on ecosystem structural patterns such as the biotic composition of the site (including sensitive and alien species) and the sizes and three-dimensional spatial arrangement of the live and dead vegetation (e.g., an old-growth pine savanna versus a dense, 20-year-old

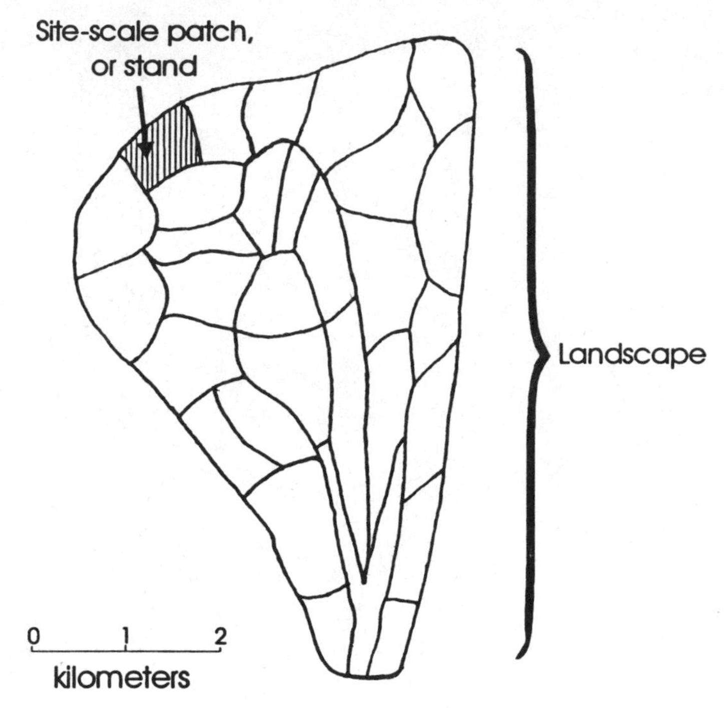

Figure 8.2 Schematic diagram of a landscape as a mosaic of site-scale patches.

plantation) is also essential because process and structure strongly interact to mold each other (e.g., the process and structural outcome of fire is quite different in a pine savanna than in a thicket). Indeed, an ecosystem perspective holistically views the ecological processes and structures of each patch as an integrated system rather than as isolated components. Overall, knowledge of site history is vital to understand observed ecological patterns and to recognize the ecological potential of any patch (Hamburg and Sanford 1986, Duffy and Meier 1992, Russell 1994). Intensive field observations are necessary to resolve this array of site-level ecological information that silviculturalists, wildlife biologists, and hydrologists require to develop sustainable management prescriptions. GIS and remote sensing applications may supplement but will never replace the primacy of field observations at this site-level scale where on-the-ground implementation occurs.

In contrast, management at the broader landscape scale is largely a planning task, implemented by designing the spatial and temporal distribution of the site-specific ecosystem patches that comprise the landscape mosaic; this requires the larger perspectives that current GIS and remote sensing approaches can provide. By managing the pattern of component patch types, sizes, and numbers in time and space the cumulative patterns and processes of the composite landscape mosaic can be managed (Franklin and Forman 1987, Rykiel et al. 1988). As at the site level, information on native landscape- (and regional-) scale structures, processes, disturbance regimes, and environmental histories is essential to develop ecology-based templates at these broader scales (Turner 1987, Urban et al. 1987). However, landscape-level information on sustainable ecological conditions is derived through somewhat different approaches (Turner and Gardner 1991) and is generally less available and less precise than more traditional site-level ecological information, in part due to the relative novelty of landscape ecology research. GIS and remote sensing methods are essential tools that will be increasingly used to develop the landscape-level spatial and temporal contexts needed for ecological management of landscapes.

The broad array of definitions and spatial scales applied to the term ecosystem necessarily makes ecosystem management a fuzzy term unless it is precisely defined. Some of the uncertainty at the field level in the U.S. Forest Service about the meaning of ecosystem management (like its "New Perspectives" progenitor [Salwasser 1992]) occurs because the different information needs and management operations of the site and landscape scales are not explicitly distinguished or outlined, as discussed above. Since ecological management must cover all aspects of ecosystem process and structure, there is a need to define ecosystem management in terms of ecological hierarchies.

From an ecological viewpoint, two core "new perspectives" may be identified in current formulations of ecosystem management: (1) increased emphasis on the use of historic, site-level, ecological processes and structures as a template for sustainable management of individual, stand-scale, ecosystem patches or sites; and (2) the importance of explicitly considering and managing the spatial and temporal patterns of the mosaic of ecosystem patches at larger landscape and regional scales. These two ideas are hierarchically linked along a continuum of spatial scales (Urban et al. 1987), and successful ecosystem management requires blending site-specific information into a landscape or regional context. Still, distinguishing the stand-scale operations at the ecosystem patch level (idea 1) as "ecological site management" and the landscape-mosaic level work (idea 2) as "ecological landscape management" would clarify understanding, and thereby improve implementation, of these new management paradigms by both field-level resource managers and the general public. Incorporating the word "ecological" into this terminology emphasizes the factors that direct management and makes it more obvious that these terms are subsets of ecosystem management. Ecosystem management may then persist as a more generalized umbrella term that covers various spatial scales as well as enhanced integration of societal purposes through cooperative, participatory relationships among land management agencies and the public.

Distinguishing ecological landscape management as an identifiable component of ecosystem management could reduce ambiguity in communications among land managers, scientists, and the public at large. Landscape is a relatively well-defined level of spatial scale that is understood by the public and professionals alike (Forman and Godron 1986). Our sense of belonging to a place is associated with the landscapes where we live and work (Tuan 1974) and which we name and often value for spiritual reasons. This public attachment to local landscapes provides a focus for long-term human attention and care of landscapes and thus the impetus for increased public involvement in land management planning. Landscapes also provide a convenient framework for the conduct of cooperative, interagency operations, an increasingly important aspect of land management.

Ecosystem management is a conservative approach to land management that basically follows Aldo Leopold's (1949) admonition to "preserve all the parts of the land mechanism" and develop "gentler and more objective criteria for its [any land management tool's] successful use." Ecosystem management elaborates on ideas expressed in Leopold's famous essay "The Land Ethic," namely: (1) "A thing is right when it tends to preserve the integrity, stability, and beauty of the biotic community"; (2) ". . . native plants and animals kept the energy circuit [of the land community] open; others may not"; (3) ". . . man-made changes are of a different order than evolutionary changes, and have effects more comprehensive than is intended or foreseen" as "man's invention of tools has enabled him to make changes of unprecedented violence, rapidity, and scope"; and (4) "The combined evidence of history and ecology seems to support one general deduction: the less violent the man-made changes, the greater the probability of successful readjustment of the [land] pyramid." Ecosystem management may be seen as a modern effort to affirmatively answer two questions Leopold raised about land management: "Can the land adjust itself to the new order? Can the desired alterations be accomplished with less violence?" Leopold (1949:274) further noted that "A science of land health needs, first of all, a base datum of normality, a picture of how healthy land maintains itself as an organism." He believed that the "most perfect norm is wilderness" while recognizing that long-sustained, cultural landscapes such as northeastern Europe could also serve as reference areas. The ecosystem management template of historic ecological conditions is an attempt to provide "more objective criteria" to the development of site-specific norms for healthy land.

Limitations to This Ecosystem Management Approach

There are a number of conceptual and practical problems with the ecological reference-template approach to ecosystem management outlined here. First of all, while ecosystem management aims to sustain the integrity of ecosystems yet still provide for human use of landscapes, the core concepts of ecosystem sustainability and integrity remain controversial (Noss 1990, Jackson 1991, Costanza and Daly 1992, Kohler 1992a, Kay 1993, Ludwig et al. 1993).

Second, the proper role of humanity in shaping ecosystem processes and patterns is subject to ongoing debate. A focus on ecological reference conditions recognizes that ecosystems do not follow the power, prestige, and monetary rules of human political, social, and economic systems, even though the entire earth is increasingly molded by these human conventions (see Cronon 1991 and Turner II et al. 1990). Environmental reconstructions often attempt to isolate the historic role of post-European-settlement land uses as "unnatural" influences, while the significant ecological effects of indigenous peoples on North American landscapes (Cronon 1983, Pyne 1982, Anderson and Nabhan 1991, Denevan 1992), including portions of the Southwest (Betancourt and Van Devender 1981, Kohler 1992b), are generally ignored or incorporated into the template descriptions of native, "natural" ecosystems. This is philosophically problematical since all people may be considered equally natural parts of landscapes. The U.S. Forest Service clearly recognizes that human purposes must be incorporated into ecosystem management through public participation in the resource management planning process (Robertson 1992, Salwasser 1992).

Third, we will always lack perfect knowledge of past and present ecosystem conditions. Relatively little ecological history work has been conducted to date and as time passes it is increasingly difficult to reconstruct bygone conditions as the evidence is lost or degraded, sometimes due to our management actions (e.g., prescribed burns can destroy the fire scars needed to develop accurate fire histories). Further, the resolution of environmental reconstruction methods is often too coarse to assemble an adequate reference template of historic ecosystem patterns and processes for practical use by managers. There are methodological limits on the quality of information we can collect on certain ecosystem parameters, such as below-ground ecological phenomena. The costs of collecting high-quality information preclude complete inventories, further constraining our knowledge of current ecological conditions. Also, more information is not always useful, as it is possible to get bogged down in trivial details or unresolvable complexities with excessive or even contradictory data. This is especially true given that our understanding of how ecosystems function is culturally mediated and continuously evolving (Botkin 1990). The necessity of working with incomplete knowledge suggests that ecosystem management will always be as much an art as a science, and that we should manage landscapes with less hubris and more humility (Kellert and Bormann 1991:210, Ehrenfeld 1993).

Fourth, the past is not a perfect guide to the future, especially in an increasingly human-altered world. When defining historic ecosystem conditions as a range to manage within, it must be recognized that environmental change is axiomatic, pervasive, and in some cases chaotic (Delcourt and Delcourt 1988, Botkin 1990, Sprugel 1991). Indeed, paleoecology shows that most modern biotic communities are not tightly coevolved and organized units but rather transitory assemblages of individual species (Hunter et al. 1990). Further, global change scenarios indicate that the use of templates based on the past several hundred to several thousand years may not be appropriate for many sites where the climate will likely be pushed out of its Holocene envelope within decades, dragging ecosystem processes and structures along to some new state (Peters 1988). Land managers will increasingly need to become facilitators of ecological changes to meet objectives such as minimizing the loss of global biodiversity. Still, by conservatively managing our landscapes to limit the rate, extent, and intensity of human interactions, many ecosystems may prove resilient enough to sustain their evolutionary potential despite the effects of climate change. For example, ecosystem resiliency is enhanced by intact plant-soil linkages (Perry et al. 1990) and by the ecological legacies of persistence and multiple possibilities inherent in biologically diverse ecosystems.

Finally, even if perfect information existed, implementation of ecosystem management would be difficult because recent human activities such as widespread habitat alteration and fragmentation (Wilcove et al. 1986), fire suppression (Covington and Moore 1992), and introduction of aggressive alien species (Westman 1990) have pervasively altered the structures and processes of many wildland ecosystems outside the ranges of historic conditions (Crosby 1986, Bahre 1991, Schwartz 1994). This is true even in our most protected southwestern landscapes (deBuys 1985, Allen 1989). The proliferation of human communities within wildland matrices increasingly constrains management options (e.g., the use of prescribed fire is limited by smoke concerns). Undertaking needed ecosystem restoration measures will be challenging from logistical, economic, and political perspectives.

Despite uncertainties in our knowledge and methodological imperfections, ecosystem management provides a conservative, reasoned approach to sustaining relatively whole, functional ecosystems in the face of accelerating global change. The increasingly apparent cumulative impacts of exponential human population growth and escalating resource consumption highlight the need to move toward such ecology-based approaches to landscape management if we are to maintain the ecosystem services and amenities we need and value.

Applications of GIS and Remote Sensing Technologies to Ecosystem Management in a Southwestern U.S. Landscape

Examples of GIS and remote sensing applications to ecosystem management are presented from the landscape of the Jemez Mountains of north central New Mexico, which includes Bandelier National Monument and portions of the Santa Fe National Forest.

Use of GIS and Remote Sensing to Provide a Landscape Context for Ecosystem Management

A landscape ecology research project (Allen 1989) used site-level fieldwork, landscape-level interpretations of aerial photographs, and GIS applications to develop ecological information at multiple spatial scales for the landscape in and around Bandelier National Monument (Figure 8.3). Bandelier, at 13,307 hectares, comprises only 2.4 percent of the land area in the Jemez Mountains. The headwater portions of most park watersheds are located on adjacent U.S. Forest Service land (Figure 8.4). Field information was collected from 969 sample points covering the Rito de los Frijoles watershed in Bandelier (Figure 8.4); this watershed provides a representative elevational transect of this landscape with a 1480-meter rise from the Rio Grande to the crest of the Jemez Mountains. At each point data were collected on numerous ecological characteristics of the

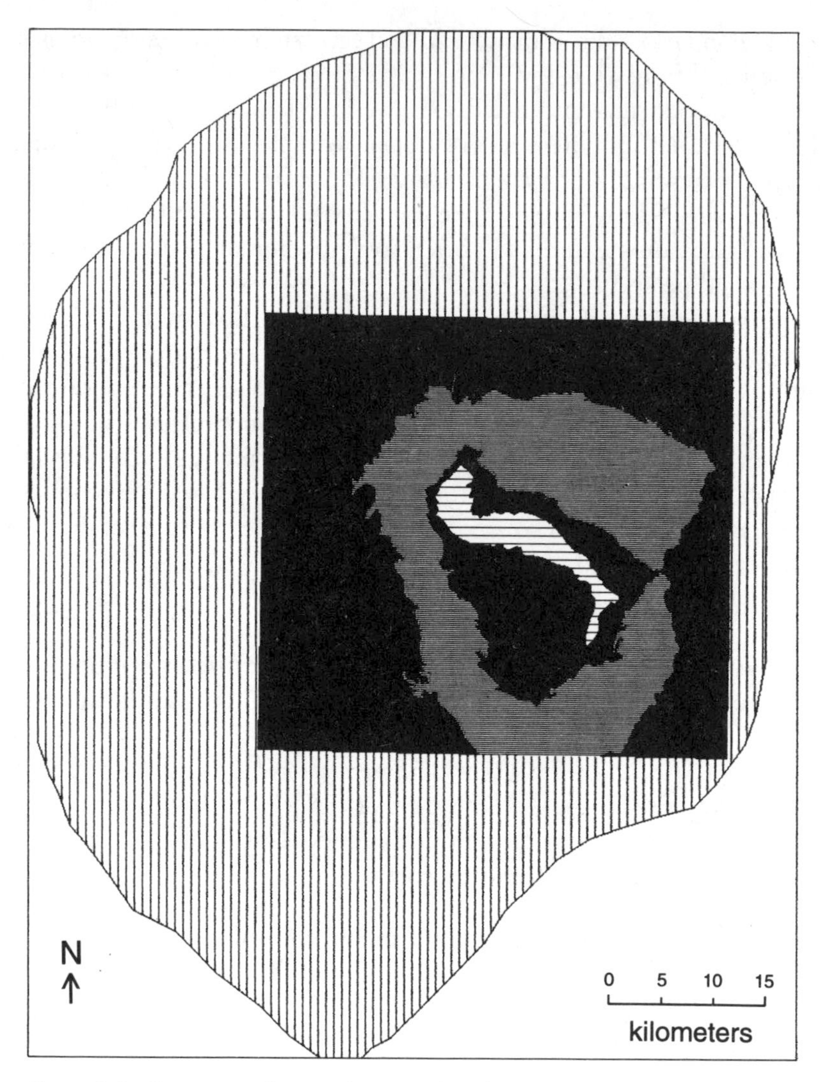

Figure 8.3 Locations of mapped areas within the 543,522-ha Jemez Mountains (vertically striped area). The black square defines the 187,858-ha area mapped from 1935 and 1981 airphotos for road networks (see Figures 8.6 and 8.7), the gray patch (85,726 ha) was mapped from 1935 and 1981 airphotos for landscape-cover types (see Figure 8.5), the irregular black area (28,684 ha) was mapped from 1981 airphotos for ecosystem patches, and the horizontally striped area is the Frijoles watershed (6500 ha), which was sampled with 969 points in the field (see Figure 8.4).

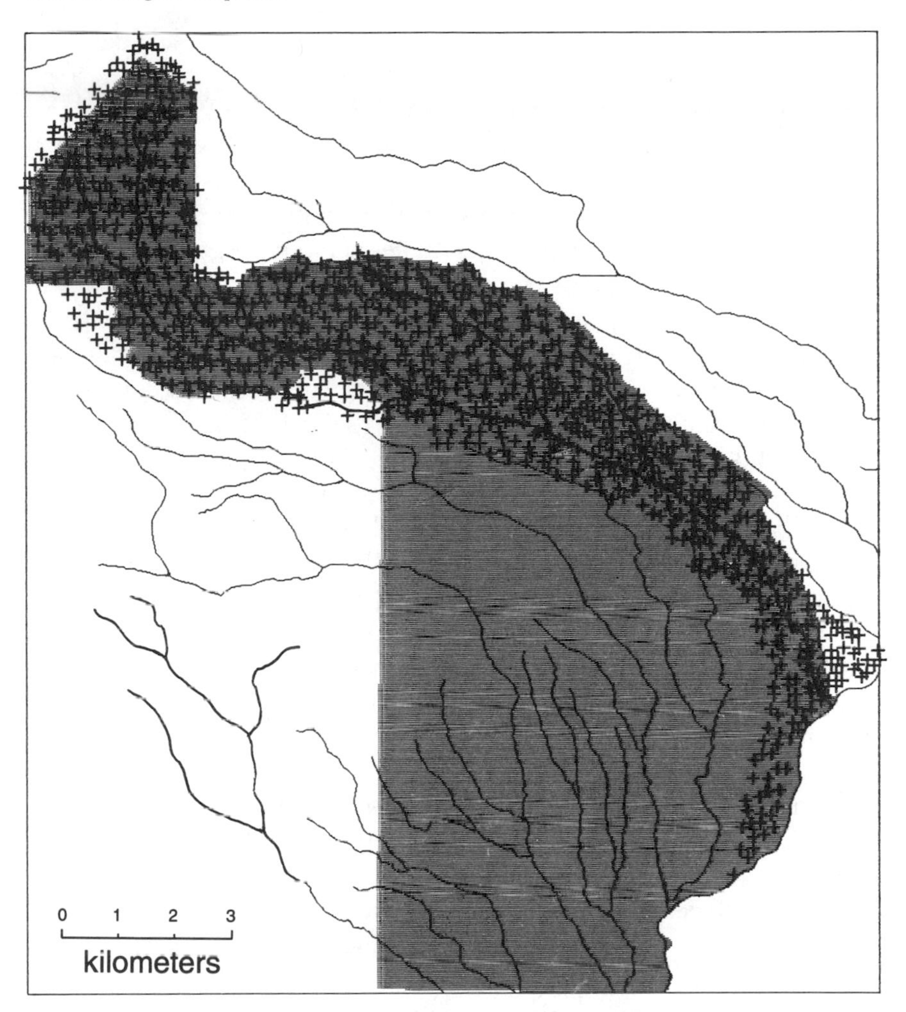

Figure 8.4 Locations of 969 field sample points in the Frijoles watershed, with the main portion of Bandelier National Monument shown in gray and canyon stream drainages as lines. The displayed headwaters area of the southern portion of Bandelier is Santa Fe National Forest land and contains the Dome Diversity Unit.

site, ranging from detailed information on vegetation and landform to evidence of processes such as erosion, fires, or floods. Changes in land cover and road networks were mapped by interpreting 1935 and 1981 aerial photographs (Figure 8.3). Combinations of vegetation, percent tree canopy cover, and landform were mapped as "ecosystem patches" at a 1.5-hectare level of resolution across Bandelier and surrounding areas by interpreting the 1981 aerial photographs (Figure 8.3). The field data provided ground truth information for the aerial photograph interpretations. These and other data were entered into a GIS (see Allen 1989 for details).

Combining this GIS information with additional field data and historical information provides multiple perspectives on the dynamics of various ecosystems that have been used to support ecosystem management in the Jemez Mountains. For example, field studies of montane grasslands (Allen 1984) had shown that these ecosystems were found in a distinctive landscape pattern on the upper slopes of most of the larger peaks in the Jemez Mountains. Several lines of evidence indicated that these grasslands have existed for millenia on these sites. Increment core dates from hundreds of trees demonstrated that tree invasion in the last 75 years has greatly reduced the extent of these grasslands, due to fire suppression and livestock grazing histories. The addition of landscape-scale information from aerial photography and GIS-generated displays and analyses of the magnitude of the grassland loss to tree invasion has triggered ecosystem management action (Figure 8.5). Tree cutting and prescribed burning have been conducted for several years in the Cerro Grande grassland in Bandelier, and a recent grassland restoration project took place on Quemazon Mountain, with other sites under consideration for treatment in the Santa Fe National Forest.

Field observations of soil erosion in Bandelier were scaled up from individual sample points to the larger landscape using the GIS, revealing extensive erosion in piñon-juniper woodlands that affects about 40 percent of Bandelier's land area. This erosion is also a cultural resource problem, as 70 percent of 1600 archeological sites surveyed between 1988 and 1991 displayed evidence of erosion damage. The GIS-supported ability to display the widespread nature of soil erosion has helped elevate this issue to the top of Bandelier's resource-management priorities, and it has fostered funding support for ongoing ecological research on the causes of the erosion and on effective treatments to restore historic ecological conditions.

GIS analyses also documented some obvious human-induced changes in this landscape. In 1935, 0.9 percent of the landscape was mapped as anthropogenic cover-type patches dominated by human land uses, primarily as dry-farmed bean fields on the mesas north of Bandelier. By 1981 the portion of this landscape converted into anthropogenic patches had increased to 6.3 percent, as the bean fields were replaced by the technical areas of Los Alamos National Laboratory and the towns of Los Alamos, White Rock, and Cochiti Lake developed. Other new cultural features include 53 pumice mines, 12 large-stock ponds, three golf courses, a ski

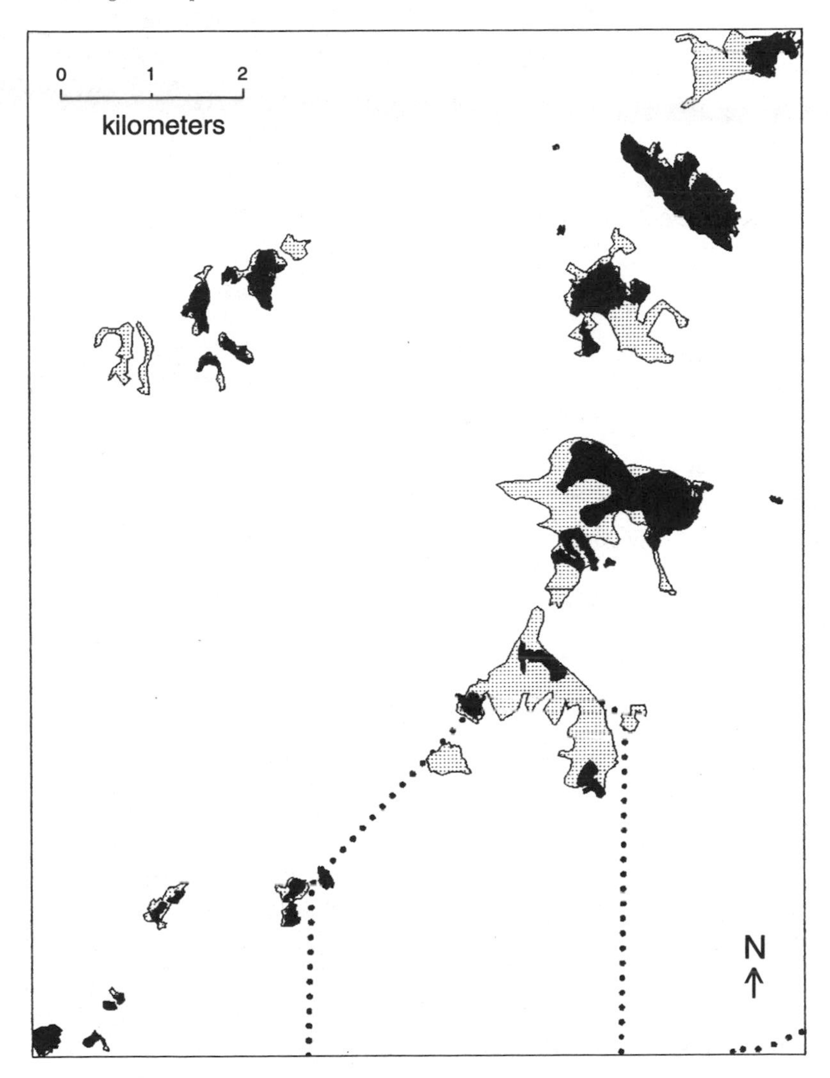

Figure 8.5 Map of reduction in montane grassland area between 1935 and 1981. Area of open grassland (with less than 10 percent tree canopy cover) is shown with a grey pattern for 1935 and as solid black patches for 1981. The dotted line encloses a portion of Bandelier National Monument.

area, and Cochiti Reservoir on the Rio Grande. Logging occurred along 11 percent of Bandelier's main unit boundaries between 1977 and 1984.

Further fragmentation of this landscape is apparent in the striking development of road networks in recent decades, as shown in Figures 8.6 and 8.7. I mapped five types of roads: railroad, paved, gravel, dirt, and

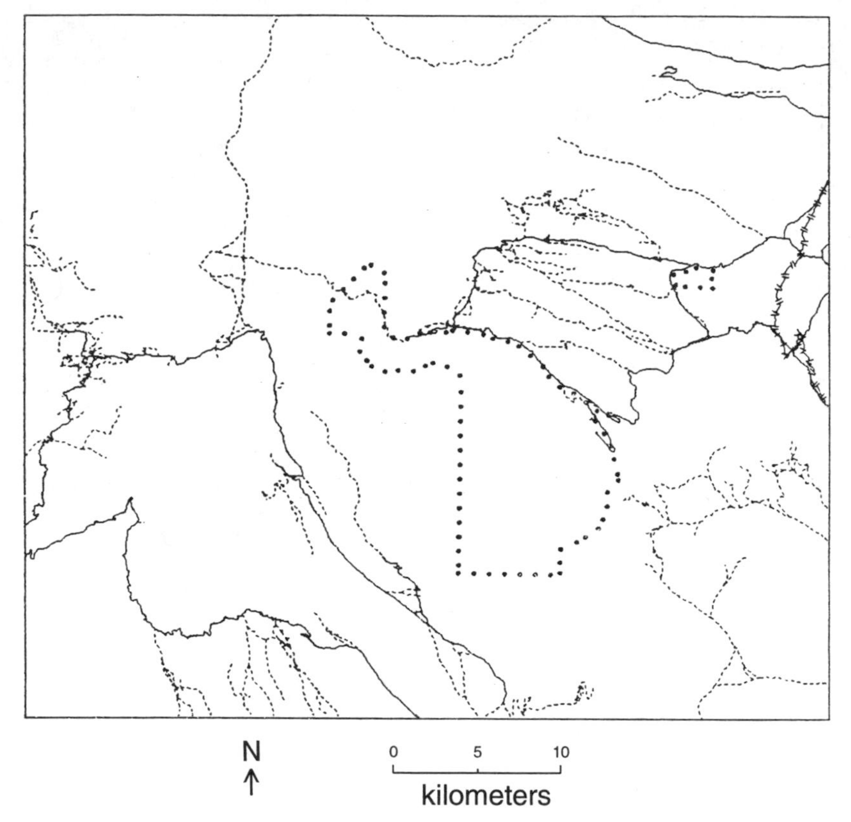

Figure 8.6 Map of all roads in 1935 across 187,858 ha of the Jemez Mountains. The crosshatched line is a railroad, the solid lines are dirt roads, the thin dashed lines are primitive roads, and the dotted lines show the current boundary of Bandelier National Monument.

primitive. Overall there was a 12-fold increase in total road length from 718 kilometers in 1935 to 8450 kilometers in 1981. Roads are indicators of direct and indirect human effects on landscapes. Examples of road impacts include: (1) roads as fire breaks, which also allow widespread access for fire suppression; (2) facilitation of motorized disturbance of wildlife; (3) introduction of edge conditions into forest interiors; and (4) roads as weedy paths whereby alien species move into new areas. In the forested portions of the Jemez Mountains roads are usually a signature of past timber harvests. By 1981 roads covered at least 1.7 percent of the mapped landscape (Allen 1989:240), with obvious implications for erosion rates, stream sediment loads, and overall landscape productivity (Maser 1988:161) in the Jemez Mountains. These GIS maps of road networks have been widely disseminated and have supported efforts by various land management

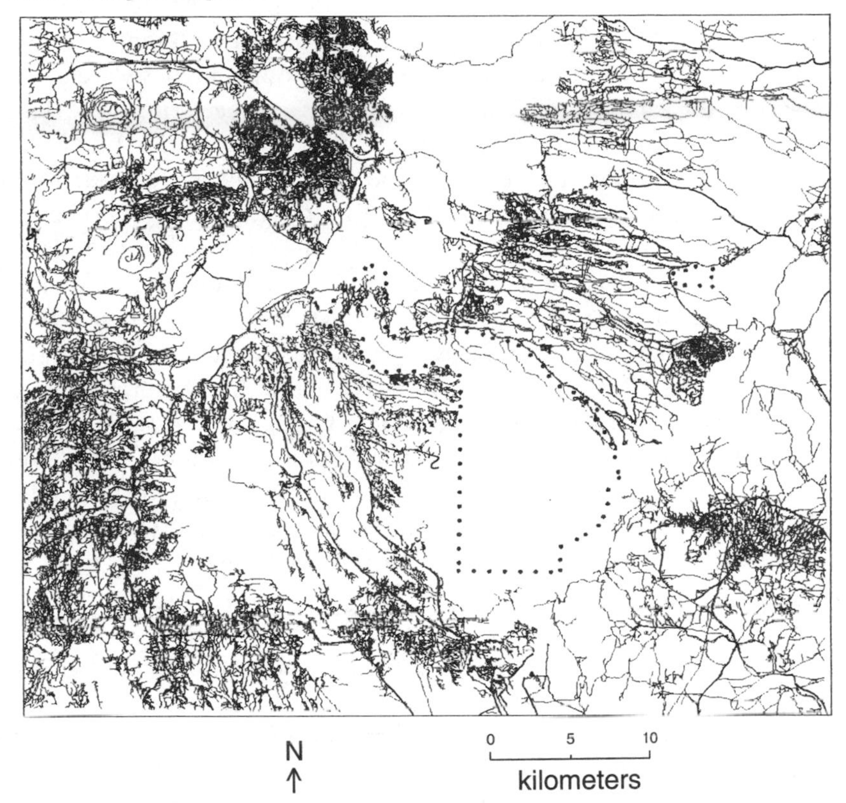

Figure 8.7 Map of all roads in 1981 across 187,858 ha of the Jemez Mountains. Distinctions between the various types of road are obscured in this black-and-white figure. The dotted lines mark the boundary of Bandelier National Monument. The roadless area of the Bandelier and adjoining Dome wildernesses is apparent. Several townsites, Los Alamos National Laboratory technical areas, powerline corridors, and many logging roads are prominent by 1981.

agencies and citizen groups to reduce open road densities and manage for less fragmented landscapes in northern New Mexico.

These landscape-scale GIS and remotely sensed data proved valuable in the most recent U.S. Forest Service planning effort in Bandelier's headwaters area, called the Dome Diversity Unit by the Forest Service (see Figure 8.4). Three timber sales had been carried out with increasing controversy in this 4630-hectare area between 1981 and 1988, and planning was underway for a fourth sale in 1989. Participation by Bandelier staff on this planning team provided an opportunity to work with the Forest Service to develop a more ecologically oriented management plan for the Dome area. Overall management direction as well as specific guidelines were derived from the Santa Fe National Forest Plan (USDA

Forest Service 1987), which provided a coarse resolution template for the Dome landscape as a cultural resource and wildlife emphasis area. GIS analysis of the canopy cover and distribution of vegetation types at a landscape level was used in consort with site-level information on vegetation structure from forest inventories and other fieldwork to assess forest structural conditions across the planning unit. These analyses demonstrated that past management actions had left Dome area resources far outside the Forest Plan standards for important structural criteria, ranging from acreages and distributions of old-growth forest stands to overall densities and distributions of snags. While the Forest Plan directs that "road use will be managed with the objective of limiting open road density to 0.3 to 1.5 miles per square mile," our GIS analysis showed 10.93 miles of road per square mile in 1981 (prior to three additional timber sales), with 2.3 percent of the Dome area in road surface (not including road shoulders and right of way). The graphic presentation of data relevant to Forest Plan standards led to substantive changes in the management plan that was developed for the Dome area (USDA Forest Service 1991), resulting in a 500-percent reduction in harvested timber volume during the current entry, designation of 27 percent of the planning area as "future old-growth forest," and widespread obliteration and closure of roads to drop open road density to 1.3 miles per square mile and provide opportunities for snags to develop in the absence of fuelwood poaching. The information produced by GIS and remote sensing methods was essential to support planning for ecological landscape management in this politically sensitive boundary area, and this success has bolstered other efforts to promote interagency cooperation in managing resources of mutual concern in this landscape.

Using GIS in Fire History Research to Support Ecosystem Management

Fire is an important disturbance process in most southwestern forests (Swetnam 1990). Recent fire history research conducted by the National Park Service and University of Arizona Tree-Ring Lab cross-dated 1180 fire scars on 74 trees within five vegetation types across a 900-meter elevation gradient in the Rito de los Frijoles watershed of Bandelier National Monument (Allen 1989, 1990). Sampled vegetation ranged from the lowest, mesatop stands of ponderosa pine (*Pinus ponderosa*) up to montane mixed-conifer forests, as well as riparian, mixed-conifer forest in Frijoles Canyon. In an ongoing cooperative project with the Santa Fe National Forest this fire history research has been extended to date fire scars from 299 trees, snags, and logs from 15 additional ponderosa pine

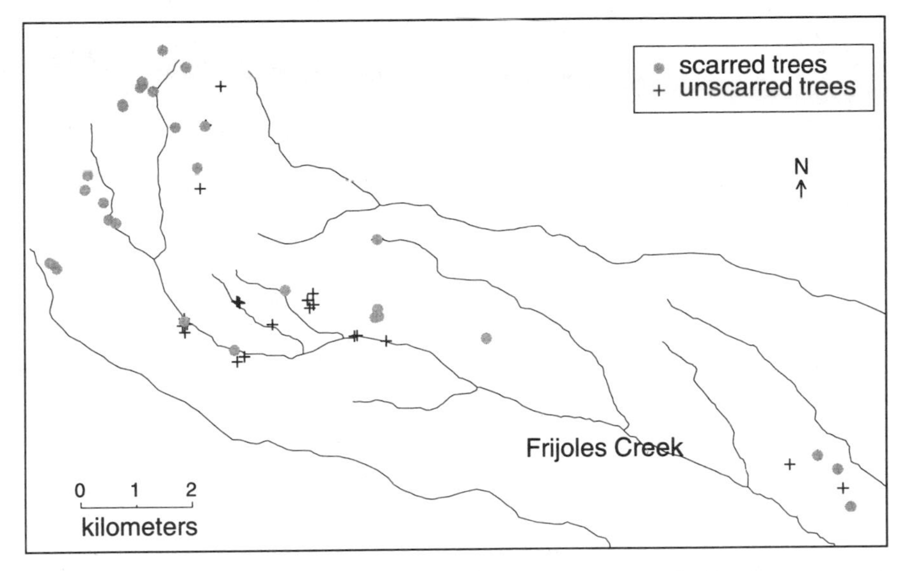

Figure 8.8 Map showing sample trees scarred by fire in 1748 in the Frijoles watershed. Similar watershed-wide fire years occurred about every 16 years, on average, between 1684 and 1851 in this watershed.

and mixed-conifer forest sites spread around the Jemez Mountains (Touchan and Swetnam 1992). Fire dates extend back to A.D. 1480, with mean fire intervals for major fires ranging from 6 to 22 years at all sites prior to the late 1800s, when fire frequencies plummeted, apparently due to the onset of heavy livestock grazing in the region (Wooton 1908, Allen 1989). GIS applications are being used to scale site-level information up to the landscape level, as in Figures 8.8 and 8.9, which display the extent of fires in the Frijoles watershed and Jemez Mountains in 1748. These data show that in many years fires burned widely across the Jemez Mountains (Allen and Touchan 1994).

Blending these fire history data with other historical and ecological information documents how fire suppression has induced major changes in the structure and function of many local ecosystems (Allen 1989). GIS analysis of the field sample-point data from Bandelier show that the distributions of many fire-sensitive species expanded markedly, for example upslope in the case of *Juniperus monosperma* and downslope in the case of *Pseudotsuga menziesii*. Formerly open ponderosa pine forests in the Jemez Mountains and throughout the Southwest have been converted into dense thickets, eliminating herbaceous understories and interrupting

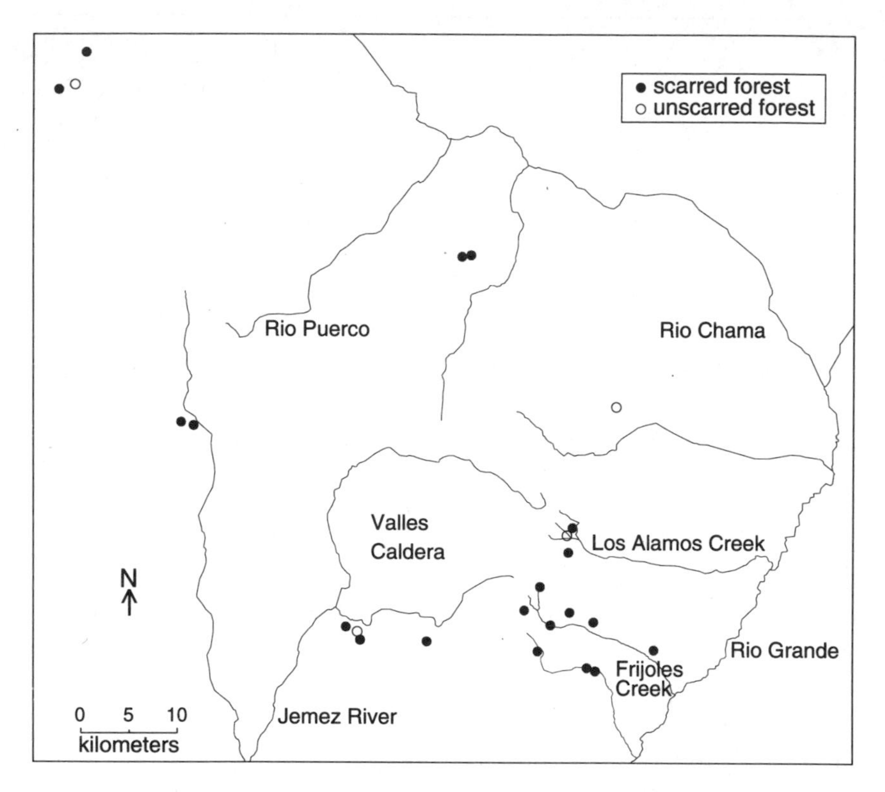

Figure 8.9 Map of the widespread 1748 fires in the Jemez Mountains. Forest sites where at least 25% of the sampled trees recorded a fire in 1748 are shown as scarred. Note that the Frijoles watershed (Figure 8.8) is located in the southeast portion of this map.

nutrient cycles (White 1991), while the potential for unprecedented high-intensity crown fires has developed (Covington and Moore 1992). The 1977 La Mesa fire concretely illustrated this structure-function relationship by converting 4000 hectares of dense ponderosa pine forest to grassland in and around Bandelier National Monument (Foxx and Potter 1984, Allen 1989).

These spatially explicit data unequivocally demonstrate the central role of fire in shaping ecological patterns at site and landscape levels in the Jemez Mountains. As a result, local land-management agencies are developing more aggressive prescribed fire programs to begin restoring historic vegetation mosaics and their associated fire regimes. The staff at Bandelier have drafted a new fire management plan that combines the ecosystem patch map and fire history data to outline the park as a mosaic of site-level patches with similar vegetation structures and fire history patterns. The patches in this landscape mosaic are being scheduled for management actions such as prescribed burns and thinning designed to restore each patch to a condition somewhere within its historic ecological envelope. This same landscape mosaic of site-level patches is also being used as the framework for an intensive fire-effects monitoring program (USDI National Park Service 1992), designed to provide information for managers on the ecological effects of management fires. Despite its small size Bandelier now has one of the most vigorous prescribed-fire programs between California and Florida in the National Park Service, with more than 500 hectares burned in 1992.

Similarly, the Santa Fe National Forest is using these high-resolution fire history data to strengthen their prescribed-fire program. For example, as I write (April 1993), smoke scents the air from a 6000-hectare prescribed burn on the west side of the Jemez Mountains, one of the ecosystem management demonstration projects on this national forest. This fire is being conducted on lands of the U.S. Forest Service, Jemez Pueblo, and Zia Pueblo by crews from the Forest Service, Jemez and Zia Pueblos, U.S. Bureau of Indian Affairs, National Park Service, U.S. Fish and Wildlife Service, and the State of New Mexico's Game and Fish Department and Forestry Division in a concrete example of cooperative management at a landscape level.

Limitations of GIS and Remote Sensing Technologies for Ecosystem Management

GIS and remote sensing approaches will become increasingly essential to implement ecological landscape management. However, many potential applications of GIS and remote sensing to ecosystem management are not yet feasible at the level of a Forest Service ranger district. Land managers must become aware of the limitations as well as the potential of these new tools (see Hoffer, Chapter 3), while proponents should avoid overselling these technologies. Otherwise, many ecosystem management applications of GIS and remote sensing will founder on unanticipated problems that may impede the adoption of these important new approaches to landscape management.

A major limitation on use of GIS and remote sensing methods is the time and expense required to develop landscape-level data and train personnel. GIS and remote sensing users must invest considerable front-end time to learn complex software packages and become familiar with the large quantities of data used in landscape-level applications. As a result land management agencies are creating positions for GIS and remote sensing specialists, but once trained these specialists are hard to retain "in the field" because they can typically command higher salaries elsewhere as their computer skills are in general demand. Rapid improvements in computer technology and reduced costs have largely removed hardware availability as a restriction on use of these methods, although systemwide procurement issues have slowed USDA Forest Service efforts to apply GIS approaches. A primary limitation on the use of these technologies for ecosystem management is the availability of trained personnel, and this will become increasingly apparent as data and hardware become ever more accessible.

A variety of data-related issues require consideration when applying GIS and remote sensing methods to ecological landscape management (Hoffer, Chapter 3). GIS applications are subject to the classic computer maxim "garbage in, garbage out." While a variety of general data layers of known resolution and quality (e.g., topography, hydrography, land ownership) can now be purchased inexpensively from sources such as the U.S. Geological Survey, the development of high-quality, ecological data to meet landscape-specific needs can be expensive. Indeed, the potential to collect data at landscape scales for low cost per unit area covered is one of the primary attractions of remote sensing approaches to data collection. However, users need to recognize that the utility and accuracy of remote sensing applications are greatly affected by the resolution of the sensor, the experience of the analyst, and the classification methodologies employed.

The data underlying land management applications of GIS continually become obsolete as forest stands and landscapes change. Thus land management agencies must plan to constantly update their GIS data in order to maintain current information over large areas, such as national forests.

Data compatibility and interpretation issues emerge when a GIS application is used to collate and integrate diverse data sets for ecological landscape management. Data resolution and accuracy may differ between data layers, which often originate from multiple sources with variable standards of quality control and documentation. Accuracy may even vary within a single layer (e.g., certain vegetation types may be classified more accurately than others). It is difficult to explicitly assess and document site-specific uncertainties in spatial data—GIS users are unable to see error bars for each pixel in a map layer. As a result, unquantified uncertainties are often layered atop one another in GIS applications, yielding outputs of unknown validity. A GIS model is subject to the limitations of all its data; therefore, the use of more data may result in a less certain output. It is already common to see naive GIS applications that overlook the limitations and uncertainties inherent in the data, leading to precise but inaccurate results or overgeneralized outputs that ignore the local complexities observed in the field (Scott et al. 1994). The flood of GIS data and rising intricacy of analytical procedures could exacerbate this problem by fostering specialized analysts who become increasingly office-bound and isolated from the field data underlying their GIS applications.

Similarly, the much-vaunted ability to easily manipulate GIS data in generalized modeling applications has a downside, as it leads to the creation of internally generated "virtual" data, which may be difficult to distinguish from "real" data that have been externally collected and physically verified (e.g., by ground-truthing). Indeed, it may be impossible to verify such model-derived "virtual" data (Oreskes et al. 1994), yet once created these data are subject to use without critical assessment of their validity.

These data issues can make the use of GIS for ecological interpretations challenging and fraught with potential for error. The illusions contained in modern television advertisements demonstrate the power that images have to deliberately or inadvertently misinform unwary consumers of visual information. The consumers of GIS outputs (including the general public, resource managers, and academics) need to critically review the validity of the "virtual worlds" presented to them, while ecosystem managers need to use the power of GIS applications with caution and awareness of the uncertainties in their spatial data. It is important to continually remind ourselves and the public that the beautiful GIS and remote sensing images we create are only simplified representations of landscapes, often removed by several interpreted steps from the external physical reality.

Finally, several data-management issues merit highlighting. Careful archiving of original data and documentation of any data alterations are required to maintain the integrity and quality of ecological data and any dependent applications. Data security will receive increasing attention as locational information on sensitive resources such as endangered species or archeological sites becomes incorporated into GIS databases, which can easily be transferred into unfriendly hands via modem links or a small diskette. GIS data management is challenging for land management agencies because of the increasingly large quantities of data used in GIS, the significant potential for deliberate and inadvertent modifications of data files without adequate documentation, difficulties in creating new positions for full-time system managers in these fiscally tight times, and the lack of continuity associated with the high turnover rates of experienced computer personnel.

Using GIS and Remote Sensing to Model Old-Growth Forests on the Santa Fe National Forest

Gonzales (Chapter 10) provides detailed information on a recent effort to use GIS and remote sensing methods to model old-growth forest locations in part of the Jemez Mountains on the Santa Fe National Forest. This project offers lessons on the use of GIS and remote sensing for ecosystem management as it concretely illustrates some of the limitations listed above.

First of all, this project depended on Landsat thematic mapper data, which lack adequate resolution to distinguish the essential structural attributes of local old-growth forests. These resolution problems apparently led the Santa Fe project to avoid consideration of the numbers of large trees and snags per acre, which are the most important structural attributes used in the Region 3 old-growth definitions and guidelines (Southwestern Region Old-Growth Core Team 1992, Popp et al. 1992). The Santa Fe model had to rely heavily on measures of canopy closure, which are not well correlated with old-growth forests in the southwest because of the historic importance of high-frequency surface fires in molding open-stand structures.

The complex and steep canyon and mountain terrain of the project area further complicated vegetation classification efforts, as did the heterogeneous structural composition of local forests. Smoothing algorithms that were applied to create polygons out of finely jumbled raster classifications likely further degraded the accuracy of the model of probable old-growth. It is clear that this model had difficulty in accurately locating old-growth stands, although the map of probable old-growth has yet to be distributed to, or reviewed by, the project's advisory group or local ecologists.

Specific goals and expected products were not clearly articulated in the early stages of the project, although the project's proposal indicated a core purpose to "spatially analyze old-growth stands within the project area" and included more than 30 "questions the project is designed to answer" (Gonzales 1991). This proposal raised expectations among some U.S. Forest Service staff and members of the outside advisory group, which led to disappointment in the outcome as few of the questions were addressed and the products of this project have not contributed significantly to the resolution of old-growth management issues on the Santa Fe National Forest. GIS and remote sensing approaches are certainly important for ecological landscape management, but managers must become aware of limitations and work closely with the analysts if they are to receive a productive return on their investments in GIS and remote sensing and see these tools achieve their potential to develop needed information products.

Conclusion: Potential Uses of GIS and Remote Sensing Approaches for Ecosystem Management

Franklin (Chapter 2) reviews the utility of a variety of GIS and remote sensing tools for ecosystem management. Several potential uses merit mention here.

Myriad possible applications of remote sensing for ecological inventory and monitoring have barely been explored, ranging from image processing of repeat photography for assessing the ground cover of small plots to satellite monitoring of global-scale phenomena such as tropical deforestation (see Skole and Tucker 1993). Remote sensing can improve the repeatability of ecological monitoring and avoid certain research impacts, such as the trampling of microphytic soil crusts in arid landscapes.

As the variety and volume of ecological data required to implement land management increases, the combined use of GIS with other computer applications will prove increasingly valuable to land managers by allowing the collation, storage, integrated analysis, and dissemination of spatial data as well as associated data, text, and graphics files. For example, one can already link the locational data for permanent vegetation plots through a GIS to computer files containing scanned images of plot photographs, raw field data, graphs of the analyzed data, and journal references relevant to the plot. One of the most important uses of GIS might emerge from its application as a focal point for tracking all land management actions and associated inventory and monitoring data, so that adaptive management can become a reality (managers cannot learn from past actions if the data from these management "experiments" is continually being lost). Also, GIS will help managers and the public keep track of the large number of promised actions contained in current land management plans (e.g., the maintenance of designated "future old-growth forest" areas in the Dome Diversity Unit).

Land managers will continue to expand their use of GIS and remote sensing approaches for assessing cumulative impacts of potential land management scenarios, as well as for envisioning desired future conditions at landscape scales. GIS and remote sensing will also be increasingly used by a diverse array of interest groups in efforts to mold public perspectives on land use issues and thereby put political pressure on land management agencies and other social institutions (e.g., the recent Project Lighthawk and Wilderness Society analyses of remaining old-growth forest in the Pacific Northwest). Indeed, many anachronistic land-use situations exist where a graphic presentation of the present ecological status, current trends, and possible future conditions under different management scenarios could serve to productively focus public attention on deadlocked issues and thereby catalyze action toward solutions.

Overall, applications of GIS and remote sensing to ecosystem management are promising means for implementing a modern land ethic. Still, these more rigorous, technologically supported approaches to land management will supplement but cannot substitute for human intuition, wisdom, and humility in decision-making, based on familiarity, respect, and appreciation for a place (Lopez 1992). Long-term intimacy with a particular landscape is essential to learn enough about the diverse verities and uncertainties of a place to develop the multiscale spatial and temporal perspectives needed to implement ecosystem management well (Jordan III 1994)—ecosystem "health is a judgment that can be made only by someone who has been intensely familiar for a long time with what is being judged" (Ehrenfeld 1993:145). Resource management agencies must provide career options that allow land managers and affiliated researchers to become knowledgeable about and attached to the landscapes in which they work—to develop a sense of place. Ultimately, sustainable land management requires more integrated, long-term relationships between people and their landscapes; this imperative may well call each of us to explore the satisfactions and obligations of "the heritage of place and the responsibility of time" (Ruch 1992).

A transformation to a more sustainable society requires changes in the ways we see and think about our relationships with landscapes. GIS and remote sensing approaches to land management have great potential to promote increased awareness and acceptance of the duty of careful stewardship implicit in the powerful, beautiful, and frightening images they produce of a living earth being rapidly changed by humans. Thus over the next decade the greatest contribution of GIS and remote sensing tools to ecosystem management may come indirectly through fostering a more universal land ethic in society at large.

Acknowledgments

Review of the manuscript by Terrell Johnson, Will Moir, James Karr, and two anonymous reviewers improved the clarity of this work. Kay Beeley of

Bandelier National Monument and Lorayne Gonzales and Milford Fletcher of the NPS Southwest GIS Support Center provided assistance with the figures. This research has been supported by the National Park Service, U.S. Forest Service, and the Engineering-2 Division of Los Alamos National Laboratory.

The perspectives expressed here are those of the author and do not represent the official views of the National Biological Survey or the National Park Service.

References

Agee, J.K. 1993. Fire Ecology of Pacific Northwest Forests. Island Press, Washington, D.C.

Agee, J.K., and D.R. Johnson (eds.). 1988. Ecosystem management for parks and wilderness. Univ. of Washington Press, Seattle, Washington.

Allen, C.D. 1984. Montane grasslands in the landscape of the Jemez Mountains, New Mexico. M.S. thesis, Univ. of Wisconsin, Madison.

Allen, C.D. 1989. Changes in the landscape of the Jemez Mountains, New Mexico. Ph.D. dissertation, Univ. of California, Berkeley.

Allen, C.D. 1990. Fire history across a landscape gradient in the Frijoles Canyon watershed, New Mexico. Bull. Ecol. Soc. Amer., Supplement to 71(2):74.

Allen, C.D., and R. Touchan. 1994. Spatial analysis of prehistoric and historic fire regimes in the Jemez Mountains, New Mexico. Poster presented at the 9th Annual U.S. Landscape Ecology Symposium, Tucson, Arizona.

Anderson, K., and G.P. Nabhan. 1991. Gardeners in Eden. Wilderness, fall 1991.

Bahre, C.J. 1991. A legacy of change: Historic human impact on vegetation of the Arizona borderlands. Univ. of Arizona Press, Tucson.

Betancourt, J.L., and T.R. Van Devender. 1981. Holocene vegetation in Chaco Canyon, New Mexico. Science 214:656–658.

Betancourt, J.L., T.R. Van Devender, and P.S. Martin (eds.). 1990. Packrat Middens: The Last 40,000 Years of Biotic Change. Univ. of Arizona Press, Tucson.

Botkin, D.B. 1990. Discordant harmonies: A new ecology for the twenty-first century. Oxford Univ. Press, New York.

Brooks, D.J., and G.E. Grant. 1992. New perspectives in forest management: Background, science issues, and research agenda. USDA Forest Service Research Paper PNW-RP-456.

Christensen, N. 1994. Landscape Processes and Ecosystem Management. Plenary address presented at the 9th Annual U.S. Landscape Ecology Symposium, Tucson, Arizona.

Costanza, R., and H.E. Daly. 1992. Natural capital and sustainable development. Conservation Biology 6(1):37–46.

Covington, W.W., and M.M. Moore. 1992. Post-settlement changes in natural fire regimes: Implications for restoration of old-growth ponderosa pine forests. pp. 81–99 In: M.R. Kaufmann, W.H. Moir, and R.L. Bassett (technical coordinators). Old-growth forests in the Southwest and Rocky Mountain regions: Proceedings of a workshop. USDA Forest Service General Technical Report RM-213.

Cronon, W. 1983. Changes in the land: Indians, colonists, and the ecology of New England. Hill and Wang, New York.

Cronon, W. 1991. Nature's metropolis: Chicago and the great West. W.W. Norton & Co., New York.

Crosby, A.W., Jr. 1986. Ecological imperialism: The biological expansion of Europe, 900–1900. Cambridge Univ. Press, New York.

Crow, T.R. 1994. Meeting review: Ecosystem management. Bull. Ecol. Soc. Amer. 75(1):33–35.

deBuys, William. 1985. Enchantment and exploitation: The life and hard times of a New Mexico mountain range. Univ. of New Mexico Press, Albuquerque.

Delcourt, H.R., and P.A. Delcourt. 1988. Quaternary landscape ecology: Relevant scales in time and space. Landscape Ecol. 2(1):23–44.

Denevan, W.M. 1992. The pristine myth: The landscape of the Americas in 1492. Annals Amer. Assoc. of Geographers 82(3):369–385.

Duffy, D.C., and A.J. Meier. 1992. Do Appalachian herbaceous understories ever recover from clearcutting? Conserv. Biol. 6(2):196–201.

Ehrenfeld, D. 1993. Beginning Again: People and Nature in the New Millennium. Oxford Univ. Press, New York.

Endangered Species Update—Special Issue. 1993. Exploring an ecosystem approach to endangered species conservation. Endangered Species Update, Vol. 10, Nos. 3 & 4. School of Natural Resources and Environment, Univ. of Michigan, Ann Arbor.

Forman, R.T.T., and M. Godron. 1986. Landscape ecology. John Wiley and Sons, New York.

Foxx, T.S., and L.D. Potter. 1984. Fire ecology at Bandelier National Monument. pp. 11–37. In: T.S. Foxx (compiler). La Mesa Fire Symposium. LA-9236-NERP. Los Alamos National Laboratory, New Mexico.

Franklin, J.F. 1989. The "new forestry." J. Soil and Water Conservation, Nov.–Dec. 1989:549.

Franklin, J.F., and R.T.T. Forman. 1987. Creating landscape patterns by forest cutting: ecological consequences and principles. Landscape Ecol. 1(1):5–18.

Gonzales, J. 1991. GIS pilot project proposal: Old-growth analysis and modeling, Santa Fe National Forest. Handout distributed to the Old-Growth Advisory Group and staff of the Santa Fe National Forest.

Grumbine, R.E. 1994. What is ecosystem management? Conserv. Biol. 8(1):27–38.

Hamburg, S.P., and R.L. Sanford. 1986. Disturbance, *homo sapiens*, and ecology. Bull. Ecol. Soc. Amer. 67(2):169–171.

Harris, L.D. 1984. The fragmented forest: Island biogeography theory and the preservation of biotic diversity. Univ. of Chicago Press, Chicago, Illinois.

Hunter, M.L., Jr., G.L. Jacobson, Jr., and T. Webb III. 1988. Paleoecology and the coarse-filter approach to maintaining biological diversity. Conserv. Biol. 2(4):375–385.

Jackson, W. 1991. Nature as the measure for a sustainable agriculture. In: F.H. Borman and S.R. Kellert (eds.). Ecology, Economics, and Ethics. Yale Univ. Press, New Haven.

Jordan III, W.R. 1994. Environmental management. Science 263:305–306.

Kay, J.J. 1993. On the nature of ecological integrity: Some closing comments, pp. 201–212. In: S. Woodley, J. Kay, and G. Francis (eds.). Ecological Integrity and the Management of Ecosystems. St. Lucie Press.

Kellert, S.R., and F.H. Bormann. 1991. Closing the circle: Weaving strands among ecology, economics, and ethics. In: F.H. Borman and S.R. Kellert (eds.). Ecology, Economics, and Ethics. Yale Univ. Press, New Haven.

Kohler, T.A. 1992a. The prehistory of sustainability. Population and Environment 13(4):237–242.

Kohler, T.A. 1992b. Prehistoric human impact on the environment in the upland North American Southwest. Population and Environment 13(4):255–268.

Leopold, A.S. 1949. A Sand County almanac, with essays on conservation from Round River. Ballantine Books edition, ninth printing of 1976, New York.

Lopez, B. 1992. The rediscovery of North America. Orion 11(3):10–16.

Ludwig, D., R. Hilborn, and C. Walters. 1993. Uncertainty, resource exploitation, and conservation: Lessons from history. Science 260:17, 36.

Maser, C. 1988. The redesigned forest. R. and E. Miles, San Pedro, California.

Norse, E.A., K.L. Rosenbaum, D.S. Wilcove, B.A. Wilcox, W.H. Romme, D.W. Johnston, and M.L. Stout. 1986. Conserving biological diversity in our national forests. The Wilderness Society, Washington, D.C.

Noss, R.F. 1983. A regional landscape approach to maintain diversity. Bioscience 33(11):700–706.

Noss, R.F. 1990. Can we maintain biological and ecological integrity? Conserv. Biol. 4(3):241–243.

Oreskes, N., K. Schrader-Frechette, and K. Belitz. 1994. Verification, validation, and confirmation of numerical models in the earth sciences. Science 263:641–646.

Perry, D.A., M.P. Amaranthus, J.G. Borchers, and S.I. Borchers. 1990. Species migration and ecosystem stability during climate change: The below-ground connection. Conserv. Biol. 4(3):266–274.

Peters, R.L., II. 1988. The effect of global climate change on natural communities. pp. 450–464 In: E.O. Wilson (ed.). Biodiversity. National Academy Press, Washington, D.C.

Pickett, S.T.A., and P.S. White (eds.). 1985. The ecology of natural disturbance and patch dynamics. Academic Press, New York.

Popp, J.B., P.D. Jackson, and R.L. Bassett. 1992. Old-growth concepts from habitat type data in the Southwest. pp. 100–105 In: M.R. Kaufmann, W.H. Moir, and R.L. Bassett (technical coords.). Old-growth forests in the Southwest and Rocky Mountain regions: Proceedings of a workshop. USDA Forest Service General Technical Report RM-213.

Pyne, S.J. 1982. Fire in America: A cultural history of wildland and rural fire. Princeton Univ. Press, Princeton, New Jersey.

Risser, P.G. 1985. Toward a holistic management perspective. Bioscience 35(7):414–418.

Robertson, 1992. Ecosystem management of the national forests and grasslands. Memo to Regional Foresters and Station Directors. USDA Forest Service, Washington, D.C.: June 4, 1992.

Ruch, J. 1992. The heritage of place and the responsibility of time. Colorado Plateau Advocate, fall 1992.

Russell, E.W.B. 1994. Meeting review: Land use history. Bull. Ecol. Soc. Amer. 75(1):35–36.

Rykiel, E.J., Jr., R.N. Coulson, P.J.H. Sharpe, T.F.H. Allen, and R.O. Flamm. 1988. Disturbance propagation by bark beetles as an episodic landscape phenomenon. Landscape Ecol. 1(3):129–140.

Salwasser, H. 1991. New perspectives for sustaining diversity in U.S. national forest ecosystems. Conserv. Biol. 5(4):567–569.

Salwasser, H. 1992. From new perspectives to ecosystem management: Response to Frissell et al. and Lawrence and Murphy. Conserv. Biol. 6(3):469–472.

Schwartz, M.W. 1994. Natural distribution and abundance of forest species and communities in northern Florida. Ecol. 75(3):687–705

Scott, T.A., J.T. Rotenberry, and M.L. Morrison. 1994. Balancing global and local conservation strategies after the GIS revolution. Poster presented at the 9th Annual U.S. Landscape Ecology Symposium, Tucson, Arizona.

Skole, D., and C. Tucker. 1993. Tropical deforestation and habitat fragmentation in the Amazon: Satellite data from 1978 to 1988. Science 260:1905–1910.

Sousa, W.P. 1984. The role of disturbance in natural communities. Ann. Rev. Ecol. and Syst. 15:353–392.

Southwestern Region Old-Growth Core Team. 1992. Recommended old-growth definitions and descriptions and old-growth allocation procedure. USDA Forest Service, Southwest Region, Albuquerque, New Mexico.

Sprugel, D.G. 1991. Disturbance, equilibrium, and environmental variability: What is "natural" vegetation in a changing environment? Biol. Conserv. 58:1–18.

Swetnam, T.W. 1990. Fire history and climate in the southwestern United States. pp. 6–17. In: Effects of fire management of southwestern natural resources. USDA Forest Service, General Technical Report RM-191.

Thomas, W.L., Jr. (ed.). 1956. Man's role in changing the face of the Earth. Univ. of Chicago Press, Chicago, Illinois.

Touchan, R., and T.W. Swetnam. 1992. Fire history of the Jemez Mountains: Fire scar chronologies from five locations. Final report submitted to the Santa Fe National Forest and Bandelier National Monument. Laboratory of Tree-Ring Research, Univ. of Arizona, Tucson.

Tuan, Yi-Fu. 1974. Topophilia: A study of environmental perception, attitudes, and values. Prentice-Hall Inc., Englewood Cliffs, New Jersey.

Turner, M.G. (ed.). 1987. Landscape heterogeneity and disturbance. Springer-Verlag, New York.

Turner, M.G., and R.H. Gardner (eds.). 1991. Quantitative methods in landscape ecology. Springer-Verlag, New York.

Turner, B.L., II, B.L., W.C. Clark, R.W. Kates, J.F. Richards, J.T. Mathews, and W.B. Meyers (eds.). 1990. The Earth as Transformed by Human Action. Cambridge Univ. Press, New York.

Urban, D.L., R.V. O'Neill, and H.H. Shugart. 1987. Landscape ecology. BioScience 37(2):119–127.

USDA Forest Service. 1987. Santa Fe National Forest Plan. USDA Forest Service, Southwest Region, Albuquerque, New Mexico.

USDA Forest Service. 1991. Environmental Assessment, Dome Diversity Unit Management Project. Santa Fe National Forest, Jemez Ranger District, Jemez Springs, New Mexico.

USDI Fish and Wildlife Service. 1994. An Ecosystem Approach to Fish and Wildlife Conservation. USDI Fish and Wildlife Service, Washington, D.C.

USDI National Park Service. 1992. Western Region fire monitoring handbook. USDI National Park Service, San Francisco, California.

Westman, W.E. 1990. Park management of exotic plant species: Problems and issues. Conserv. Biol. 4(3):251–260.

White, C.S. 1990. The role of monoterpenes in soil nitrogen cycling processes in ponderosa pine. Biogeochemistry 12:43–68.

Wilcove, D.S., C.H. McLellan, and A.P. Dobson. 1986. Habitat fragmentation in the temperate zone. pp. 237–56. In: M.E. Soulé (ed.). Conservation Biology. Sinaurer Associates Inc., Sunderland, Massachusetts.

Wooton, E.O. 1908. The range problem in New Mexico. Agriculture Experiment Station Bulletin No. 66, New Mexico College of Agriculture and Mechanic Arts. Albuquerque Morning Journal, Albuquerque, New Mexico.

9

Resource Management Perspective: Practical Considerations for Using GIS and Remote Sensing at the Field Level

Douglas P. Schleusner

Ecosystem *management*, by definition, involves human interaction with natural systems. It is not a "hands off" philosophy, nor is it merely a detached examination of the world to satisfy one's intellectual curiosity. Lessons learned are intended for application through a variety of management practices. The principles of ecosystem management do not change the statutory multiple-use mission of the USDA Forest Service, but they do change *how* that mission is implemented.

Discussion of ecosystem management has spread rapidly throughout the Forest Service during the past several years. Sprouting from the seeds of New Forestry and New Perspectives, ecosystem management has been the subject of brochures, papers, and workshops. Pilot projects and demonstration areas have been implemented on districts and forests around the country—many of them quite successfully. Even so, a large gap still remains between theory and practice. If ecosystem management is ever to be a success, GS-7 forestry technicians, GS-9 district biologists, and GS-5 seasonal employees will be the reason.

One of the biggest frustrations voiced at the field level is the lack of common understanding of what it means to implement ecosystem management "on the ground." As one Southwestern Region (Arizona and New Mexico) employee recently stated, "I can't tell you what ecosystem management is, but you'll know it when you see it." Others view it as "just another initiative" that causes more work, but will soon fade away (Senn and Crawford 1993).

This chapter describes how advanced technologies, such as geographic information systems (GIS) and remote sensing, can be used to help incorporate concepts of ecosystem management into the field-level organizations of the USDA Forest Service. With the experience gained on more than 190 million acres of public land, actions taken by this agency may have relevance for other groups and organizations. This chapter traces the path from national principles to regional strategies to field level applications on the Santa Fe National Forest.

The concepts of increased public involvement, greater interaction with the research community, use of advanced technology, and integration of inventories and databases are widely known and generally accepted. It is the idea of "taking an ecological approach" that seems to be least understood and most widely debated. Forest and district employees are searching for examples and practical guidelines to use in project planning and implementation. While creativity and innovation should be encouraged at this early stage—and rigid prescriptions avoided—some minimum level of consistency would be valuable.

It has been said that the clearest picture of any organization's core values can be seen in their budget sheets. Starting with the four areas of emphasis in the 1990 RPA (Forest and Rangeland Renewable Resources Planning Act) Program, there have been significant program shifts in the past several years (USDA 1990). It has only been in the past several months, however, that both Congress and the Forest Service have formally recognized ecosystem management in the budget allocation process. It is this recognition that promises to add an element of consistency to field level applications.

In the final allocation for fiscal year 1993, two new line items dealing with ecosystem management were added to the budget structure—one dealing with operations and maintenance items, and the other with construction of structural improvements. While these line items may not be fully integrated into the budget until 1994, their intent is already being implemented. The Southwestern Region has supplemented national program direction by establishing specific activity codes that identify actions which implement some aspect of ecosystem management.

For purposes of this discussion, two groups of factors influence ecosystem management at the field level—natural and organizational. The first category includes the type of landscapes commonly associated with National Forest lands (i.e., intensity of management) and the scale of those landscapes. The main emphasis of this chapter is on the second category—organizational factors—which includes policies at various levels, the planning structure established by law and regulation, and budget direction. The remainder of this chapter describes the influence of these factors on eventual field-level applications, provides examples of projects that are currently in progress, identifies opportunities for use of GIS and remote sensing, and describes a number of practical guidelines that could minimize potential barriers to acceptance.

National Forest Landscapes

Before examining the progression of ecosystem management through the Forest Service organization, it is useful to briefly review the kinds of lands that this agency administers, and the various scales at which landscape ecology is being applied. Both factors ultimately influence the type of data needed for resource management and the potential roles of GIS and remote sensing.

It has been proposed that three main categories of ecosystems exist on National Forest lands: (1) the artificial, devoted to high-input intensive resource "farming"; (2) the natural, which is preserved in wilderness areas; and (3) the seminatural, which can be managed for multiple resources in perpetuity (Rowe 1992). Until very recently, these categories could be accurately portrayed as a bell-shaped curve, with the artificial and natural systems occupying the extremes of the curve and most lands being "seminatural" or "multiple use."

As an agency average, the natural type of ecosystem (i.e., wilderness) comprises 17.6 percent of the land base (USDA 1992a). The artificial (i.e., intensive timber management) end of the spectrum has occupied about the same percentage. On the Santa Fe National Forest these two categories account for 18.7 percent and 17.2 percent, respectively, although clearcutting has never been extensively applied. The remaining 64 percent of the Forest is considered seminatural (USDA 1987).

Given the recent direction issued by the Chief of the Forest Service to sharply reduce clearcutting as a routine form of stand treatment (Robertson 1992), this single emphasis end of the bell curve will shift over time. With the current move to uneven age management, the number of acres devoted to "farming" will decrease as they are treated with a new set of silvicultural prescriptions, and the "artificial" label will lose its relevance in this classification scheme.

Acreage managed in seminatural conditions, either due to actual land allocations or the application of less intensive practices on a forestwide basis, will be the focus of most ecosystem management efforts. It is here where the greatest diversity of human uses meets the largest land area that the "ecological approach to multiple-use management" will be most evident. This is not to deny the importance of wilderness ecosystems, which possess the largest contiguous blocks of undisturbed land in the National Forest system. However, they also have fewer public demands and lower restoration priorities.

Ecosystem Scales

As evidenced by the body of literature that has accumulated in just the past several years, "ecosystems" are being examined at many scales. These range from the neighborhood stream studied by a sixth grade science class to the entire planet in global warming models. While both of the above examples are certainly interesting, neither is particularly relevant to most situations encountered on a typical ranger district or national forest. To help determine the most likely opportunities for using GIS and remote sensing at the field level, the following scheme provides some context for various levels of analysis (Table 9.1).

Prior to the current emphasis on ecosystem management, project analysis has most often focused on relatively homogeneous land units referred to as sites. One of the most familiar examples is the timber stand, usually encompassing several tens to several hundreds of acres. When a series of these sites are linked together in a patchwork, they create a landscape mosaic. In northern New Mexico, the combination of canyons and mesa tops is a classic example. When a variety of landscape mosaics are viewed in a broader context, a region can then be identified (Bailey 1988).

This hierarchy has value to planners and resource managers for a variety of reasons, including the design of GIS support systems for ecosystem management. Two of the most important factors involve predictability and diversity. Regions, by definition, are large areas within which local ecosystems recur in a more or less predictable fashion. Assuming that these landscape mosaics have been identified and mapped, it is certainly possible to observe and record their response to a variety of resource management activities. With sufficient monitoring of these "managed" areas, it is also possible to predict the behavior of a previously unmanaged area. It is also possible, based on experience, to develop and apply management direction over a broader area, rather than start over again at each and every site.

With adequate knowledge of the landscape mosaics present within a region, it is possible to measure, monitor, and mitigate relative diversity and potential impacts resulting from management activities. It is here that GIS and remote sensing can play a vital role.

Table 9.1 Relative scales of national forest activities

Ecoscale	Primary influence	NFS level	Map scale	Areal extent
Macro (region)	Climate	National/Regional	1:250,000 and smaller	Thousands of square miles
Meso (landscape)	Landform	Forest	1:100,000–1:24,000	10,000 ± to 500 acres
Micro (site)	Microclimate and site conditions	District-project	1:24,000 and larger	10–500 acres

Source: Modified from Bailey (1988) and USDA (1993)

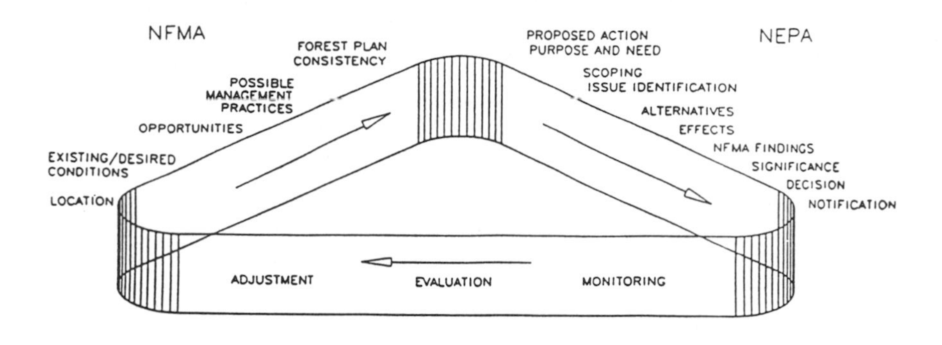

Figure 9.1 Forest Plan Implementation Process

Planning Structure

Application of ecosystem management practices does not occur apart from other existing requirements, such as the National Forest Management Act (NFMA) and National Environmental Policy Act (NEPA). These two pieces of legislation and their implementing regulations have directly influenced the Forest Service planning structure now in effect. Forest plan implementation courses conducted within the agency have emphasized a threefold approach: (1) identification of existing and desired conditions, opportunities, and possible management practices consistent with the forest plan; (2) identification of specific project proposals with purpose and need, alternatives, and consequences as required by NEPA; and (3) monitoring, evaluation, and adjustment (USDA 1991a). Somewhat coincidentally, this model (Figure 9.1) lends itself very well to many of the key concepts of ecosystem management.

Each of these phases has somewhat different data and analysis needs. The "NFMA side," as it is known, looks at the big picture. Questions of diversity and pattern can be examined in the context of current conditions—and what is desired for the future. Issues and opportunities then suggest the types of management practices that can move a given ecosystem from the present situation to its desired future state.

The "NEPA side" examines the details. With the use of more intensive data, alternatives are considered for each project (or set of management practices) that when taken together move the ecosystem in the desired direction.

In the Southwestern Region there is an additional bridge between the forest plan and project implementation known as Integrated Resource Management (IRM). Incorporating elements of plan consistency, public involvement, budget linkage, data collection, alternative development, documentation, and monitoring, this process is also the subject of a GIS guidebook that identifies a number of GIS products for each of the 13 stages of implementation (USDA 1991b). As a general rule, the amount and detail of information required increases with each step of the IRM process.

Budget Direction

As previously mentioned, the final budget for fiscal year 1993 not only includes direction for ecosystem management, but sets the stage for modification of the budget structure to increase accountability of the allocated funds. This in itself will have a positive influence on the role of GIS in ecosystem management.

Two broad categories of funds are identified for ecosystem management. One is for operation and maintenance activities. The second category is for various types of facility construction (e.g., water developments, fencing, road relocation or reconstruction). Within the Southwestern Region these budget line items are earmarked for use as follows: (1) training and interpretation; (2) demonstration planning areas; and (3) current year project implementation (Henson 1993).

Budget direction further specifies that these funds should be used to emphasize landscapes with piñon-juniper ecosystems, riparian areas, and/or areas associated with improving forest health. Within these emphasis areas and budget line items there are seven activity codes that track expenditures for such activities as research collaboration, external partnerships, establishment of regional database and inventory standards, inventories and database development for landscape analysis, definition of desired future conditions, scheduling and implementation of treatments, and monitoring.

A major portion of this year's funding is directed at demonstration planning areas, where the primary emphasis is on automation of an integrated database that allows spatial display and analysis. To accomplish this, the Southwestern Region has identified nine GIS map layers (as a minimum) that should be acquired for ecosystem management planning.

These include the seven Primary Base Series: hydrology, transportation, boundaries, land net, land status, cultural features, and topography. In addition, an Ecological Unit layer derived from the Terrestrial Ecosystem Survey, and a current vegetation layer are also required. Completion of these planning efforts will set the stage for future project implementation.

National Forest Policies

In 1990 the Forest Service initiated the program of "New Perspectives," which emphasized increased public involvement, conservation partnerships, and greater collaboration among land managers and research scientists. This was directly linked to the four themes of the 1990 RPA Recommended Program (USDA 1990). They included: (1) enhancing recreation, wildlife, and fisheries resources; (2) ensuring that commodity production is environmentally acceptable; (3) improving scientific knowledge about natural resources; and (4) responding to global resource issues.

Forest Service Chief Dale Robertson announced the formal transition to Ecosystem Management in June 1992 (USDA 1992b). At that time he required each regional forester to submit a strategy for implementation within 90 days. As a result, in the Southwestern Region four primary documents have been used as guidance by the field: (1) Chief Robertson's Working Guidelines for Ecosystem Management (Robertson 1992); (2) Deputy Chief Overbay's speech on Ecosystem Management (Overbay 1992); (3) the statement of Ecology-Based Multiple-Use Management (Henson and Montrey 1992a); and (4) the Ecology-Based Multiple-Use Management Strategy (Henson and Montrey 1992b). These last two documents were jointly prepared by employees of the Southwestern Region and the Rocky Mountain Forest and Range Experiment Station, and represent the commitment of the Regional Forester and Station Director to increased cooperation between managers and researchers.

To more clearly define the nature of ecosystem management at the field level (i.e., Forest and Ranger District), and the potential roles of GIS and remote sensing technologies, Table 9.2 outlines a logical progression from national principles to regional strategies to forest applications.

Demonstration Areas

To illustrate some ecosystem management ideas currently being considered on the Santa Fe National Forest, the following section briefly

TABLE 9.2 Ecosystem management policy by organizational level

National principles[1]	Regional strategies[2]	Forest applications
(1) Sustain diversity, health, and productivity of ecosystems.	Mimic intensity, frequency, and areal extent of naturally occurring disturbances in the Southwest.	Reintroduce fire to recreate vegetative conditions that existed prior to settlement.
	Emphasie ecosystem management principles for the following Southwestern Region priorities: piñon-juniper woodland, riparian, and forest health.	Focus on improved grazing practices that protect riparian zones and reduce erosion.
(2) Recognize that future conditions are not perfectly predictable and that changes in ecosystems cannot be absolutely controlled.	Focus on what should be accomplished rather than spelling out precisely how.	Use GIS capabilities to keep data on continuously variable features in raw form until needed.
	Allow applications of ecosystem management to grow and evolve over time as more information and experience is acquired.	Share the results of various practices with other units, incorporate what works, modify what does not.
	Encourage flexibility, creativity, and innovation.	Monitor effects of forestwide activities on diversity, and share with districts in early planning.

(3) Integrate ecological, economic, and social considerations in description of Desired Future Conditions.	Move from single-species emphasis to ecosystem approach in all aspects of management.	Identify and develop data that reflect minimum criteria for species' habitat needs, and that can be used to model "what if" questions through GIS.
	Incorporate ecosystem management principles into Forest Plan amendments and revisions, e.g. Mexican spotted owl and goshawk direction.	Emphasize use of Forest Plan Implementation process in identifying existing conditions, desired future conditions, and the means to accomplish the change.
	Recognize the unique and diverse cultures of the Southwest, and how traditional uses interact with Desired Future Conditions.	Incorporate traditional use patterns in the design of fuelwood harvest for overstocked piñon-juniper stands.
(4) Coordinate management strategies for terrestrial, riparian, and aquatic systems at various scales across multiple ownerships.	Recognize regional, national, and international ecological systems, e.g. Colorado and Rio Grande Rivers, or Chihuahuah Desert.	Incorporate data relating to adjacent ownerships (e.g. Park Service, BLM, other forests, and Pueblos) when developing layers for GIS landscape analysis.
	Establish a network of demonstration areas on each forest around the Southwestern Region.	Identify at least one demonstration area on the forest that emphasizes one or more of the following: riparian, piñon-juniper woodland, and forest health.

continued

Table 9.2 *continued*

National principles[1]	Regional strategies[2]	Forest applications
(5) Integrate ecological inventories, classifications, data management, and analytical tools for land and resource management.	Emphasize completion of Terrestrial Ecosystem Surveys on all forests.	Use completed TES surveys and other data to identify basic ecological units for the entire forest as the basis for future planning.
	Work with other organizations to identify and conserve critical and/or unique ecosystems.	Develop memorandum of understanding with the Forest Conservation Council to outline procedures for sharing information related to old growth.
	Develop/implement a uniform system for existing vegetative cover.	Review/analyze stand database, aerial photo inventory, Landsat imagery, TES survey used in old-growth study to determine reasons for disagreement and potential best applications.
(6) Integrate monitoring and research activities with land and resource management.	Share funding in emphasis areas (e.g., ecological classifications) between the Southwestern Region and Rocky Mountain Station.	Coordinate GIS and remote-sensing efforts with TERRA, Rocky Mountain Station, and the Southwest Forestry Consortium (NAU).
	Continue to work with universities in development of Decision Support System (DSS).	Identify ways of using GIS and/or remote sensing to develop multiresource monitoring methods, i.e. "proxies."
		Use satellite imagery to view and measure vegetative diversity and changes in that diversity over time.

[1]From Robertson (1992) and Overbay (1992)
[2]From Henson and Montrey (1992a, 1992b)

describes several demonstration area proposals submitted for fiscal year 1993. While the Forest has acquired the seven Primary Base Series layers previously mentioned, very little of the data has been field verified for any of the proposals. The Terrestrial Ecosystem Survey has also been completed for all areas of the forest, but the basic ecological units have yet to be determined for the demonstration areas. These descriptions provide a more detailed idea of the type of data needed and possible GIS and/or remote sensing applications.

Mesa del Medio

Located on the Coyote Ranger District, management emphasis within this 17,000-acre area is primarily on restoration of riparian habitat that has suffered due to excessive and/or uncontrolled grazing and poor road locations. Other related resource concerns include habitat condition of the Rio Grande Cutthroat Trout (sensitive species), competition for forage between livestock and the expanding elk herds in the area, and a dwarf mistletoe infestation. Potential management practices that could be applied include prescribed fire, salvage harvest of infected timber stands, fencing, development of alternate water sources away from the riparian zones, shorter livestock use periods, and road relocation and/or obliteration.

Data needs for this area will probably include fish habitat condition by stream segment, allotment boundaries and fence locations, locations of mistletoe infestations and risk values for other stands, primary elk use areas, transportation net and condition by segment. Some of this is already available in the nine standard layers, or can be interpreted from them. Other layers will have to be obtained from aerial photo interpretation and field investigations.

Since GIS is more than just a system for collecting and displaying data, a classic opportunity exists in this planning area for analysis of elk/livestock grazing conflicts. Location of elk summer and winter ranges and their migration paths is known, as well as the location of livestock pasture fencing and rotation dates, and existing range vegetation condition. This information will allow managers to identify areas of greatest overlap, types of forage at greatest risk, and opportunities to relocate use.

El Pueblo

This area is located almost entirely within the piñon-juniper woodlands of the Pecos/Las Vegas Ranger District. As with other areas on the forest, this 26,000-acre proposal also has a completed TES survey and the seven primary base series layers. Adjacent to a five-mile stretch of the Pecos

River, overgrazing and exclusion of fire from the piñon-juniper type has contributed to excessive soil erosion. Presence of numerous cultural resource sites and high local demand for fuelwood harvest are social issues that must be addressed in the overall approach to ecosystem management.

Management practices that could contribute to restoration and enhancement of the area include carefully planned fuelwood harvest to thin overstocked P-J stands, fencing, prescribed fire, erosion control structures, road obliteration, and streambank stabilization. In addition to the data layers common to most projects on the Santa Fe, this area will need information on cultural resource sites and traditional fuelwood areas.

Operational Constraints

While not ecosystem management projects in the fullest sense, other efforts during the past several years have yielded some insights into GIS and remote sensing applications at the field level. During 1991–1992 a pilot project on the Jemez Ranger District examined the utility of a supervised classification of remotely sensed data for identification and modeling of old-growth stands within a 230,000-acre study area.

The pilot project, described in more detail by Gonzales et al. (1992), illustrated that satellite imagery can play a valuable role in data collection and analysis. The accuracy of several component layers (cover type and crown closure) meets or exceeds that of any other available source for forestwide coverage at this time. The resultant probable-old-growth model not only provides an initial indication of where planners and managers should direct field investigations, it can also be adjusted to accommodate and test other combinations of structural attributes.

In spite of these advantages and opportunities, barriers to acceptance also developed during the project, some of which still exist. They can be lumped into three categories: philosophical, technical, and operational. In the first case, some key members of the forest management team disagreed with the assumptions underlying the old-growth structural attribute definitions. Others voiced serious doubts about the utility of satellite imagery from the very outset based on previous experience.

Technical concerns revolved around integration of the pilot project's results with the existing timber stand database. Key individuals were unwilling to spend additional time and money on something that would not assist in completing and updating an existing data structure. Operational barriers included the transition from a pilot project, with its specifically assigned employees, to day-to-day use and maintenance by employees who are already working 50-hour weeks. This last factor may be the most critical to any project's success in the long run.

As other projects have explored the use of GIS on the forest, it is apparent that there are other barriers that limit the effectiveness of advanced technology at the field level—whether used for ecosystem management or other more traditional projects. In addition to a basic resistance to change, lack of understanding and a resultant lack of ownership or commitment can also be a hindrance.

In a recent IRM Activity Review, several variations of this were found. Many district employees believe that ecosystem management is not necessary, since it is already viewed as taking place through Integrated Resource Management and the multiple-use mission of the Forest Service. Combined with this is a definite lack of common understanding about ecological principles, scale of analysis, and terminology. Other barriers were identified that would hinder an integration of GIS and ecosystem management from working smoothly, including remnants of a perceived overemphasis on commodity production targets, the budgeting process, functional organizational structure (e.g., range, wildlife, timber, recreation), lack of hardware, software, and inventories, and current emphasis on single-species management such as the Mexican spotted owl (Senn and Crawford 1993).

Summary and Recommendations

This chapter identifies several factors that influence the type of information needed for ecosystem management at the field level within the Forest Service, and potential roles of GIS and remote sensing in planning and implementation. They include: (1) types of landscapes typically found at the forest and district level; (2) the scale of ecosystems common to field-level applications; (3) ecosystem management policies at national and regional levels; (4) desired future conditions for various resources and geographic locations; (5) planning structures required by law or regulation; (6) budget direction for ecosystem management; and (7) operational constraints. Within the context of these factors, a number of items should guide the use of GIS and remote sensing in ecosystem management at the field level:

- Most of the lands that will be considered under ecosystem management are "seminatural" in character. While the wilderness end of the spectrum should not be ignored, an era of limited budgets suggests that the areas with the greatest need of restoration and broadest mix of competing uses should be emphasized.

- Management emphasis will be shifting from intensive, localized activities to more extensive, dispersed practices that lend themselves to analysis at a landscape mosaic scale.
- Use of GIS and remote sensing techniques will be increasingly valuable for predictive models of effects and for monitoring changes in diversity.
- Data should be maintained in "raw" form until needed for a specific purpose, and not interpreted too early in the analysis process.
- To the extent possible, comparable data layers for adjacent and intermingled non-National Forest ownerships should be acquired.
- For the near future, emphasize data collection and analysis tools that yield the greatest benefit for the Southwestern Region's emphasis areas of piñon-juniper woodlands, riparian areas, and forest health.
- Identify basic ecological units using the Terrestrial Ecosystem Survey for the entire Santa Fe National Forest at the earliest possible date.
- Develop agreements with local organizations such as the Forest Conservation Council to obtain new and/or improved data.
- Coordinate current efforts by the Rocky Mountain Experiment Station, TERRA, and the Southwest Forestry Consortium to eliminate redundancies and focus on realistic and achievable tools for end users.
- Encourage use of GIS and remote sensing to identify multiresource monitoring activities that can be adopted at the field level.
- Refine and validate the nine data layers identified in the Southwestern Region's GIS Information Structure for the entire forest: hydrology, transportation, boundaries, land net, land status, cultural features, topography, ecological units, and existing vegetation.

- When considering "pilot projects," explore ways in which the same results could be achieved with existing inventories, procedures, and staffing in order to minimize future transition difficulties.
- Actively work with the end users (district staff and resource specialists) *and their supervisors* to make sure that ownership and commitment exist for new methods and applications.

References

Bailey, R.G. 1988. Ecogeographic analysis: A guide to the ecological division of land for resource management. USDA Forest Service Misc. Publ. 1465. 16 pp.

Gonzales, J., M. Barry, H. Lachowski, P. Maus, J. Johnson, and V. Landrum. 1992. Vegetation classification and old-growth modeling in the Jemez Mountains. USDA Forest Service Remote Sensing Steering Committee. 32 pp.

Henson, L. 1993. Ecosystem management direction and FY93 allocations. Letter to forest supervisors and staff directors. January 21, file designation 6520.

Henson, L., and H.M. Montrey. 1992a. Ecology-based multiple-use management. USDA Forest Service, Southwestern Region and Rocky Mountain Forest and Range Experiment Station. June 15. 7 pp.

Henson, L., and H.M. Montrey. 1992b. Ecology-based multiple-use management strategy. USDA Forest Service, Southwestern Region and Rocky Mountain Forest and Range Experiment Station. September 1. 17 pp.

Overbay, J.C. 1992. Ecosystem management. Unpublished speech at national workshop Taking an Ecological Approach to Management. Salt Lake City, UT. April 27.

Robertson, F.D. 1992. Ecosystem management of the national forests and grasslands. Letter to regional foresters and station directors. June 4, file designation 1330-1.

Rowe, J. Stan. 1992. The ecosystem approach to forestland management. Forestry Chronicle. April, 68(2).

Senn, R., and J. Crawford. 1993. IRM activity review: Final report of findings. Southwestern Region, USDA Forest Service. 41 pp.

USDA Forest Service. 1993. National hierarchical framework of ecological units. Watershed and Air Management Staff, Washington, D.C. (in press).

USDA Forest Service. 1992a. Report of the Forest Service: Fiscal year 1991. pp. T-14, T-23.

USDA Forest Service. 1992b. Memorandum from USDA Forest Service Chief F. Dale Robertson to Regional Foresters and Station Directors. Reply to: 1330-1. June 4, 1992.

USDA Forest Service. 1991a. Forest plan implementation: Course 1900-1. Land Management Planning Staff, Washington, D.C.

USDA Forest Service. 1991b. A geographic information systems guidebook: For use in the integrated resource management process. Southwestern Region. 73 pp.

USDA Forest Service. 1990. The Forest Service program for forest and rangeland resources: A long-term strategic plan. Recommended 1990 RPA Program.

USDA Forest Service. 1987. Santa Fe National Forest land and resource management plan, Southwestern Region. p. 47.

10

GIS Applications Perspective: Using Remote Sensing and GIS for Modeling Old-Growth Forests

Jessica Gonzales

Introduction

The U.S. Forest Service has begun to integrate the use of digital remotely sensed data and geographical information system (GIS) data into its resource data collection and analysis methods for planning and management on the national forests. Since the difficulty and complexity of today's ecological management questions require Forest Service managers to use the best available analytical tools, the Forest Service is acquiring the GIS hardware and software to help address these difficult questions. Forest Service managers recognized that acquiring the technology, namely remotely sensed digital imagery and a GIS, and effectively using it were two different actions requiring simultaneous pursuit. Concurrent testing and acquisition of remotely sensed digital data and GIS have created opportunities for Forest Service personnel to explore the use of these technologies through pilot projects. Pilot projects provide a means for personnel to gain skills and knowledge in using upcoming technology through practical application of new equipment and data. New and more technical ways of doing business bring with them their own set of issues to resolve. Each initial use of technology by an organization is surrounded by its own set of circumstances and environment. However, some of what is learned and experienced in pilot projects can be very useful and relevant to others who are using similar technology and conducting similar projects.

Significant land management, organizational, and social issues; project management experiences; data analysis techniques; and results of a remote sensing and GIS pilot project will be discussed as a case study.

A demonstration project was conducted in the southwest to use digital satellite data and existing ecological information in a geographic information system (GIS) to model probable old-growth forest. In this cooperative project conducted by the Santa Fe National Forest and the Forest Service's Nationwide Forestry Applications Program (NFAP), digital Landsat Thematic Mapper (TM) data were used to produce a GIS database of vegetation characteristics for a portion of the Jemez Mountains in northern New Mexico. Using a combination of computer- and user-driven classification techniques, along with field and photo interpreted data, forest vegetation was classified by cover type, crown closure, and tree size. Map accuracies of 77 percent, 82 percent, and 76 percent, respectively, were attained. Probable old-growth was modeled, using crown closure and tree size data from the TM classification and existing forest type and site productivity data from an ecological mapping inventory. The distribution and abundance of probable old-growth were generated from the model.

This demonstration/pilot project serves as a case study for implementing large-area remote sensing/GIS pilot projects within the Forest Service. Technical findings acquired through the Landsat TM classifications and GIS modeling are discussed, as well as the land management, organizational, and social issues associated with the project; project management experiences and constraints; and accuracies of the derived vegetation layers.

Project Scope and Objectives

The pilot project was generated from land management planning issues and data needs that are typical of many national forests. The study area contains a wild and scenic river, proposed expansion of pumice mining, a proposed national recreation area, ongoing timber harvest, and key wildlife habitat. Old-growth forest components were known to be significant habitat requirements for several rare wildlife species occurring in the area, including the northern goshawk, Mexican spotted owl, and Jemez salamander. The majority of timber harvest activity had occurred and was continuing in old-growth areas. Consequently, concern for the management of old-growth was building, along with public opposition to timber harvest. The Santa Fe National Forest's land management plan was under appeal and lacking the spatial analysis of resources. Recreation, timber, mineral, and wildlife management plans needed preparation and required old-growth data. The Santa Fe National Forest did not have an adequate inventory of old-growth, a field-verified definition of old-growth, nor an operational GIS.

The Santa Fe National Forest's data needs were combined with regional and national objectives for the use of remote sensing to develop the project. The Southwestern Region of the Forest Service (Arizona and New Mexico) sought to conduct a pilot project that used digital satellite data and GIS to address an important regional and national issue. The combined use of supervised (user driven) and unsupervised (computer driven) image classification techniques had not been fully explored in the southwest and the Forest Service was interested in testing satellite imagery for large-area forest inventory. NFAP, a Forest Service Washington Office engineering unit, provided technical assistance in the project. The project was part of NFAP's mission to integrate the use of remote sensing into resource data collection, test the use of Landsat TM data as a vegetation data source for GIS, and transfer remote sensing and GIS skills to Forest Service personnel. The Forest's existing vegetation database lacked old-growth attributes and was fragmented. Also, the accuracy of some data elements was questionable. Therefore, the data were not entirely suitable for old-growth analysis on a landscape level. The Santa Fe National Forest personnel wanted any newly acquired vegetation data to be useful for old-growth inventory as well as other resource analyses, to fill existing data gaps in wilderness and privately owned land, and to be GIS compatible. Santa Fe National Forest data needs and Forest Service regional objectives led to the use of digital satellite imagery.

In 1990, a pilot project designed with forest, region, and national objectives was proposed and jointly funded by all three levels of the Forest Service. The project's four major objectives were to:

- Determine the spatial distribution and abundance of probable old-growth in the study area, using key criteria in the regional old-growth definition.
- Develop a practical vegetation classification scheme that could be used with Landsat TM data and for a variety of resource analyses.
- Test the use of Landsat TM to derive vegetation data in the southwest.
- Transfer remote sensing skills to Forest Service personnel.
- Exchange information and ideas about the project with an old-growth inventory citizen's advisory group.

Information gained from the pilot project was to be used to determine whether further vegetation inventories should be conducted using Landsat TM, to determine future use of digital satellite imagery and classification methodologies in the Region, and to develop a practical definition of old-growth for use in Forest land management planning.

Developing an Ecological Classification Scheme

Mapping old-growth using all the structural requirements in the southwestern regional definition would be difficult and expensive; therefore, a more practical inventory approach was taken. A few key characteristics of old-growth derived from remotely sensed data were used to model probable old-growth. Probable old-growth areas were those that most likely met southwestern old-growth structural requirements, but whose structural characteristics had not been field verified. The Region's field verification requirement for old-growth made stratification and prioritization of probable old-growth areas very important. Key vegetation characteristics could be used to explore more practical definitions of old-growth, conduct spatial analyses for forest land management planning, and develop more efficient ground surveys for old-growth verification.

A practical vegetation classification scheme that included old-growth attributes derivable from Landsat TM and aerial photographs was developed. The regional definition of old-growth contains specific parameters for live and dead trees, which are categorized by site potential and forest type. Dead tree characteristics and some live tree characteristics, such as basal area and tree age, could not be derived from satellite imagery. However, several key characteristics for live trees could be derived from Landsat TM. Site potential or index data was to be taken from another data source. A practical vegetation classification scheme was designed to accommodate key old-growth criteria, be usable with remotely sensed data, and remain flexible enough for general Forest use. The interpretive vegetation classification scheme included the following:

- The number of large trees per acre was converted to percent crown closure of large trees, which could be determined from field data and Landsat TM.

- Total percent canopy cover was categorized into four classes, which could be identified on aerial photographs and from Landsat TM. The use of Landsat TM was a determinant factor for setting the maximum crown closure of low-density vegetation at 20 percent.

- Five broad tree size categories were created to facilitate field data collection and photo interpretation of tree size.

- Dead wood, stand age, and average stand basal area data were collected at 75 percent of the field training sites; but these stand characteristics were not mapped. The data were used to ensure that old-growth training sites were well represented in the classifications.

The vegetation classification scheme used to classify the Landsat TM data is shown in Figure 10.1. Tree size categories contain both multiple- and single-storied stands and are labeled according to the size category having the highest percent crown closure. For example, areas in the small size class contain mostly small sized trees, but may contain trees of one or more sizes. Areas with less than 20 percent crown closure are considered low-density vegetation.

Location and Characteristics of the Study Area

The study area is located in the Jemez Mountains of northern New Mexico and contains vegetation types representative of those on the Santa Fe National Forest. The study area covers a six 7.5-minute quadrangle map

VEGETATION COVER CLASSES

Class	Description
SF	Spruce-Fir
MC	Mixed Conifer
PP	Ponderosa Pine
P-J	Pinyon-Juniper
AS	Aspen
OAK	Oak
G-F	Grass-Forbs
BA	Bare & Low Density Vegetation
CA	Cottonwood/Agriculture

CROWN CLOSURE CLASSES

Class	Description
1	0 - 20%
2	21 - 40%
3	41 - 70%
4	71 - 100%

TREE SIZE CLASSES

Class	Description
Seed/Sap/Pole	Seedling/Sapling + Pole ≥ 20% and Small + Medium + Large < 20%
Small	Seedling/Sapling + Pole < 20% and Small + Medium + Large ≥
Medium	Medium + Large ≥ 20%
Large	Large ≥ 20%
Nonforest	Bare (< 20% Crown Cover), Grass/Forb, Shrub & Oak

Figure 10.1 The vegetation classification scheme

area (approximately 231,000 acres) containing the piñon/juniper, ponderosa pine, mixed-conifer, aspen, oak, cottonwood, and willow forest types. The mixed-conifer type includes a characteristic mixture of fir, spruce, hardwoods, and pine species found in northern New Mexico.

Methods

Major activities in the study are discussed below, including classification of the Landsat TM imagery, accuracy assessment of the resulting vegetation data layers, use of existing GIS data, and developing the old-growth model.

Landsat Image Analysis

Field data collection, computerized image processing, and GIS activities were used to derive vegetation data layers from Landsat TM. A GIS was used with an image analysis system to use other data layers in the image analysis, provide storage for layers generated during analysis, and perform modeling and GIS operations. Figure 10.2 illustrates the basic steps used to create vegetation data layers from Landsat TM data.

Classification of the TM data was an interactive process requiring both process and experimentation conducted by an experienced image analyst and resource specialist. Image analysis began with a geocoded (referenced to a map coordinate system) and terrain corrected (topographic relief correctly positioned) image prepared by the vendor. In an effort to fully use both spectral and vegetation information, a combination of supervised and unsupervised (maximum likelihood) classifications was used. In unsupervised classifications, the image is classified using only the spectral data. Supervised classifications use field data from training sites to aid in the classification. The vegetation cover-type layer was derived from a combination of two unsupervised classifications, one to separate forest from nonforest areas and hardwoods and one to separate forest types. Nonforest areas resulting from the first classification were masked out of the image. Thirty-five classes resulting from the second classification were aggregated into six forest types, using the field crew's knowledge of the study area and aerial photos. The nonforest and hardwood classes were then added to the vegetation cover-type classification, resulting in a total of nine classes. Crown closure and tree size were classified using a combination of supervised and unsupervised classifications. Nonforest and low-density vegetation areas were obtained from an unsupervised classification. Using supervised classification techniques, the statistics for four

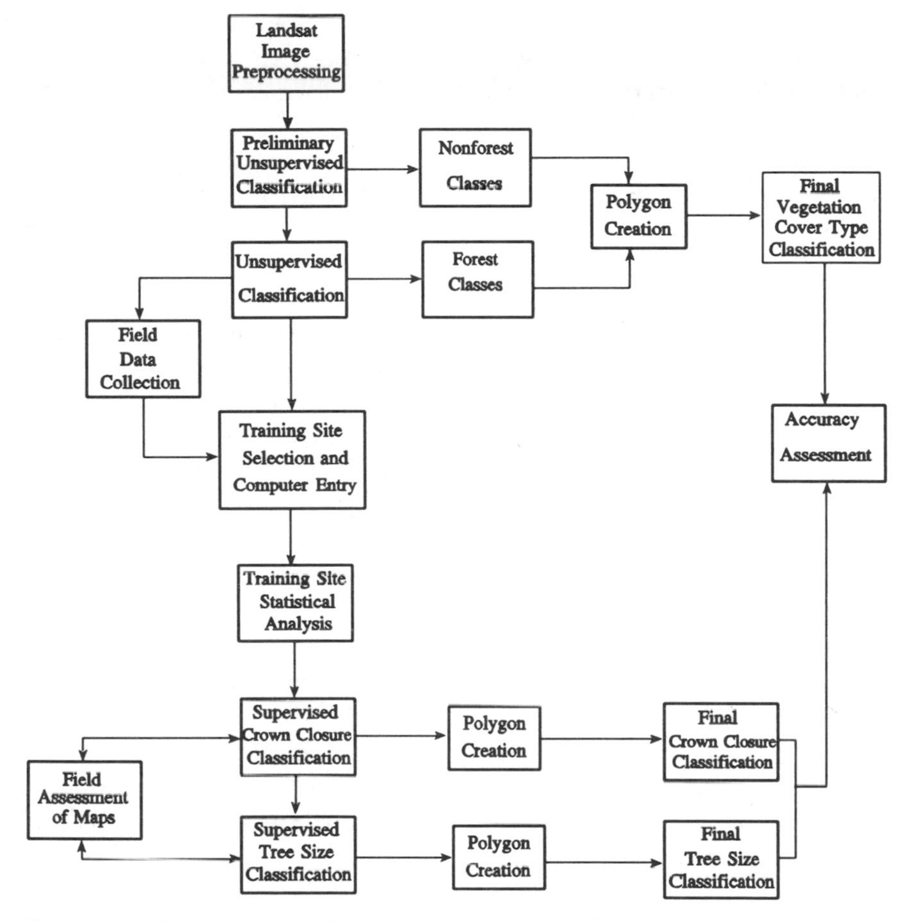

Figure 10.2 Basic process used to create vegetation data layers

bands of TM data (bands 3, 4, 5, and 7) and the vegetation characteristics associated with 112 training sites were used to classify crown closure and tree size in forested areas.

Training Site Data Collection

Training sites for the supervised classifications were selected from the vegetation cover-type classification, from aerial photographs (1:24,000 scale, color), and in the field. Training sites selected from various data sources resulted in a representative sample of training sites in all vegetation classification categories.

Vegetation characteristics for 168 training sites were recorded in the field (75 field sites) or interpreted from aerial photos (37 photo sites). In all of these sites, estimates of percent crown closure for each unique combination of tree species and diameter class were made from aerial photos viewed in stereo. A matrix of percent crown cover by tree species and size class, as well as the percent cover of shrubs, forbs, grasses, and rock/bare ground was constructed for each training site. The matrix allowed the data to be categorized or summarized differently for each classification.

Statistical Analysis

Statistics collected from four spectral bands of TM digital data were produced for each training site and cluster analysis was performed on the means of the training site spectral values. Analysis of the spectral and vegetation characteristics of each training site assisted in determining the optimum combination of training sites to use in the final classifications of crown closure and tree size. Cluster analysis also indicated where spectral variability occurring in the image had not been accounted for and where confusion between training sites occurred.

Polygon Creation

The raster or pixel (picture element) data for the final classifications were generalized by systematically aggregating similar pixels into larger groups. A series of raster and vector filters was applied to each raster data layer to create homogeneous groups of pixels with similar vegetation characteristics. These pixel groups were then converted into polygons (vectorized) to facilitate use of the data.

Ancillary Data

Cartographic Feature Files (CFFs) and Terrestrial Ecosystem Survey (TES) data were two sets of ancillary data used in the project. Ancillary data aids in the classification and/or modeling process, but is not derived from the satellite data. The CFFs are planimetric digital data files containing geographic features found on the U.S. Geological Survey 7.5-minute topographic maps (Forest Service Primary Base Series Maps). CFFs were overlayed on the Landsat image to aid in orientation and land feature display. Site indices derived from TES data provided a means of incorporating different old-growth structural requirements for high- and low-potential sites into the model. TES data contains ecological mapping units that are classified on the basis of soil, vegetation, and climate characteristics (USDA-FS 1986). Each TES polygon was assigned a site index and forest type attribute, based on the vegetation, soil composition, and interpreted site indices within each polygon. Site indices and values used to separate high- and low-capability sites in each forest type were: 50 for piñon/juniper (Howell 1940), 55 for ponderosa pine (Minor 1964), and 50 for mixed conifer (Edminster and Jump 1976). Site indices for piñon/juniper areas were estimated based on the TES interpreted values for fuelwood production. Fuelwood production, expressed in cords per acre, was converted to site index by applying a conversion developed by Howell (1940).

Probable Old-Growth Model

A GIS model was developed to illustrate the integrated use of vegetation data derived from Landsat TM and existing GIS data. Probable old-growth areas were located by selectively applying modeling constraints to four raster data layers, forest type, and site index layers from ancillary data, and crown closure and tree size layers derived from Landsat TM. The probable old-growth model is illustrated in Figure 10.3. Forest type and site index layers obtained from TES data were used in the model, to keep interpretations of site index and forest type consistent within an area and to use existing ecological data. Probable old-growth polygons were required to be within a minimum tree size class. Size requirements for each forest type were held constant between high- and low-productivity sites. To qualify as probable old-growth in the model, 20 percent or more of the total crown closure was required in one or more of the allowable size classes.

Results

Vegetation Classifications

Differentiating vegetation cover types, using unsupervised classifications, worked well for most cover types. The resulting spectral classes were interpreted by the field crew with relative ease. Class interpretation was done without the full season of field knowledge; therefore, some classification errors may have resulted from mislabeling classes. Knowledge of the area's vegetation, topography, soil, and ecological relationships obtained from field visits continues to be very useful for data interpretation. The spruce/fir forest type was not well represented in the study area; therefore, it is not present in the vegetation cover-type classification. Hardwood forest types, consisting of oak, aspen, and cottonwood, appear to be separable in most cases. However, some editing based on field observations was necessary. Differences in aspect and elevation do not easily distinguish hardwoods in the Jemez Mountains. Cottonwood was grouped with agri-

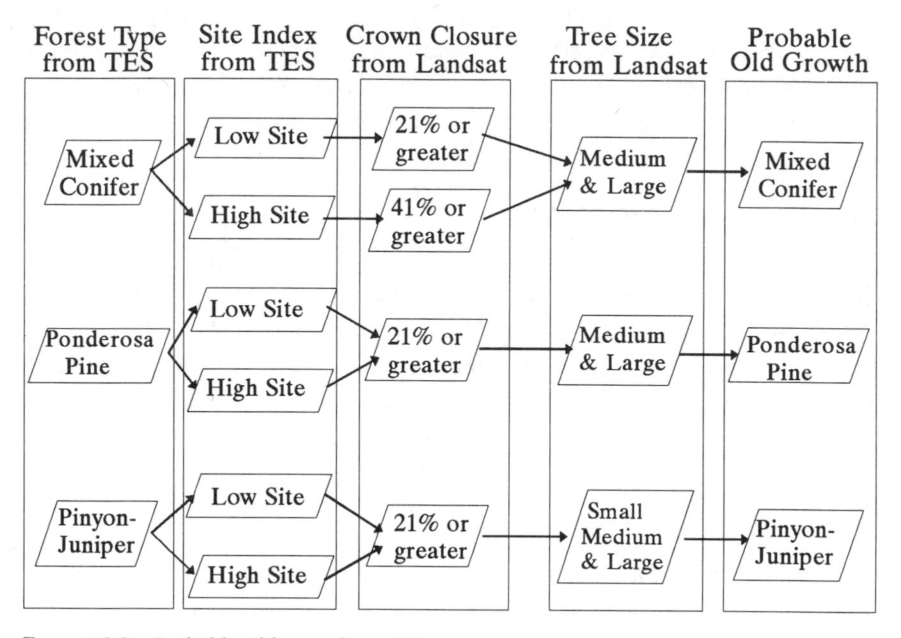

Figure 10.3 Probable-old-growth model

cultural lands, as they occur adjacent to one another in the Cañon de San Diego and are spectrally similar.

The combination of classification techniques used to derive crown closure worked well, partly due to the utility of the classification scheme. The derived crown closure polygons correspond well with crown closure changes seen on orthophotos and with reference data. The few oak and shrub areas were included in the 0–20 percent crown closure class. A finer separation of crown closure may be possible; however, the identification of fine breaks in crown closure may be more limited by our ability to interpret available aerial photos than by the ability of TM data to sense these differences. Further refinement of crown closure in areas of low tree density (0–20 percent class) remains a difficult task using Landsat TM data.

Size was the most difficult vegetation characteristic to map, due to the ecological conditions of the area and the limitations of using remotely sensed data. Developing a practical classification scheme for use in the southwest was difficult because of the size variability that occurs within stands. Forest management practices and the exclusion of wildfires have created many multistoried, uneven-aged stands that have a wide variety of tree sizes and indistinct canopy layers, particularly in the ponderosa pine and mixed-conifer forest types. Even in undisturbed areas, biotic and abiotic factors and microclimate effects make it difficult to classify or model tree size. A few small single-storied stands in the seedling/sapling/pole and small size classes were identified from the TM data and placed in their corresponding multistoried size class, but most stands are multistoried. Variable tree-size conditions also made converting the pixel groups into polygons a challenge. It was apparent that, at least for some types of analysis, a less generalized map containing many small polygons, the five size categories, and/or a different approach in characterizing variable tree-size conditions may be more useful than the generalized maps familiar to most data users. In fact, some Forest data users preferred the less generalized versions of the maps on file.

Accuracy Assessment

Both qualitative and quantitative accuracy assessments were completed for the polygon maps of each classification derived from Landsat TM. A qualitative assessment of accuracy was done on the preliminary classifications by plotting them at 1:24,000 scale, ground checking the maps, and comparing them to orthophotos and/or aerial photos. The final polygon maps were assessed quantitatively for accuracy by comparing them (the classified data) to field/photo collected data (the reference data) for the same site. Polygon maps with a minimum mapping unit size of 5 acres were assessed for accuracy. A systematic sample of 126 accuracy assessment sites was taken across the entire study area and used to assess the accuracy of all three classifications.

The same vegetation data collected for training sites were collected in the accuracy assessment sites. Accuracy of the Landsat TM classifications is expressed in the form of an error matrix that quantifies classification errors between classes. In the matrix shown in Figure 10.4, the reference data from the accuracy assessment sites are in columns and the classified data are in rows. Each row represents a unique category or class in the classification and each column represents that same class for the reference data. The cell values in the major diagonal (where the corresponding categories intersect) represent the number of sites correctly classified. Values in the off-diagonal positions represent errors or disagreement between the classified and reference data.

Accuracy was quantified in terms of the "overall," "user's," and "producer's" accuracy, according to assessment techniques discussed by Story and Congalton (1986). Overall accuracy is generated from the number of correctly classified samples divided by the total number of reference sam-

VEGETATION TYPE

OVERALL ACCURACY = 96/125 = 77%

CLASSIFIED DATA (rows) × REFERENCE DATA (columns)

Class	MC	PP	P–J	AS	OA	BA	TOTAL
MC	34	2	1	1	0	0	38
PP	2	24	16	0	1	1	44
P–J	0	0	16	0	0	0	16
AS	1	0	0	4	0	1	6
OAK	0	0	0	2	3	0	5
BARE	0	0	1	0	0	15	16
TOTAL	37	26	34	7	4	17	125

PRODUCER'S ACCURACY	USER'S ACCURACY
92%	90%
92%	55%
47%	100%
57%	67%
75%	60%
88%	94%

CROWN CLOSURE

OVERALL ACCURACY = 103/126 = 82%

CLASSIFIED DATA (rows) × REFERENCE DATA (columns)

Class	0–20%	21–40%	41–70%	71–100%	TOTAL
0–20%	19	4	0	0	23
21–40%	3	25	1	0	29
41–70%	1	10	21	0	32
71–100%	0	0	4	38	42
TOTAL	23	39	26	38	126

PRODUCER'S ACCURACY	USER'S ACCURACY
83%	83%
86%	64%
66%	81%
91%	100%

TREE SIZE

OVERALL ACCURACY = 96/127 = 76%

CLASSIFIED DATA (rows) × REFERENCE DATA (columns)

Class	1	2	3	4	5	TOTAL
1	59	4	2	0	3	68
2	14	15	2	0	0	31
3	3	1	3	0	0	7
4	0	1	0	0	0	1
5	0	1	0	0	19	20
TOTAL	76	22	7	0	22	127

PRODUCER'S ACCURACY	USER'S ACCURACY
78%	87%
68%	48%
43%	43%
0%	0%
95%	95%

TREE SIZE CATEGORIES

Class 1 Seedling/Sapling/Pole
Class 2 Small
Class 3 Medium
Class 4 Large
Class 5 Nonforest and low density vegetation

Figure 10.4 Error matrices for vegetation classifications

ples. The overall accuracy indicates the probability of any class being correctly classified. "User's accuracy" quantifies how reliable the map is from the user's standpoint or how well the map represents what is actually on the ground. Errors quantified in the "user's accuracy" are sites assigned or committed to the wrong class. "Producer's accuracy" quantifies how well a particular characteristic or class can be mapped. Sites that have been omitted from the correct class are quantified in the "producer's accuracy." "User's" and "producer's" accuracies are shown in Figure 10.4, along with the error matrix values used to calculate them.

The overall accuracies for the classifications were 76 percent for tree size, 77 percent for vegetation cover type, and 82 percent for crown closure. Crown closure was the most accurately classified vegetation characteristic, due to the aerial collection of the Landsat data and because slight changes in crown closure create distinct changes in spectral reflectance. As predicted, tree size was the most difficult vegetation characteristic to classify and was the least accurate classification. The variability of tree size within and between stands was difficult to characterize from the spectral data and accurately interpret from aerial photos. The classification accuracies obtained in this project indicate the difficulty in classifying southwestern vegetation characteristics using remotely sensed data.

Old-Growth Model

The spatial distribution and abundance of probable old-growth generated from the model was generally as expected; very few probable old-growth areas exist. A total of 5598 acres of probable old-growth were generated within the main boundaries of the Forest (3 percent of 169,763 acres). Acres of probable old-growth found in each forest type were as follows: 2840 acres in mixed conifer (51 percent), 288 acres in ponderosa pine (5 percent), and 2470 acres in piñon/juniper (44 percent). Probable old-growth polygons generated from the model occurred primarily in the northern half of the study area and were relatively small and scattered. Old-growth distribution patterns in ponderosa pine and mixed-conifer forest types result from both tree harvest activities and the natural distribution of these forest types.

The model did not adequately estimate probable old-growth in two areas, land outside the main forest boundary and in the piñon-juniper forest type. The probable old-growth model was restricted by the input data layer having the smallest area, which was the forest type and site index data from TES. TES data did not include areas outside the main boundary of the forest; therefore, the model did not generate probable old-growth on large tracts of land not administered by the Forest Service. Yet, these areas are known to contain substantial amounts of probable old-growth. More probable old-growth was generated in the piñon/juniper type than in the ponderosa pine type as a result of using TES forest type data. The TES data identifies large mapping units in the ecotone of piñon/juniper and ponderosa pine as being the piñon/juniper type. In the model, these piñon/juniper areas meet the requirements for piñon/juniper old-growth. These piñon/juniper areas in TES are most likely small-sized ponderosa and would not meet model requirements for ponderosa pine. Therefore, the model overestimates old-growth in the piñon/juniper type.

Cost-Benefit Analysis

The cost and benefits of using remote sensing for forest inventory are greatly affected by the size of the project area, the labor force used (e.g., contractors or in-house personnel), the knowledge and expertise of the work force, and other factors inherent in a particular project. The case study was conducted as a demonstration project and covered a relatively small area for the use of satellite imagery. Project costs presented here are meant to give the reader a general idea of the time and fixed costs involved in conducting similar classification projects.

Project costs were analyzed in terms of the time and people needed for training and accuracy-assessment data collection, and for specific project activities, as well as the cost of acquiring remotely sensed imagery. The total cost of one geocoded and terrain corrected Landsat TM scene, a digital elevation model, and aerial photographs was approximately $6350. Collection of all training and accuracy-assessment site data for 261 sites required 69 person/days to complete. (A person/day is equivalent to one person working eight hours.) Landsat TM image analysis and classification took 211 person/days. Post-classification computer work, which included filtering and creating polygon coverages, took 61 person/days. Accuracy assessment analysis of the three coverages took 104 person/days. Project development and coordination and GIS integration and analysis took 27 and 11 person/days, respectively. The entire Landsat classification and old-growth modeling project required approximately 483 person/days to complete. Potential old-growth was photo interpreted within the same area, using 1:24,000 natural color aerial photos, and took approximately 33 person/days to complete at a cost of approximately $2700.

Although the aerial photo interpretation project was less expensive in terms of time and money, the Landsat TM classification does have significant advantages. It is important to consider the differences between the resulting GIS data layers in terms of their use under changing old-growth definitions and management requirements, the objectivity with which polygons are delineated, the extent of data coverage, the age of the data captured, and the data's accuracy.

Conclusions

Landsat TM data can provide useful vegetation data for GIS when used in heterogeneous southwestern vegetation conditions. Unsupervised classification techniques were successfully used to separate vegetation cover types in the study area. Crown cover and tree size classifications were also accurately produced from Landsat TM data, using a combination of unsupervised and supervised classification techniques. However, some vegetation characteristics were easier to derive than others. Tree size was difficult to derive, due to the limitations of remotely sensed data and the variability of tree size conditions in the study area. Differentiation between the spruce/fir and mixed-conifer forest types was not fully investigated because very few spruce/fir areas exist in the study area. Other remote sensing techniques should be explored to further refine and classify tree size in the Southwestern Region. Based on the results of this project, the combined use of supervised and unsupervised classification techniques to derive vegetation characteristics in the southwest from Landsat TM data is recommended.

A well-designed classification scheme is critical to deriving a classification that meets the data user's needs. Vegetation classification schemes are highly dependent on the data source(s) and vegetation variability. The cost of doing a classification and determining its accuracy are also affected by the classification scheme. Most important, a well-designed classification scheme is a flexible one. It is expensive to maintain flexibility in the classification; but, compared to the cost of trying to improve poorly classified data or not being able to use the data effectively, the cost is worth it. Therefore, the time and money necessary to design a good classification scheme should be available in projects using Landsat TM data.

Several barriers to integrating the use of Landsat TM at the Forest level were observed and should be considered in future projects. Although an effort was made to educate forest managers and the public about satellite imagery and its potential use, people did not embrace the technology for its potential. It was clear that the success of past applications of satellite imagery continue to play a major role in formulating opinions about its use today. Most people were not familiar with current advances in satellite imagery, image processing software, or even GIS capabilities. Therefore, a person's optimism for how well Landsat TM could classify vegetation in the southwest was generally based on preconceived notions, with little or no direct knowledge of the technology. There is a clear need for training and awareness by Forest Service personnel in the use of remote sensing and GIS, as well as increased remote sensing awareness and information exchange with the public. There is also the need to continue technology transfer to forest personnel and to maintain local expertise in remote sensing and GIS. People were often skeptical of using satellite imagery because they believed maps produced from satellite imagery were too inaccurate or broad for their needs and the cost of using the data was too high. Although both conditions can exist for some projects, the true cost and accuracy of existing data and traditional data collection must also be considered and compared to alternative sources of data. The Forest Service has chosen to emphasize the use of existing data, but has not yet taken a comprehensive look at the adequacy of existing data in relation to the decisions they support (Evanisko 1990). Existing data need to be critically evaluated before they are used, and more effort should be directed toward integrating relational data. Budget and time constraints continue to have a significant impact on the use of satellite imagery for small-area applications.

The case study's time constraints required that all field data be collected in one field season. In addition, the study required assistance in field data collection due to inadequate funding. The purchase of satellite imagery required a large portion of the project's budget. Finally, the resistance to change within the organization can be a formidable barrier. It was difficult to foster ownership in the data and the desire to use the data among different disciplines. Full support and commitment of management and a means of transferring that commitment throughout the organization is vital to the successful use of new technology.

Landsat TM has several desirable qualities as a data source for GIS. Landsat TM entirely covers a large area; therefore, it fills the gaps that may exist in other databases. Vegetation data for areas such as wilderness and privately owned land may contain information that can significantly affect the distribution and abundance of old-growth. Collection of Landsat TM data is repeatable and consistent through time, which provides for both current and future data needs. Not only can current old-growth conditions be assessed—these conditions can be monitored through time by detecting vegetation changes from satellite imagery obtained at a later date. Because the data are already in digital form, TM provides an accessible and flexible data source for GIS.

When considering the potential uses of GIS, it is important to remember that GIS outputs are dependent on database characteristics. The area extent and positional accuracy of the data both significantly affect GIS outputs. The detail and flexibility of the data, which is inherent in the classification scheme, determines how effectively the data can be used. Finally, knowing the accuracy of each data layer allows the user to appropriately apply the data in project analyses and allows others to critically evaluate the data's use. Data characteristics, including accuracy, will need to be documented and made available to data users and the public.

The larger job of using the classified data and continuing to build on what was learned in the case study is still to be done. The Santa Fe National Forest must determine appropriate uses of the data in forest and project-level planning, develop additional models of old-growth, conduct spatial analyses on old-growth model outputs, and determine how to use the Landsat TM data with existing databases.

The Southwestern Region of the Forest Service has established a regional remote sensing program and continues to use satellite imagery throughout the region (Krausmann 1993). Vegetation mapping, using supervised and unsupervised classification techniques, are being conducted on the Kaibab and Carson, as well as the Santa Fe National Forest. The Prescott National Forest is integrating the use of Landsat imagery into its ecosystem inventory. Coarse vegetation and landscape diversity maps are being developed from Landsat and used to design and refine field sampling for TES. The Cibola National Forest is using Landsat TM to examine vegetation changes resulting from logging, grazing, roading, and recreation in the Zuni Mountains. The Coronado and Tonto National Forests are also beginning to use Landsat imagery for small-area resource analyses. Forests will be able to use remote sensing technology more effectively with remote sensing technical assistance now available in the Region.

Landsat TM imagery can provide GIS data layers needed for forest management activities. A GIS with high-quality vegetation data is essential for forest management planning and monitoring on a landscape level. Classified Landsat TM data can provide vegetation data layers useful for stratification of intensive field vegetation sampling. The data could be used to develop efficient vegetation sampling schemes for forest stand examinations and ecological land unit surveys. Also, computer modeling with GIS data interpreted from Landsat TM is a valuable tool for evaluating particular ecological conditions or wildlife habitats. "When the increasing availability and sophistication of geographic information systems are coupled with the increasing availability of high-resolution multispectral data, the outlook for utilizing remote sensing data as an integral component of the land management process is indeed a bright one" (Lillisand and Kiefer 1987).

References

Edminster, C.B., and L.H. Jump. 1976. Site index curves for Douglas-fir in New Mexico. U.S. Department of Agriculture, Forest Service, Rocky Mountain Forest and Range Experiment Station Res. Note RM-326.

Evanisko, F. 1990. "Ecosystem mapping: Possibilities and impediments for remote sensing." Paper presented at Third Forest Service Remote Sensing Applications Conference, American Society for Photogrammetry and Remote Sensing, Tucson, Arizona.

Howel, J., Jr. 1940. Pinyon and juniper: A preliminary study of volume, growth and yield. U.S. Department of Interior, Soil Conservation Service Reg. Bull. No. 71.

Krausmann, William. Southwestern Regional Remote Sensing Scientist. Personal communication. January 5, 1993.

Lillisand, T.M., and R.W. Kiefer. 1987. Remote Sensing and Image Interpretation. John Wiley and Sons, New York.

Minor, C.O. 1964. Site-index curves for young-growth ponderosa pine in Northern Arizona. U.S. Department of Agriculture, Forest Service, Rocky Mountain Forest and Range Experiment Station Res. Note RM-37.

Story, M., and R. Congalton. 1986. Accuracy assessment: A user's perspective. Photogramm. Eng. Remote Sens. Vol. 52, No. 3, pp. 397–99.

USDA-FS. 1986. Terrestrial ecosystem survey handbook. U.S. Department of Agriculture, Forest Service, Southwestern Region, Albuquerque, N.M. 125 pp.

Ecosystem Management in Southern Appalachian Hardwood Forests: Minimizing Fragmentation

Introduction

Charles C. Van Sickle

Federal agencies in the southern Appalachian region have been experimenting with many different approaches to ecosystem management. Each separate venture is influenced by the social and economic context, by political and organizational considerations, and by the surrounding geography and physiography. The chapters in this section illustrate some of the approaches used.

GIS and remote sensing have become essential tools for ecosystem management. Thus, the demand for geographic analysis capabilities is growing rapidly. A recent meeting of GIS specialists within the southern Appalachians included more than 30 specialists representing about 20 organizations. Those who attended were eager to share information and experience.

This sense of cooperation has been fostered by the Southern Appalachian Man and Biosphere (SAMAB) program. SAMAB is a cooperative program that brings government agencies (National Park Service, USDA Forest Service, Tennessee Valley Authority, U.S. Fish & Wildlife Service, U.S. Department of Energy, and U.S. Economic Development Administration), academic institutions, and citizens together to address the region's natural resource problems. As many parts of the world today, the southern Appalachians are threatened by economic and population growth. But rather than trying to restrict growth, which is probably unrealistic in the short term, SAMAB is seeking ways to manage growth and resource use in ways that minimize the impact on a treasured environment.

The southern Appalachians have long been recognized as a region with special attractions and special problems. It is one of the most biologically diverse regions in the world. Its mountains are old and beautiful, inviting settlement and exploitation early in our country's history.

But the region has also suffered neglect and overuse—of both its resources and its people. From the story of the Cherokee relocation, to the farmers who tried to grow crops on steep slopes, to the poverty of the coal fields, the people of the southern Appalachians have struggled for survival. For many residents the latest wave of immigration and development is welcome. Today, the region is experiencing rapid development, especially in recreation and in construction of retirement homes. This development, while adding to the region's economic base, is also straining its infrastructure and placing great demand on its water resources and quality, its forests, its ability to offer scenic vistas, and its plant and animal habitat.

The region's national forests and national parks often serve as a buffer for mitigating the influence of land use decisions on adjacent private lands. But even as this function becomes critical, it is increasingly obvious that these federal lands are again showing signs of overuse, and that the functions they serve are becoming more precious as the lands surrounding them are altered.

The land management equation consists of federal and private lands in a mosaic of ownership patterns. The patterns reflect the creation of national parks and forests from the abandonment of private lands. The resulting edge effect is advantageous for some species while disadvantaging to others.

This, then, is the context for ecosystem management in the southern Appalachians: an area where actions taken on one ownership often affect the character and quality of lands belonging to adjacent owners; where federal lands are becoming increasingly important in preserving the character of the region; and where environmental quality depends on the ability of federal agencies and public interests working together to plan future development.

There are numerous illustrations of ecosystem management to choose from. One project worth mentioning is the Chattooga River Basin Demonstration Project. The Chattooga River Basin consists of about 122,000 acres of national forest and 58,000 acres of private lands. The Forest Service has been asked to develop a separate plan for the Chattooga River Basin based on its special set of geographic and social features. These include a wild and scenic river, close proximity to the expanding Atlanta metropolitan area, a planning unit involving three separate national forests in three separate states, and the important influence of the included private lands within the planning unit. This project will be a real test of GIS and remote sensing technology. Part of the challenge will be to combine digital data from several sources into a framework that will furnish information for private as well as public lands, allow shifts in scale and resolution, and provide easy access for a variety of users.

Ecosystem management has organizational implications as well as technological ones. Many resource management organizations, such as the Forest Service, are organized along easily defined functional lines—timber management, wildlife management, recreation, fire, water, and so on. These functions must become more integrated to accomplish ecosystem management. The transition to ecosystem management will require more emphasis on the development of interdisciplinary teams. Similarly, the culture and training of scientists is oriented toward specialized, independent, individual efforts. Teamwork doesn't come easy. Sometimes it is useful to create an integrating framework to help organize separate pieces of information. One form of integrating mechanism is GIS technology. It allows us to bring information together in an analytical mode and to display the outcome of our decisions. It also helps us understand relationships that are not easy to grasp in any other way. Although complex relationships are integrated, ultimately, by the human minds of the managers and workers, GIS and remote sensing technology can be an important catalyst and tool in this process.

11

Ecological Perspective: Understanding the Impacts of Forest Fragmentation

Scott M. Pearson

Southern Appalachian forests are rich in biological diversity. To understand the impacts of habitat fragmentation, the sensitivity of ecosystems, communities, and species to fragmentation must be determined. Recognizing the natural patterns of heterogeneity in these forests and the importance of this heterogeneity to ecological processes will promote our understanding of fragmentation. The impact of a specific forest use (economic development, forest harvesting, recreation) depends on the type of habitat modification and the spatial extent and pattern of this use. Information on ecological processes, maps of natural communities, and projections about activities that modify forests are needed in order to implement management strategies that will minimize forest fragmentation. Information on ecological processes is needed for communities and species that are sensitive to fragmentation. This information includes the importance of a metapopulation structure, impact of modified habitats on dispersal abilities, edge effects, and sensitivity to fragmentation. Accurate maps about ecologically important variables are also needed. The forest stand and condition maps provide an important database, but this information may not be useful for mapping suitable habitat for many species. While public lands represent a significant proportion of this region, much of the modification of forests is taking place on private lands. Therefore, conservation strategies must be coordinated across the region. Forest use must be coordinated over a range of spatial and temporal scales, recognizing that public lands are part of a larger landscape.

Introduction

Southern Appalachian forests contain many species adapted to conditions ranging from mesic coves to xeric ridges to cold, windswept summits. The major plant communities include wetlands and balds as well as hardwood and coniferous forests (Braun 1950). It is one of the most floristically diverse regions in the eastern United States, hosting 1200 or more species of trees, shrubs, and forbs (White 1982).

While some species are found in several community types, each community type hosts a set of unique plant and animal species. For example, rock outcrops and grassy balds support populations of rare sedges, grasses, and forbs like the Blue Ridge goldenrod (*Solidago spithamaia*). Mesic cove forests contain wildflower species such as ginseng (*Panax quinquefolium*) and golden seal (*Hydrastis canadensis*), and invertebrates such as the noonday snail (*Mesodon clarki nantahala*). The endangered Carolina northern flying squirrel (*Glaucomys sabrinus coloratus*) inhabits high-elevation spruce-fir and northern hardwood forests. Streams and rivers are inhabited by endemic fishes such as the yellowfin madtom (*Noturus flavipinnis*). Bog turtles (*Clemmys muhlenbergi*) and green pitcher plants (*Sarracenia oreophila*) are found in montane wetlands. A high proportion of the southern Appalachians' avifauna are neotropical migrant birds that are specialized to specific plant communities (MacArthur 1959, Willson 1976).

Preserving the biological diversity of this region will depend on the preservation of the diversity of biological systems. One of the greatest threats to these natural systems is the modification of native habitats and the associated fragmentation of landscapes. Habitat loss and forest fragmentation is recognized as one of the greatest threats to biological diversity. This chapter will discuss patterns of heterogeneity and connectivity in the southern Appalachians, impacts and causes of fragmentation, and information needed to avoid further fragmentation. This discussion should assist land managers in assessing the effects of fragmentation and developing management strategies that minimize impact on these biologically rich ecosystems of the southern Appalachians.

Natural Patterns of Environmental Heterogeneity in the Southern Appalachians

Environmental heterogeneity is a natural part of ecosystems. Heterogeneity is variation in physical and biotic conditions in space and time. The phrase "patterns of heterogeneity" refers to the amount and types of variation associated with a given ecosystem. All systems have some characteristic level of environmental heterogeneity. Some systems exist in highly variable environments and others function in environments with relatively less variation (i.e., low heterogeneity). Heterogeneity in space and time has been important in the evolution of biological diversity and remains important in its maintenance. Indeed, spatial heterogeneity due to changes in soil type, elevation gradients, and topography, and temporal variability in climate since the Pleistocene are responsible for the biological diversity of the southern Appalachians.

Spatial patterns in ecological communities are hierarchical. At the broadest scale, the southern Appalachian landscape appears as one that is dominated by deciduous forest with interspersed patches of coniferous forest and other habitats such as balds, rock outcrops, wetlands, and riparian zones (Figure 11.1). Deciduous forests cover most of the middle and lower elevations of this mountainous region. However, the dominant species of deciduous trees vary with soil type, elevation, and topographic exposure within a given watershed. In any particular portion of a watershed, the density, age structure, and seral status of trees vary due to the disturbance history of that site. Variation in the size and density of trees influences light levels reaching the forest floor and, therefore, the herb community. Thus, variation at multiple scales influences the spatial distribution of plant species.

An examination of the spatial distribution of the plant communities will illustrate the hierarchical pattern of ecological communities in the southern Appalachians and some of the major factors that influence their spatial relationships. The spatial distribution of the major plant communities in this region are largely determined by interactions between elevation and site moisture (Figure 11.2, Whittaker 1956). Site moisture is affected by topographic features like aspect, shape and steepness of slope, and relative position on the slope. For example, the oak-chestnut forest type extends to elevations as high as 1500 meters on slopes with southerly aspects where it is replaced by the northern hardwoods forest type (Braun 1950). However, the upper extent of oak-chestnut forest is limited to approximately 1400 meters on north-facing slopes. Mesophytic cove forests are also more common on north-facing slopes that receive less solar radiation and retain more soil moisture. Thus, topography and elevation place constraints on the spatial distribution of forest types. These constraints ultimately determine the spatial pattern of community types on the landscape, but other factors like natural disturbance and human land uses can affect the species compositions of these communities.

Within the bounds of any specific community type, there is variation due to disturbance history. Tree falls are the most common natural disturbance in this region, affecting approximately 1 percent of the deciduous-forested landscape annually (Trimble and Tryon 1966; Runkle 1981, 1982). Recovery (canopy closure) from these small isolated disturbances is relatively quick. Tree falls do, however, provide a continuous source of

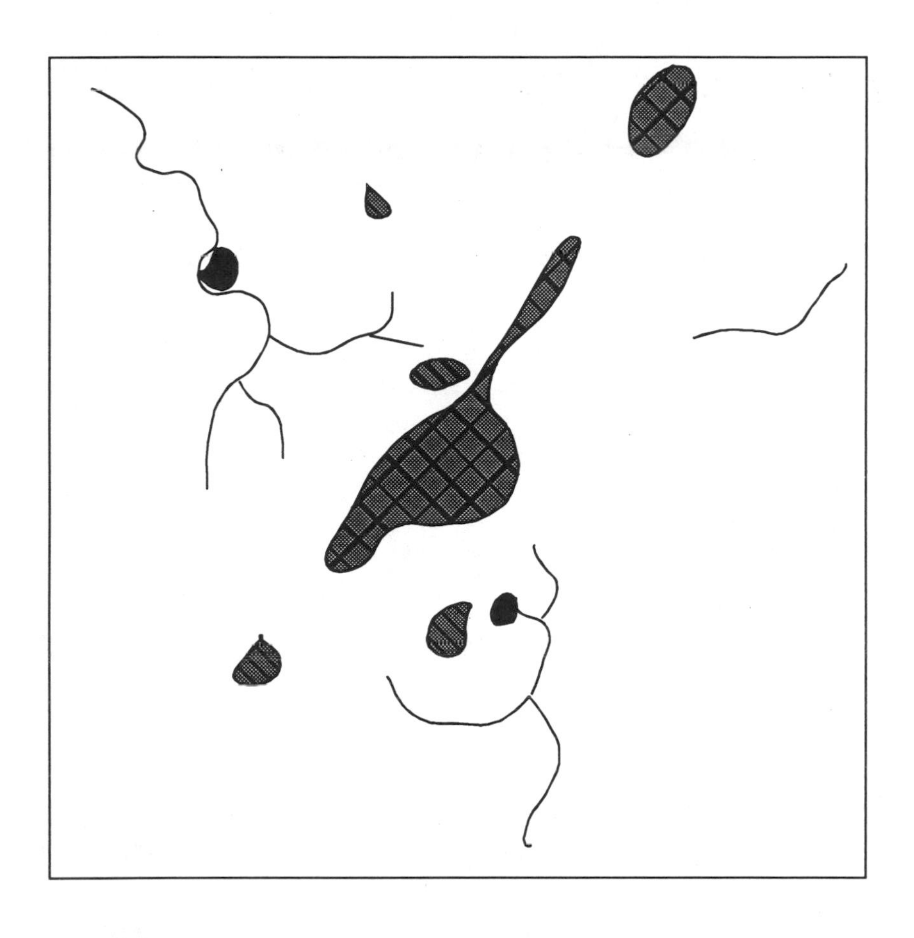

Figure 11.1 An idealized map of a southern Appalachian forest. Wetlands, streams, conifer forest, and grassy and heath balds appear as islands or strips of habitat surrounded by a "sea" of deciduous forest.

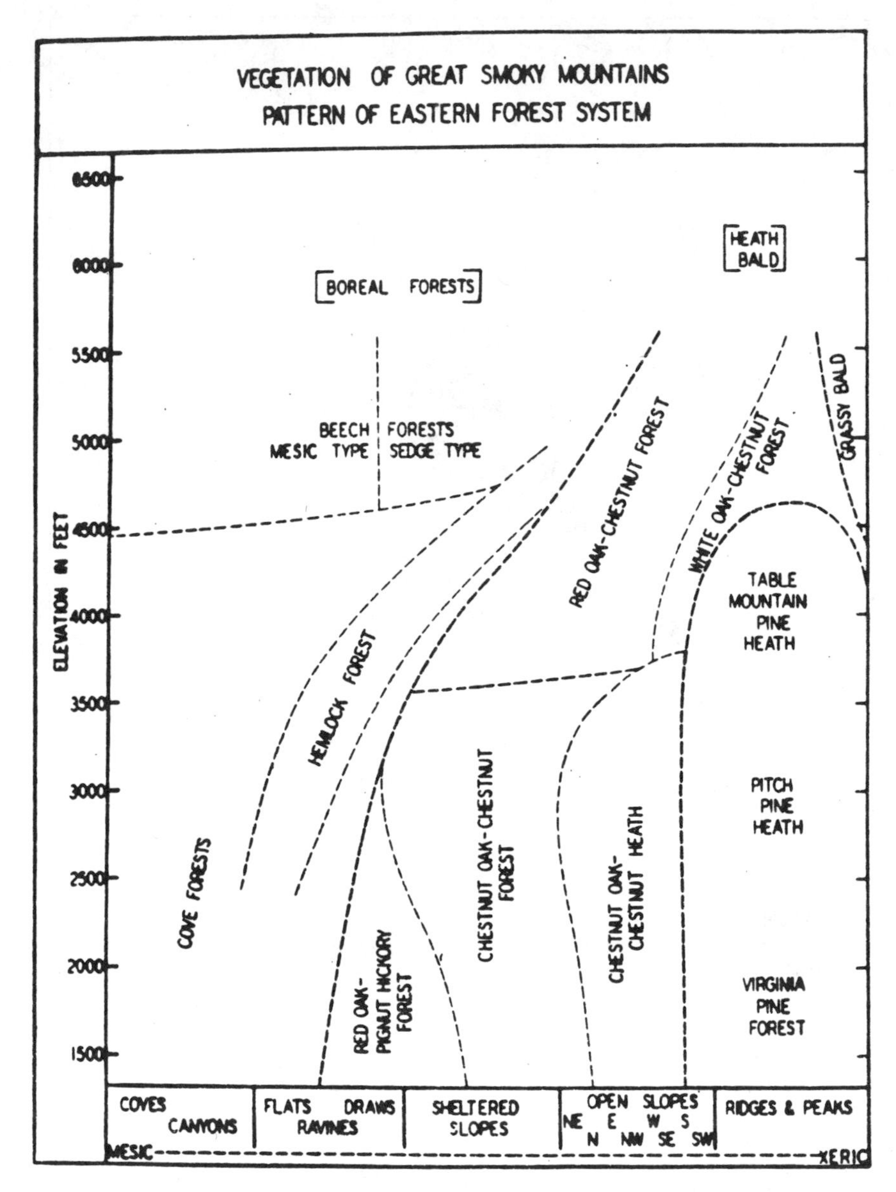

Figure 11.2 The spatial arrangement of plant community types of the Great Smoky Mountains relative to elevation and site moisture. From "Vegetation of the Great Smoky Mountains" by R.H. Whittaker, Ecological Monographs, 1956, 21, 1–80. Copyright © 1956 by The Ecological Society of America. Reprinted by permission.

habitat for species that cannot persist at the low light levels under closed-canopy forest. Human-caused fire might have been an important factor in the maintenance of grassy balds (Lindsay and Bratton 1979) and extensive stands of pine (*Pinus* spp.) forest (Harmon 1982). Natural fire is a relatively rare source of disturbance for southern Appalachian forests compared to forests in other regions of North America. The major fires affecting forests now contained in the Great Smoky Mountains National Park within the past 100 years followed a dramatic increase in down and dead wood after extensive logging activities conducted before park establishment (Pyle 1988).

At broad scales, patterns in vegetation in the southern Appalachians are determined by topography and elevation. At finer scales, variation in tree species composition, tree density, and age is due to disturbance and species interactions. Microbe, fungi, and animal communities respond to some of the same factors as the vascular plants. These species may be affected directly by factors like microclimate or indirectly by the effect of these factors on the plants on which these species depend. These factors, which influence vegetation at multiple scales, generate a hierarchical pattern of environmental heterogeneity (O'Neill et al. 1986) that is reflected in the spatial distribution of species and communities.

Understanding the Impact of Forest Fragmentation

Recognizing patterns of environmental heterogeneity is important because they reveal the conditions under which the current species, communities, and ecosystems have developed. The spatial patterns of heterogeneity also reveal the patterns of connectivity present within this system. Whether or not the current patterns of heterogeneity are desirable or will foster the persistence of native ecosystems in the southern Appalachians is an important question. If management goals for this region include protection of its rich biological diversity, then a landscape-level perspective on the region's ecological relationships and potential impacts of human land use is necessary. A landscape-level perspective is essential for understanding the impacts of the effects of a broad-scale phenomenon like forest fragmentation because this perspective is holistic in its approach to ecological systems, and it recognizes spatial and temporal patterns of heterogeneity at different scales.

Landscape connectivity is important for many ecological processes. Connectivity can exert strong influences on ecological processes such as the movement and dispersal of organisms (Gardner et al. 1989, 1991), the use of resources by animals (O'Neill et al. 1988, Pearson et al. in press), gene flow (Gilpin and Soule 1986), and the spread of disturbance (Turner et al. 1991). Changing the patterns of connectivity can disrupt ecological processes that depend on movement within the landscape. For example, the persistence of a metapopulation of small mammals may depend on the ability of dispersing young to reach other populations. If the overall connectivity of the landscape is altered by the creation of habitats that act as barriers to dispersing young, the isolated populations of this small mammal could become extinct due to their own demographic instability (Brown and Kodric-Brown 1977, Beuchner 1989) or competitive interactions (Nee and May 1992).

Fragmentation occurs when natural patterns of heterogeneity and connectivity are modified to the extent that normal ecological processes are disrupted and/or new processes are initiated. An example of a normal process that could be disrupted is the small mammal dispersal explained above. Examples of new processes that could be initiated are increased light levels in forests due to an increase in edge, the invasion of exotic plants, and increased predator densities in forest fragments with high edge-to-area ratios. The specific changes in natural heterogeneity that result in the disruption of an ecological process depend on the process in question and the natural histories of the species involved (Pearson et al. in press).

All species are probably sensitive to habitat fragmentation if it occurs at a scale that interferes with their life histories. Species that are sensitive to fragmentation at the scales of human land use decisions (1–1000 hectares) often share similar life history traits. These species often (1) have large home ranges, (2) require uninterrupted areas of suitable habitat, (3) are habitat specialist, (4) are edge sensitive, (5) exist in metapopulations, or (6) have limited ability to disperse or move across modified habitats (cf. Terborgh's [1974] extinction-prone species).

Communities susceptible to fragmentation include (1) communities whose keystone species have one or more of the qualities listed above, and (2) communities occurring in small isolated patches whose component species exist as small populations and are dependent on metapopulation structure (e.g., Thomas and Harrison 1992). Communities exhibiting this latter condition are particularly susceptible if these metapopulations depend on natural corridors, such as streams or ridge tops (e.g., Szacki and Liro 1991). For example, populations in wetlands may depend on riparian strips as dispersal corridors. Similarly, high-elevation islands of spruce-fir forest may be connected by strips of conifers on intervening ridges. Such habitats could become fragmented if the connecting corridors were eliminated.

What Activities Increase Fragmentation?

Humans tend to modify landscapes to suit their own needs. Within the southern Appalachians, examples of intensive human land uses include agricultural, industrial, and residential or urban. These land uses dramatically alter the ecological character of the land where they occur, and they tend to be permanent modifications at ecological time scales. Both ridges and riparian zones on private lands are being modified by economic development in the southern Appalachians. Ridges are attractive for home sites because they offer scenic vistas. Agricultural, industrial, urban, and recreational (golf courses) development has concentrated in valleys, threatening to eliminate most of the bottomland and riparian forests of the region. Flamm and Turner (submitted) found that during the period 1976–1990 the majority of land use change in one southern Appalachian watershed occurred on private lands on sites with shallow slopes at lower elevations. Public lands, which account for a large proportion of the land area in this region, are largely used for forest products, recreation, and conservation. These land uses can occur with less dramatic alterations of the current "natural" conditions; however, exploitative activities cannot be conducted without some impact on natural patterns of heterogeneity.

The impact of any land use activity depends on (1) the degree of habitat modification and (2) the spatial extent and pattern of modification. The degree of habitat modification is the extent to which a modified land unit differs from the set of seral stages likely for that land unit under a natural disturbance regime. For example, a paved parking lot in a valley does not resemble any of the set of possible land covers (seral stages) expected for that site. Thus, constructing a parking lot, like other forms of urban land use, represents an impact with a large degree of habitat modification. Moreover, heavily modified habitats show a high degree of contrast (sensu Kotliar and Wiens 1990) to native habitats. The suitability of these modified habitats for native organisms is roughly related to the habitats' contrast to naturally occurring habitats. Harvesting timber products represents an impact with a lesser degree of habitat modification (and lower contrast) because natural disturbance regimes can produce seral stages where trees have been killed, resetting the site to an earlier successional stage. Harvest methods, of course, do vary with their degree of impact because they differ in the amount of tree biomass removed (selective cut versus clearcut) and in the amount of disturbance to soil and water. Some long-term effects of clearcutting have been recorded (Leopold and Parker 1985, Duffy and Meier 1992).

As the degree of habitat modification increases, the likelihood of that modified habitat supporting native species decreases. Succession will likely restore a facsimile of the original community to a site that has been clearcut at a rate similar to the natural rate of recovery from disturbance (but see Duffy and Meier 1992). However, urban and agricultural activities ensure a much longer time to recovery, because they remove practically all native flora and fauna as well as seed sources and replace the former community with material and organisms that inhibit the reestablishment of native species. Thus, these intensive "disturbances" with a high degree of habitat modification do not resemble any of the possibilities in natural disturbance-recovery regimes.

The spatial extent and pattern of habitat modification must be examined from a broad, landscape-scale perspective. Spatial extent is the amount of the landscape modified. Spatial pattern refers to the spatial arrangement of this modification. Pattern and extent are not independent in their relevance to landscape-level ecosystem function. The following discussion of pattern and extent assumes that the degree of habitat modification is sufficiently severe that modified sites become unable to support the ecological processes of interest.

If the area affected by habitat modification is a small proportion of the landscape (e.g., <35 percent), then broad-scale properties of the landscape, such as connectivity, likely will not be affected, and the spatial pattern of modification is likely not important (Pearson et al. in press). This statement assumes that the landscape in question is relatively uniform spatially in terms of ecological importance. That is, there are no "special places" of higher value in the landscape that are to be modified. Examples of such special places include demographic "sources" for metapopulations (sensu Pulliam 1988) or corridors important for connecting specific habitats. If the landscape does contain special places, then the modification of one or more of those places could be particularly detrimental.

If a large proportion of the landscape is modified (>75 percent), then ecological processes, such as metapopulation dynamics, that depend on a minimum proportion of the landscape remaining suitable and unfragmented will not continue. Spatial pattern does not matter in this situation either. When intermediate proportions of the landscape (35–75 percent) are modified, the spatial arrangement of unmodified habitat can dra-

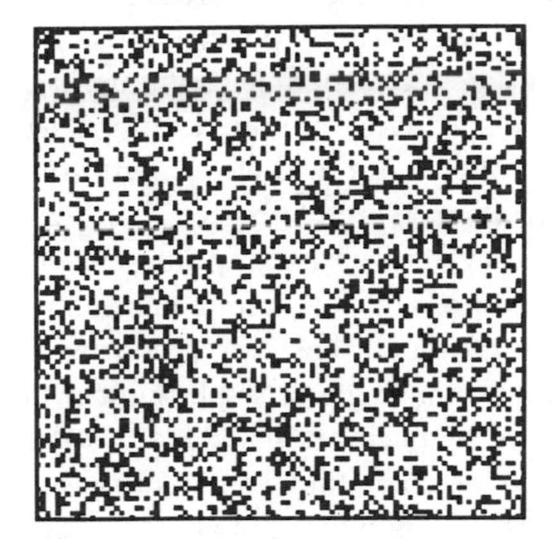

Figure 11.3 The spatial arrangement of habitat loss (open) can affect the connectivity of the remaining habitat (shaded). Both of these maps are 100 × 100 cells in extent and have 3280 cells (approximately 33%) of habitat remaining. Habitat loss that is random in space can result in a highly fragmented landscape consisting of many small habitat patches. When habitat loss is spatially organized (aggregated), the remaining habitat is highly connected. (After Pearson et al. in press).

matically affect the connectivity or fragmentation of the landscape. Random, fine-scale patterns of habitat loss result in greater degrees of landscape fragmentation than coarse-scale patterns in which destructive activities are clumped in space (Figure 11.3; Franklin and Forman 1987, Pearson et al. in press).

The greatest opportunities for using spatial pattern to positively affect landscape connectivity is at intermediate levels of habitat abundance. Connectivity can be maintained by organizing impacts in space such that regions which become impacted and those that remain unaffected are contiguous (i.e., clumped in space, Figure 11.3). For example, if home construction threatens to fragment the landscape, fragmentation could be averted by restricting home construction to some subregion of the landscape, like a city, rather than randomly or uniformly placing home sites across the landscape. However, managers and researchers should keep in mind that the effects of spatial pattern and extent could be specific to the ecological processes and species affected by a particular type of habitat modification. For example, a population of wind-dispersed plants may persist indefinitely on a small archipelago of suitable sites, but a viable population of black bear (*Ursus americanus*) may need large areas of unfragmented, undisturbed habitat in order to remain viable.

The importance of evaluating landscape change from an organism-based perspective cannot be understated (Wiens 1989). Species and communities differ in their resource needs, tolerance of modified habitats, and effects of spatial pattern. Adopting the perspective of an individual species or group of species can reveal landscape patterns not readily apparent from a human perspective (e.g., Wiens and Milne 1989, Flather et al. 1992). Biological diversity, by definition, includes species with many different life history strategies. Differences in life histories and resource requirements occur both within and among community types. Therefore, devising a plan to protect biological diversity based on a single or small number of taxa may not address the resource needs of species not included in the plan. Indeed, single-species approaches have come under sharp criticism. This criticism has, in part, encouraged the development of an ecosystem management approach being adopted by the U.S. Forest Service.

Information Needs for Managing Landscapes to Minimize Fragmentation

The southern Appalachians with its diversity of communities presents an exceptional challenge for conserving natural systems in the face of an increasing need to utilize the landscape for providing services for a growing human population. To meet this challenge, managers need knowledge of the current patterns of biological diversity, a scientific understanding of the ecology of these diverse communities, and the ability to coordinate and regulate land use so as to not disrupt the ecological processes necessary for the persistence of these natural ecosystems.

First, the current patterns of biological diversity must be available for identifying communities and areas that are vulnerable to fragmentation. While a generalized habitat map may provide a starting point, it may not be useful for designing strategies to protect specific species and communities. Ideally, habitats should be mapped from the perspective of the species of interest. Habitat maps of the same landscape but produced for different species may differ, because each species may be sensitive to different ecological parameters. For example, salamanders might respond to soil moisture and water quality, whereas squirrels respond to mast production and the availability of tree cavities. Remotely sensed data, in conjunction with collateral data like digital elevation models and soils maps, provide the means to produce habitat maps over large areas. Being able to tailor these maps for specific communities or species depends on the scale and interpretation of the remotely sensed and ground-survey data. Vegetation maps produced in gap analysis (Scott et al. 1987) is a good first step toward developing a spatially explicit database on biological diversity.

Second, a general understanding of the ecology of these communities is needed in order to estimate their sensitivity to habitat modification. Management and protection of any ecological system is dependent on understanding the important processes shaping and maintaining the system. These processes include external factors such as climate as well as internal factors such as essential interspecies dependencies. While knowing the details of each species' natural history and its interspecific relationships would assist in the development of precise management plans, such detailed information is not available for the vast majority of the species in the southern Appalachians. Moreover, many of these details may not be needed if the general ecological requirements (e.g., activities of key species, soil type, topographic position, tolerance to modification) of the community types are known and connectivity with similar communities is maintained.

Ecological characteristics to consider while attempting to limit the effects of fragmentation include (1) natural disturbance regimes and their role in creating heterogeneity in the landscape (e.g., fire, flooding, insect outbreaks, windstorms); (2) effect of habitat modification on species' dispersal and persistence; and (3) species' sensitivities to edge habitats and fragmentation of suitable habitat.

Finally, temporal and spatial variability in land use must be coordinated at multiple scales. Natural ecosystems can be better protected if habitat modification in one area is balanced by preservation in other areas. By staggering habitat modification in time, most sites can be utilized over a long period. This period of time depends on the type of disturbance and time to recovery. It may be hundreds of years for some intensive land uses or decades for land uses with less habitat modification. This temporal, spatial balancing is least likely to threaten the persistence of native ecosystems if the spatial extent and frequency of habitat modification resembles the natural disturbance regime. Such coordination will be challenging in the southern Appalachians because much of the landscape is in private ownership, where social and economic factors largely determine patterns of land use.

Acknowledgments

L.K. Mann, R.O. Flamm, R.L. Kroodsma, and two anonymous reviewers provided suggestions for improving previous drafts of this chapter. This work received support from an Alexander Hollaender Distinguished Post-doctoral Fellowship sponsored by the U.S. Department of Energy administered through Oak Ridge Institute for Science and Education and from the Office of Health and Environmental Research, Department of Energy under contract No. DE-AC05-840R21400 with Martin Marietta Energy Systems, Inc. This is publication number 4094 of the Environmental Sciences Division of Oak Ridge National Laboratory.

References

Beuchner, M. 1989. Are small-scale landscape features important factors for field studies of small mammal dispersal sinks? Landscape Ecol. 2:191–199.

Braun, E.L. 1950. Deciduous forests of eastern North America. New York: Hafner Press.

Brown, J.H., and A. Kodric-Brown. 1977. Turnover rates in insular biogeography: effects of immigration on extinction. Ecology 58:445–449.

Duffy, D.C., and A.J. Meier. 1992. Do Appalachian herbaceous understories ever recover from clearcutting? Conserv. Biol. 6:196–201.

Flamm, R.O., and M.G. Turner. Land cover changes in rural Western North Carolina. Submitted to Sistema Terra.

Flather, C.H., S.J. Brady, and D.B. Inkley. 1992. Regional habitat appraisals of wildlife communities: a landscape-level evaluation of a resource planning model using avian distribution data. Landscape Ecol. 7:137–147.

Franklin, J.R., and R.T.T. Forman. 1987. Creating landscape patterns by forest cutting: ecological consequences and principles. Landscape Ecol. 1:5–18.

Gardner, R.H., R.V. O'Neill, M.G. Turner, and V.H. Dale. 1989. Quantifying scale-dependent effects of animal movement with simple percolation models. Landscape Ecol. 1:19–28.

Gilpin, M E., and M.E. Soule. 1986. Minimum viable populations: processes of species extinctions. In: M. E. Soule (ed.). Conservation biology: the science of scarcity and diversity. pp. 19–34. Sunderland, Mass.: Sinauer Associates, Inc.

Harmon, M. 1982. Fire history of the westernmost portion of Great Smoky Mountains National Park. Bull. Torreya Bot. Club 109:74–79.

Kotliar, N.B., and J.A. Wiens. 1990. Multiple scales of patchiness and patch structure: a hierarchical framework for the study of heterogeneity. Oikos 59:253–260.

Lee, S., and R.M. May. 1992. Dynamics of metapopulations: habitat destruction and competitive coexistence. J. Animal Ecol. 61:37–40.

Leopold, D.J., and G.R. Parker. 1985. Vegetation patterns on a Southern Appalachian watershed after successive clearcuts. Castanea 50:164–186.

Lindsay, M.M., and S.P. Bratton. 1979. Grassy balds of the Great Smoky Mountains National Park: their history and flora in relation to potential management. Environ. Manage. 3:417–430.

MacArthur, R.H. 1959. On the breeding distribution pattern of North American birds. Auk 76:318–325.

O'Neill, R.V., D.L. DeAngelis, J.B. Waide, and T.F.H. Allen. 1986. A hierarchical concept of ecosystems. Princeton, N.J.: Princeton University Press.

O'Neill, R.V., B.T. Milne, M.G. Turner, and R.H. Gardner. 1988. Resource utilization scales and landscape pattern. Landscape Ecol. 2:63–69.

Pearson, S.M., M.G. Turner, R.H. Gardner, and R.V. O'Neill. In press. An organism-based perspective on habitat fragmentation. In: R.C. Szaro (ed.). Biodiversity in Managed Landscapes: theory and practice. Cary, N.C.: Oxford University Press.

Pulliam, H.R. 1988. Sources, sinks, and population regulation. Am. Nat. 132:652–661.

Pyle, C. 1988. The type and extent of anthropogenic vegetation disturbance in the Great Smoky Mountains before National Park Service acquisition. Castanea 53:183–196.

Runkle, J.R. 1981. Gap regeneration in some old-growth forests of the eastern United States. Ecology 62:1041–1051.

Runkle, J.R. 1982. Patterns of disturbance in some old-growth mesic forests of the eastern North America. Ecology 63:1533–1546.

Scott, J.M., B. Csuti, K. Smith, J.E. Estes, and S. Caicco. 1988. Beyond endangered species: a geographic approach to protecting future biological diversity. Bioscience 37:782–788.

Szacki, J., and A. Liro. 1991. Movements of small mammals in the heterogeneous landscape. Landscape Ecol. 5:219–224.

Terborgh, J. 1974. Preservation of natural diversity: the problem of extinction-prone species. BioScience 24:715–722.

Thomas, C.D., and S. Harrison. 1992. Spatial dynamics of a patchily distributed butterfly species. J. Animal Ecol. 61:437–446.

Trimble, G.R., Jr., and E.H. Tryon. 1966. Crown encroachment into openings cut in Appalachian hardwood stands. J. Forest. 64:104–108.

Wiens, J.A. 1989. Spatial scaling in ecology. Functional Ecol. 3:385–397.

Wiens, J.A., and B.T. Milne. 1989. Scaling of landscapes in landscape ecology, or, landscape ecology from a beetle's perspective. Landscape Ecol. 3:87–96.

White, P.S. 1982. The flora of Great Smoky Mountains National Park: an annotated checklist of the vascular plants and a review of previous floristic work. USDI, National Park Service, Southeast Regional Office, Research/Resource Management Report SER-55.

Whittaker, R.H. 1956. Vegetation of the Great Smoky Mountains. Ecol. Mongr. 26:1–80.

Willson, M.F. 1976. The breeding distribution of North American migrant birds: A critique of MacArthur (1959). Wilson Bull. 88:582–587.

12

Resource Management Perspective: Managing to Minimize Fragmentation of Native Hardwood Forests

David Cawrse

The southern Appalachian forests provide a challenge for ecosystem management. This area, which extends from Virginia to northern Georgia, is one of the most biologically diverse regions in the world. The close proximity to major population centers makes ecosystem management on public lands very complex. Geographic information systems (GIS) have become an important tool. Three important aspects of ecosystem management and fragmentation in the southern Appalachians are discussed. They are:

- Ecosystem management on national forest lands in the southern Appalachians, including a discussion on the importance of scale in making decisions
- Forest fragmentation in that area
- Strategies to minimize forest fragmentation

The George Washington National Forest in Virginia and the Nantahala National Forest in western North Carolina are described as examples. One of the examples used GIS and remote sensing tools, the other did not.

Ecosystem Management in the Southern Appalachians on National Forest System Lands

In June 1992 the Southern Region of the USDA Forest Service developed its "Strategies for the Southern Region" in response to the policy issued by the Chief of the Forest Service. The Chief's statement outlined the following principles:

- Participation of the public and partners
- Integration of management
- Sustainability of values and uses in the landscape
- Collaboration of scientists and partners in applying state-of-the-art thinking and technologies

Ecosystem management is now the basis for developing forest plans and for forest plan implementation of projects such as timber harvesting. Moving from the forest plan to a project has three important parts. Managers must first define the desired condition, determine the existing condition, and then determine the actions that will achieve the desired condition. In the second part, the alternatives and environmental effects are disclosed through the NEPA document. Finally, monitoring and evaluation are designed to provide feedback and to adjust the plan as necessary.

Of the steps mentioned above, defining the desired condition is the most important because that determines the landscape goal. Use of remote sensing and GIS can be valuable techniques for defining both the desired conditions and existing conditions over different spatial scales. Scale is important—all ecological processes and types of ecological structure are multiscaled—and so are the related decisions. GIS and remote sensing techniques aid in modeling and decision-making at these different scales and levels.

How we classify ecological systems has a parallel with how we make decisions. Roughly five levels of decision-making exist regarding National Forest System lands (Salwasser 1992). The national hierarchy for levels of classification (McNabb 1993) and the relationship to Forest Service planning and analysis are shown in Table 12.1.

Yet, our decisions on National Forest System lands usually focus only on a few of these levels and are often within a context of another level. Decisions are made at readily observable levels. Sometimes these decisions appear as answers to simple dichotomous questions at the ground level. Does this tree get marked to cut or not? Does this stand get prescribed burned or not? Does this trail get built or not? But complexity increases when context and scale are considered through time and space. Changing scale can change relationships that may affect the decision. Does this tree get cut next to this old-growth stand? Does this trail get built when it is in a known bear corridor? Does this stand get prescribed burned when water yields are critically high due to past cutting on private land? Decisions made at one scale usually influence at least one level

TABLE 12.1 National hierarchy for levels of classification and the relationship to Forest Service planning and analysis

Organizational level	Land division	Related laws or regulations
National or Congressional level	Domain, Division, or Province	Environmental laws, where our national parks, forests, and wildernesses occur
Regional level	Section	Regional guidelines, recovery plans
Forest level	Landtype association	Forestwide guidelines and standards, objective for the area and suitable lands
District level	Landtype (watershed)	When and where harvesting occurs, delineation of old growth, road closure
Project	Landtype phase or site	What type of harvesting, marking guidelines, whether a stand is old growth

below and one above; therefore, a decision-maker should take these other levels into consideration.

The grain (detail or precision) of information needed at each level varies. Decisions over wide extent (which includes spatial and temporal considerations) may be confused by too many fine-grain observations, so a coarse-grain approach would be more appropriate. For example, at the stand level, a forester might need information on the species, density, vigor, and age of a stand. A district ranger making a decision on a timber sale would be overwhelmed by information on each tree or even a stand—the information is too fine grained. More useful information might be: How much volume is removed? Are viable populations maintained for all indicator species? Are visual quality objectives met? At the forest level, a forest supervisor would be overwhelmed by such information on individual sales. Again a coarser grain of information is needed. The grain needs to be appropriate to the information needed in decision-making. Understanding the relationships between all the levels will help the manager to find a satisfactory answer.

Forest Fragmentation of Native Hardwood Stands in the Southern Appalachians

National Forest System lands in the southern Appalachians are all purchased lands. Nearly all these lands were farmed and logged over and characterized by fragmented ownership. Much has been written about these "lands that nobody wanted" (Shands 1977). A quote from a 1905 report characterizes this region (Ayers 1905): "At present, trees of choicest timber are killed to make fields on which corn costs $1 a bushel or to be grazed until worn out and gullied by rapid erosion. On these clearings the mountaineers make only a miserable living. The markets are distant, the once abundant game is gone, the population is sparse, and the roads wretched. The material prosperity of these people depends on the development of the one important natural resource—the forest."

In 1911, the Weeks Act gave the federal government authority to purchase lands. Purchase units were formed, and by 1918 the first national forest in the southern Appalachians was proclaimed. Today, within the southern Appalachians, 4.4 million acres are contained in eight national forests. These lands form the largest single ownership in the southern Appalachians. In the early years, gaining access to the area and flood prevention were the most important management objectives. Today, the issues are different—restoration of old-growth habitat, population viability of rare plant and animal species, forest fragmentation, invasion of nonnative species, visual quality, chronic air pollution, global warming, and ozone depletion.

Forest fragmentation is a major concern in the southern Appalachians for two reasons (Harris 1984). First is the loss of habitat for interior species that are dependent on contiguous forest. Second is isolation of remaining habitat in islands surrounded by hostile environments. Further loss of habitat may occur due to edge effects. Isolation reduces gene flow between populations and thus threatens genetic diversity. Many endangered, threatened, and sensitive species are rare because human activities have greatly diminished their populations. Ginseng, for example, is a plant that was once widely distributed but is now rather rare, with population viability a concern in some areas.

Forest fragmentation varies both spatially and temporally, and its effects vary by species. Fragmentation should be analyzed across the landscape encompassing both public and private lands. On public lands, particularly growth and variety of life give the southern Appalachians resiliency to changes in vegetation from harvesting, although some researchers believe that secondary forests may be less diverse biologically than old-growth forests (Duffy and Meier 1992). But with approximately 50 percent of the area of the national forests in the southern Appalachians not suitable for harvest, forest fragmentation on private lands is more likely to threaten species' viability and biodiversity. On private lands, fragmentation tends to be permanent if pasture is developed for residential use. It is likely to remain in this condition for several generations. An aerial flight over Highlands Ranger District in western North Carolina reveals that nine golf courses and the associated second-home developments are the greatest source of fragmentation there. Social patterns also change in the landscape. Road densities are greater, pets (particularly dogs) are common in the area, and poaching may be common. Black bear is usually the first species to be displaced from the area. Also the effects of forest fragmentation vary as to the matrix within which they occur. A 30-acre clearcut in a 60-acre woodlot surrounded by agricultural fields has a much different effect than a 30-acre clearcut within a 10,000-acre forested landscape.

For U.S. Forest Service managers, fragmentation must be managed to ensure population viability of all species across the forest. Management indicator species are used to monitor the effects. Many forest plans now include species that are sensitive to forest fragmentation. In North Carolina, for example, these species include black bear, oven bird, and cerulean warbler. Interior obligates, such as the oven bird, need a minimum patch size to be successful, while species with large home ranges, such as the black bear, need connecting corridors. Although the indicator species concept has been criticized as an inadequate approach to ecosystem management (some species' habitats may be missed), the manager does have discretion to analyze other indicators of ecosystem health at the project level. As fragmentation becomes more critical, three factors need to be analyzed. They are isolation, patch size, and forest reserve (Hillman 1990). Analyses should also include cumulative impacts on the entire landscape for public and, as much as possible, private lands. This is required in NEPA (40 CFR 1508.7).

National forests in the southern Appalachians hold the best hope for minimizing forest fragmentation and maintaining species viability in that region. The following illustrations show what is being done to analyze and minimize forest fragmentation. GIS was used to construct the George Washington National Forest plan but was not available for the Nantahala National Forest.

Nantahala National Forest

The headwaters of the upper Chattooga River contain 180,000 acres, of which 67 percent is in the National Forest System. The National Forest System lands within this watershed are managed by three different districts on three different national forests in three different states. The Chattooga River, which flows through this tri-state area, is a Wild and Scenic River and is considered by some to be the flagship of the Wild and Scenic River System. In 1990 a timber sale was planned on the Highlands Ranger District in North Carolina in a tributary of the Chattooga River. The analysis process that was followed for this sale was described earlier. The desired condition was defined over a 10,000-acre area and compared with the existing condition. The area was already fragmented from past harvesting, particularly by clearcutting; also, 40 percent of the area was developed private land. One of the desired conditions was to maintain a continuous canopy cover. This condition was formulated in response to visual quality objectives as well as concerns on forest fragmentation. Key issues related to forest fragmentation were the effects on old growth, salamanders, and neotropical migratory birds.

The first step in meeting that desired condition was to allocate and map old growth. Since many interior forest birds have an aversion to large openings, stand size was maximized as much as possible. Rather than have two 25-acre stands, one 50-acre stand would be designated in an area. (It is recognized that even this stand size is not large enough for many forest interior obligates; however, the stands do not occur as isolated patches but rather as old stands within second-growth forests of the same tree height.) A diversity index was developed (Carlson 1990) to measure how different the old-growth candidate stands were. This was done to ensure that a variety of sites would have adequate representation in the landscape—from cove sites to upland hardwood sites. Also, distances between old-growth candidate stands and connecting corridors were a concern. Riparian zones were used as connecting corridors. How effective these corridors are as dispersal mechanisms remains to be seen, as each organism encounters a corridor differently. The final step was hand mapping these stands and analyzing alternatives that would minimize fragmentation.

A timber sale designed for the area involved harvesting using thinning, two-age shelterwood (where about 40 square feet of basal area is retained, with no additional removal planned), and group selection. Groups are small (1/3 acre to an acre) and vigorous pole timber was maintained within a group, so they do not appear as miniclearcuts. Use of group-selection harvesting received the most criticism in the project. Although the analysis characterized group selection as appearing as freckles across the stand, one person commented that it was more like smallpox infecting the skin of the earth. Possibly, however, the use of small groups may mimic the successional patterns caused when large trees fall over and young trees fill in the openings. The desired conditions chosen attempt to replicate natural disturbance patterns in the landscape.

Although the effect this sale would have on neotropical migratory birds is not known, research is underway by the Southeastern Forest Experiment Station. It is interesting to note that initial research findings show that no cowbirds (a negative indicator of habitat for interior obligates) were observed on any of the research plots, which may be because few feeding sites are present for the birds. Also interesting on the Highlands Ranger District is that the greatest diversity of interior birds was not in old-growth stands but on 60-year-old stands that were thinned 10 years ago. These stands had an overstory of white pine, a midstory of hardwoods, and an understory of rhododendron and mountain laurel (Hawkinson 1991).

With rainfall sometimes exceeding 100 inches per year and mild summers and winters, the southern Appalachians are home to a wide variety of salamanders. The endangered green salamander (*Aneides aeneus*) finds optimal habitat in rock outcrops, which are currently being mapped by the district. No harvesting will occur within 100 feet of potential sites. Harvesting techniques that meet the desired condition of continuous canopy cover would maintain diffuse shade across the ground. Fragmentation effects should be minimal.

Although GIS and remote sensing were not used in the Big Creek analysis, there were several potential applications: old-growth inventory and mapping; mapping of habitat for indicator species, such as salamanders and neotropical migratory birds; defining existing and future disturbance patterns on the landscape on both private lands and National Forest System lands; inventory and coordination of data across state boundaries and private lands; and delineation of past harvest patterns. A map of areas developed or likely to be developed versus lands that are preserved or protected (wilderness, state parks, unsuitable national forest lands) would have aided in identifying fragmentation patterns. Edge effects could be modeled that would map effective, functioning habitat for certain species. Patch size and isolation relative to the forested area could be analyzed. Forest successional patterns could be modeled to see if desired conditions are met; perhaps the spatial models could be linked with silvicultural expert systems that will aid in recommending to the manager those systems that would best achieve desired conditions. Although black bear was not analyzed in this project, GIS and remote sensing could aid in giving representation of the different criteria for black bear management. This includes calculating road densities, identifying early- and late-successional habitat, and mapping potential population corridors between known populations. This would need to be done at a large scale (50,000 to 200,000 acres).

George Washington National Forest

GIS was a key component in the recent update of the George Washington National Forest Plan. Primary concerns were management of habitat for the Cow Knob Salamander (*Plephodon punctatus*) and inventory and analysis of effects on old growth (Croy 1993).

The Cow Knob Salamander occurs only on Shenandoah Mountain, an area approximately 24 miles long and one mile wide. The Forest Service, in cooperation with the U.S. Fish and Wildlife Service and Virginia Division of Natural Heritage, developed a prelisting recovery plan for this salamander. Based on past inventory, it was found the salamander preferred habitat above 3000 feet with a forest type of late-successional or old-growth mixed hardwood stands dominated by oak. By using forest stand inventory data stored in an Oracle database, GIS was used with ArcInfo software to develop a map of this habitat. Managers looked for bottlenecks, or connecting corridors, of suitable habitat. In these critical areas, vegetative removal and rock collecting were not allowed. This approach should minimize fragmentation and enhance recovery of the salamander.

With the help of the Nature Conservancy, the Forest Service identified 10 different old-growth forest types on the forest. These types developed old-growth characteristics at different ages (e.g., 80 years for yellow pine types and 180 years for cover types). By using forest stand inventory data and GIS, a map was produced of old-age stands by old-growth type. This resulted in about 18 percent of the forest, or more than 180,000 acres, inventoried as having old-growth characteristics. Spatial arrangement was analyzed, along with effects on old-age stands from various plan alternatives and management area prescriptions. Connecting areas were developed using designated wilderness, backcountry areas, special biological areas, riparian areas, and unsuitable areas. Because most mountains run northeast to southwest, riparian areas were used as connectors across valleys, although private lands made it difficult to connect some of the areas. This approach of using coarse-grain data to minimize fragmentation effects on old-age stands appeared successful.

In summary, GIS and remote sensing should be an integral part of the plan-to-project implementation requiring defining the desired and existing conditions; GIS can assist the manager in assessing environmental impacts of proposed projects; and GIS can also aid the manager in monitoring so that proper adjustments may be made to achieve desired conditions on the landscape. Using GIS and remote sensing to integrate our thinking in all these areas will significantly aid forest managers in implementing ecosystem management.

References

Allen, Timothy F.H., Thomas W. Hoekstra. 1992. Toward a unified ecology. New York: Columbia University Press.

Ayres, H.B., W.W. Ashe. 1902. The Southern Appalachian Forests. Washington, D.C.: U.S. Government Printing Office.

Carlson, Paul. 1990. Analysis of the Upper Big Creek and Clear Creek drainages of the Highlands Ranger District for identification of Old-growth Restoration Areas. Unpublished. Highlands, NC, December 31.

Duffy, David C., and Albert J. Meier. 1992. Do Appalachian herbaceous understories ever recover from clearcutting? Conservation Biology, June, 1992.

Harris, L.D. 1984. The fragmented forest. University of Chicago Press, Chicago, IL.

Hawkinson, Dr. Margaret. 1991. Personal communication. Highlands, NC, July.

Hillman, Lauren L. 1990. Forest fragmentation: Implications for national forest system management in the Southern Region. Asheville, National Forests in North Carolina, March 1.

McNabb, Henry. 1993. Personal communication. Bent Creek Experimental Forest, Asheville, NC, February.

Salwasser, Hal. 1992. Personal communication. Southern Appalachian Man and the Biosphere Conference, Gatlinburg, TN, November.

Shands, William E., and Robert G. Healy. 1977. The lands nobody wanted. The Conservation Foundation, Washington, DC.

13

GIS Applications Perspective: Multidisciplinary Modeling and GIS for Landscape Management

Richard O. Flamm and Monica G. Turner

Ecological dynamics in human-influenced landscapes are strongly affected by the socioeconomic factors that influence land-use decisions. Incorporating these factors into a spatially explicit landscape-change model requires the integration of multidisciplinary data. A model was developed to simulate the effects of land use on landscape structure in the Little Tennessee River Basin in western North Carolina. This model uses a variety of data, including interpreted remotely sensed imagery, census and ownership maps, topography, and spatial representations of social and economic data. Data are integrated by using a geographic information system and translated into a common format—maps. Simulations generate new maps of land cover representing the amount of land-cover change that occurs. With spatially explicit projections of landscape change, issues such as biodiversity conservation, the importance of specific landscape elements to conservation goals, and long-term landscape integrity can be addressed. For management to use the model to address these issues, a computer-based landscape-management decision aid is being developed. This tool, called the Land Use Change Analysis System (LUCAS), integrates the models, associated databases, and a geographic information system to facilitate the evaluation of land-use decisions and management plans. This system will estimate landscape-level consequences of alternative actions and will serve to focus coordination among different landowners and land-use interests in managing the regional landscape.

The southern Appalachian landscape is a product of the interaction between ecological and socioeconomic processes. Effective landscape management in this region requires (1) an understanding of how these processes are linked in time and space to influence landscape dynamics, and (2) a methodology that takes this knowledge about the landscape and makes it available to managers. One approach for linking the processes is to develop a spatially explicit multidisciplinary model that can simulate landscape change induced by land use and then evaluate its impacts on ecological and resource supply variables. The advent of geographic information systems (GIS) and remote sensing makes construction of this model feasible. Applying this model in a landscape management program involves "packaging" it in a form that is desirable to the decision-makers. Designing this "package" requires a cooperative effort among computer scientists, research scientists, and potential users of the model.

In this chapter, we present an approach for integrating ecological and socioeconometric information for application in a landscape management program. First, we discuss how one can integrate information for use in landscape-level conservation planning. This discussion is placed in the context of a landscape-change simulation model being developed for the southern Appalachians and the Olympic Peninsula. Second, we present a methodology of how this model can be applied to address landscape management and conservation questions. This methodology is discussed in terms of the landscape-management decision aid being developed called the Land Use Change and Analysis System (LUCAS).

Linking Ecology and Socioeconomics for Simulating Landscape Change

Landscapes traditionally are viewed ecologically as a mosaic of land cover types (Forman and Godron 1986). For example, in the southern Appalachians, the landscape can be characterized as forest with interspersed patches of agriculture, range or brushy lands, urban areas, and wetlands. Landscapes can also be viewed as a mosaic of socioeconomic units called ownership tracts. Individual tracts can be categorized as a subset of either public or private land (e.g., state, USDA Forest Service, residential, commercial, industrial). As management goals differ among the ownership categories, so too may the land use differ. Consequently, the structure and function of the landscape is directly related to the abundance and arrangement of land in each ownership category, in addition to the mosaic of ecologically defined patches. Combining these ecologic and socioeconomic views is necessary for a more complete understanding of landscape-scale processes and, therefore, more informed conservation-management decisions.

In the southern Appalachians, a model is being developed that integrates the socioeconomic and ecologic views of the landscape for the purpose of simulating the influence of land use on landscape change and its impacts. Model development revolves around two considerations. First, can we represent information derived from the socioeconometric and biological disciplines as a common data structure to facilitate integration? Second, can solutions to landscape management and conservation problems be generalized? In other words, can a single solution serve as a standard approach for addressing a broad array of landscape management issues?

Integrating Ecology and Socioeconomics

In the landscape-change model, integration is accomplished through the database, and the unifying data structure is the raster map. All data used and produced by the model are represented as a map. Maps created by the model can then be evaluated for specific purposes, such as an analysis of changes in the landscape's biodiversity.

The data layers relate to land cover, land use, access or transportation costs, and land-use potential. The data used to create these layers originate in many forms, including remote imagery, digital elevation models, census tract data, TIGER/Linc™ census files, county tax assessor maps of private ownership boundaries, and federal ownership maps. Some of the information is used directly, like the road network maps from the TIGER data. Other maps must be created, such as land cover (e.g., interpretations of remote imagery), land use (e.g., combination of county tax assessor maps, interpreted remote imagery, and TIGER data), land-use potential (e.g., composite of elevation and slope), and maps of access or transportation costs (e.g., the distance between each patch and a road or the nearest market or cultural center).

Land-cover maps were created by interpreting 80-meter resolution multispectral scanner (MSS) images. Four MSS images from 1975, 1980, 1986, and 1990 were used to create a time series of land-cover change for the Little Tennessee basin. This time series was used to derive probabilities of land-cover change for input into the landscape-change model. The land-cover classifications included forest, disturbed/unvegetated (includes urban and recently cleared areas), agricultural/grassy/brushy (includes row crops, rangeland, lawns, young regrowth, etc.), water, bare rock, and balds.

Land cover is an expression of land use. Different land uses, however, might occur within the same land-cover class. For example, an area classified from a remote image as hardwood forest might be used as an unmanaged woodland, recreation area, plantation, or a wooded residential area. Although these areas might appear identical on the remote image, they probably function differently within the landscape. In the landscape-change model, a land-use map is being constructed from county tax assessor maps and TIGER data. Land-use classifications include commercial, residential, industrial, agricultural, cleared, other forested, and transportation. Land-use potential was represented as an overlay of elevation and slope maps derived from USGS 1:24,000 DEMs.

Access and transportation cost maps were created using TIGER files of road networks. Access measures the cost associated with movement away from paved roads. This cost was estimated by creating a map of the distance that each pixel is located from a paved road. Transportation measures the costs associated with distance along roads to specific points like market or cultural centers. In this map, each cell was assigned a value representing the shortest distance from a point on a road nearest to each pixel to the center of Franklin, North Carolina, the major market and cultural center in the Little Tennessee watershed.

Of the data layers discussed above, the most fundamental to the model are the land-cover maps derived from the remote images. The selection of the type of imagery and the time periods determine the spatial and temporal scale by which land-cover changes are measured by the model. Improvements in remote imagery and the software available for their interpretation have greatly enhanced our versatility to create land-cover maps at a variety of scales and detail.

Linking the land-cover maps with the other data layers is accomplished using a GIS. The GIS is used to overlay the data to construct a composite map. The composite map is represented as an ASCII file for input into the model. "Cells" in the composite map are categorized by a string of characters called a landscape-condition label. Each character of this label is a category value from one of the original maps. For example, if the label for a cell from the composite map is 3264, the first position (4, moving from right to left) might be a land-cover category, the second position (6) land use, the third position (2) distance-to-the-nearest-road, and so on. Because the landscape-condition label is a character string, its length (e.g., the number of data layers) is essentially not limiting. A time series of the composite maps can then be used to estimate transition probabilities for land-cover change based on a wide variety of spatial information. They can also be applied in the landscape-change simulation or used to assess the environmental and socioeconomic impacts of change.

A General Solution

The southern Appalachians is a landscape dominated by humans. Much of the landscape remains unchanged from year to year, however, because decision-making on the large public ownerships (Great Smokies National Park and several national forests) is oriented toward preservation, wildlife management, recreation, water resources, and forestry. Regardless, every hectare is under the stewardship of humans, and consequently landscape properties such as fragmentation, connectivity, and the degree of dominance of habitat types are influenced by market processes, human institutions, and landowner knowledge in addition to ecological processes. With this consideration, the following general solution was proposed for the landscape-change model (Lee et al. 1992) (Figure 13.1). First, an econometric analysis for estimating the propensity for ecological processes, market forces, and social factors to influence land use in the region is conducted. Second, the results of the econometric analysis are passed to a landscape-change simulation model and changes in landscape structure based on land use are estimated. Third, the results of the landscape-change simulation are passed to models that estimate impacts of change on environmental integrity and resource supplies. Fourth, results of the environmental- and socioeconomic-impact models are then analyzed for their influence on landscape sustainability. We will discuss the first three steps of the general solution as they are being applied in the southern Appalachian landscape.

The econometric analysis involves estimating probabilities of land-cover change as a function of selected socioeconometric driving variables. Presently, these variables include transportation networks (access and transportation costs); slope and elevation (indicators of land-use potential); ownership (land-holder characteristics); and land cover.

Preliminary analysis of the Little Tennessee river basin revealed that the watershed remained primarily forested, although there was a net loss of 4580 hectares of forest between 1975 and 1986. Areas categorized as disturbed/unvegetated increased almost fourfold, and the grassy/brushy cover class increased by about 70 percent. Although only 5 to 7 percent of the forested sites changed cover, the landscape was indeed dynamic. Less than 20 percent of the disturbed/unvegetated and grassy/brushy cover types remained in these categories between time intervals.

Probabilities of land-cover change were directly related to whether the parcel was privately or federally owned. Between 9 and 10 percent of all private lands experienced a land-cover change during the study period,

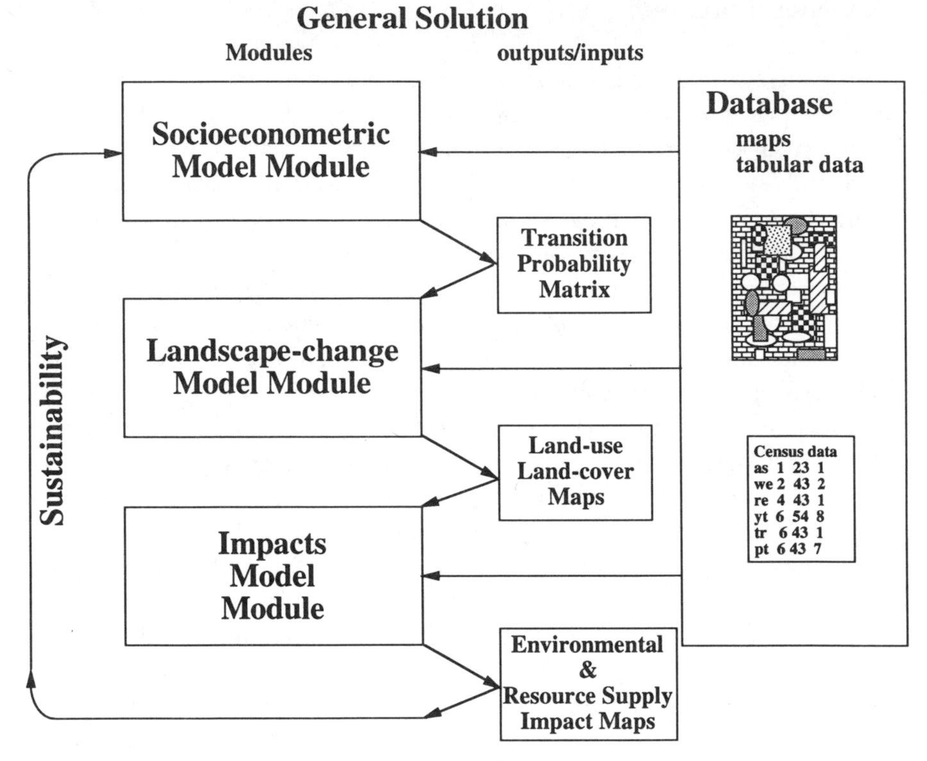

Figure 13.1 The three principal modules of the general solution being applied in the landscape-change model in the southern Appalachians. The output of each module serves as the input to the subsequent module. The database is shared by all modules.

compared to only 0.8 to 1.6 percent of federal lands. Thus, a given parcel of land was 5 to 11 times more likely to undergo land-cover transition if it was in private ownership.

A statistical analysis indicated that physical and spatial attributes of a site influenced land-cover change. In particular, the probability of forested land being harvested or developed is influenced significantly by the slope and elevation of a site, as well as the distance between the site and a paved road, and between the road location and the center of Franklin, North Carolina, the major market and cultural center in the watershed.

In general, land-cover change is most likely to occur on private land, near a paved road, at low elevations, with mild slope, and close to Franklin. Most of the transitions in land cover are forest converting to grassy/brushy and disturbed/unvegetated cover types.

Impacts of land use on landscape structure are estimated by applying the transition probabilities of land-cover change in a Markov model. Each grid cell in the map is evaluated for land-cover change based on probabilities associated with its landscape condition label. Several simulations are run to produce a set of possible land-cover maps for a future time. Spatial measures are then calculated for each map and represented as frequency histograms. Land cover change can then be evaluated in terms of each spatial measure's mean, median, mode, maximum, minimum, percentiles, distribution shape, and so on.

For example, simulations of land-cover change were run using a transition matrix derived from a coincidence analysis for the period beginning in 1975 and ending in 1986 to estimate land cover for the Little Tennessee River basin for 1997. Two frequency histograms of forest change were derived from the simulated maps, proportion cover and patch size distribution (Figure 13.2). Evaluation of the histogram for proportion cover suggests that insignificant change in proportion forest is most likely from 1986 to 1997. However, the structure of the landscape is expected to change significantly, with the forest becoming increasingly fragmented. The increase in forest patches 1 pixel in size indicates that conversions from forest changes are expected to occur near highways and close to Franklin.

The simulated maps can also be evaluated for changes in specific environmental and resource supply variables. For example, risk to an endangered species can be examined by comparing the amount of habitat or the number, size, and spatial arrangement of habitat patches that are in the initial and simulated landscapes. Fundamental to this type of evaluation is designating the rules that define a patch for a specific species. For example, habitat could be defined as simply as forest or as complex as interior hardwood forest with no less than 100 meters to an edge, at least 7 hectares in size, containing at least one perennial stream, and no closer than 3 kilometers from commercial, industrial, or residential land uses.

Landscape Management Application: The Knowledge System Environment

Packaging the model so that it is useful to natural resource decision- and policymakers is a significant issue. Failure to address this issue will result in the model having limited utility and applicability in a management setting. Fortunately, techniques derived from artificial intelligence (AI) concepts and object-oriented programming fundamentals can be used to make contemporary modeling technology available to landscape man-

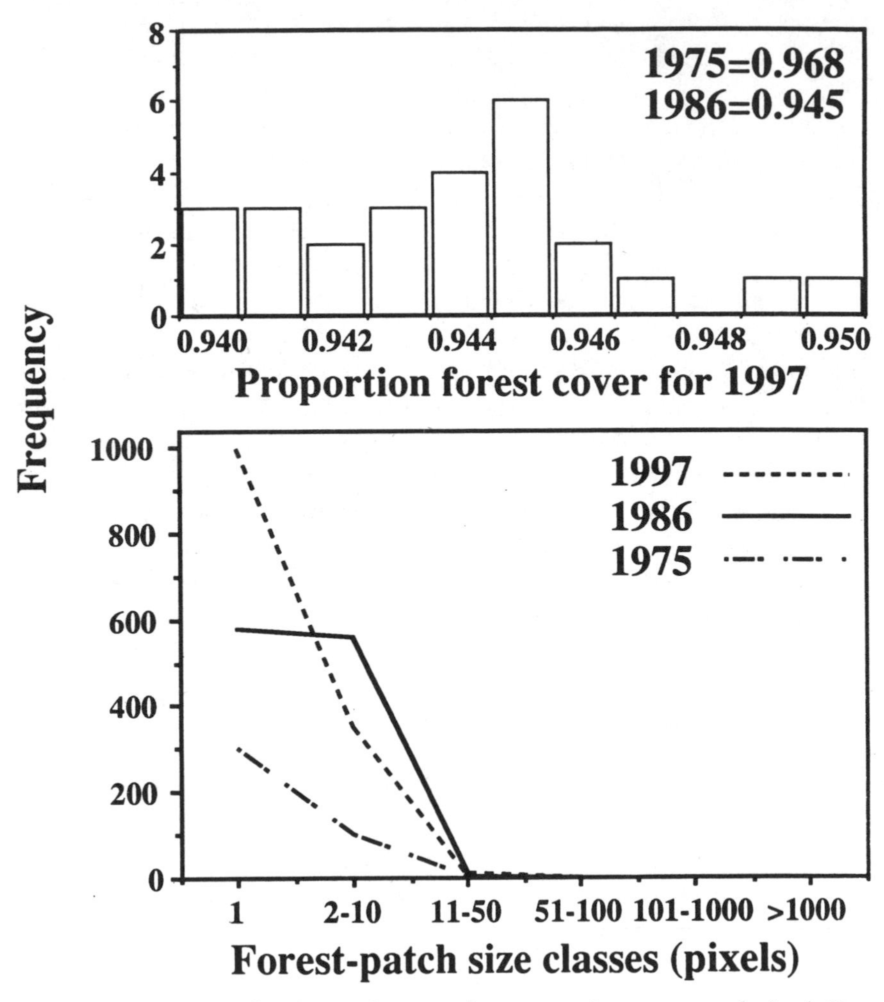

Figure 13.2 Frequency histogram of estimated proportion forest cover in the Little Tennessee River basin for 1997 that was derived from 30 simulations using the landscape-change model and forest-patch size distributions for 1975, 1986, and the estimated mean for 1997.

agers (Tanimoto 1987, Saarenmaa et al. 1988, Folse et al. 1989, Flamm et al. 1991).

A modeling environment that employs AI and object-oriented programming techniques is called a Knowledge System Environment (KSE) (Coulson et al. 1989). A KSE is a computer-based methodology developed to address issues of integration and application of different forms of information to solve unstructured problems. A KSE being designed for address-

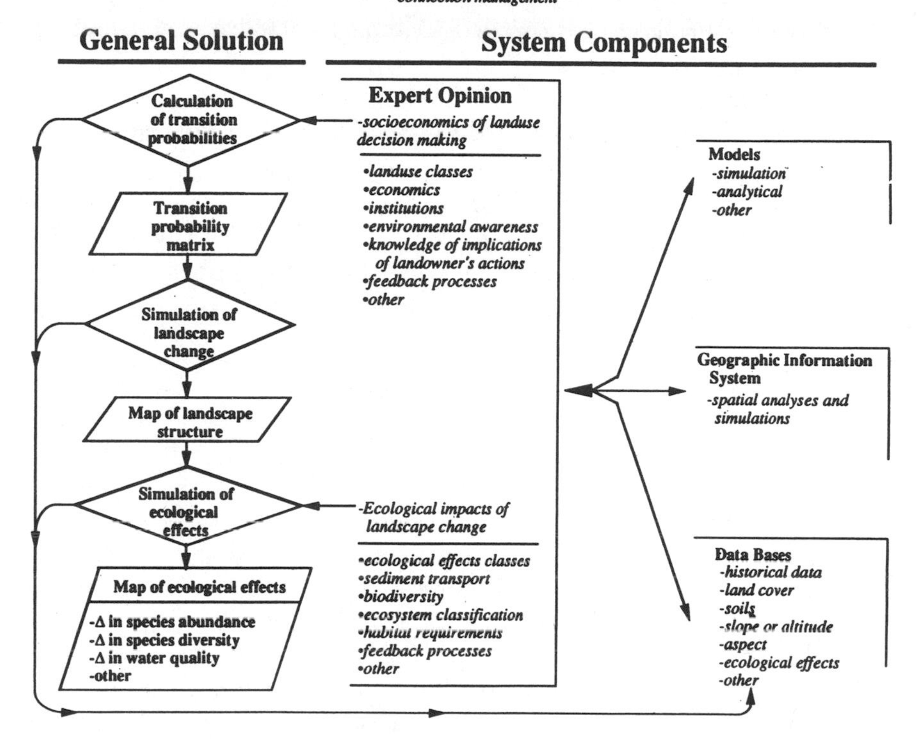

Figure 13.3 Schematic of the Land-Use Change and Analysis System. In the general solution, diamonds represent actions and parallelograms symbolize products. The entire system is being developed within a graphic user interface.

ing problems in landscape management for the southern Appalachians is called the Land Use Change and Analysis System (LUCAS) (Figure 13.3). LUCAS has four distinct modules: a model base, GIS, database, and a graphic user interface (GUI). The model base houses the quantitative models. The GIS manipulates the spatial data. The database serves as the reservoir for nonspatial data. The graphical user interface in LUCAS serves several functions: (1) it is how users communicate with the system; (2) it is used to connect system modules; and (3) it contains the expert opinion represented as the contents of windows, the order that windows "pop-up" on the screen, and the interpretation of quantitative model results.

LUCAS has many attributes, all of them a function of object-oriented methodologies and AI concepts. First, recently acquired data or new technologies such as a simulation model can be incorporated with little impact on those components of the system not being changed. As such, LUCAS can serve as a warehouse for knowledge gained from research and development. Second, event-driven, in addition to time-driven, land-use scenarios can be evaluated. This feature is particularly beneficial to landscape management, because it allows for the examination of the impacts of a specific event or series of events, such as the construction of a road or the expansion of an urban area. Third, LUCAS can provide facilities for documenting the logic behind a decision as well as help guide a regional research and development program. Such documentation may be necessary in the advent of litigation or meetings with concerned citizens.

KSE as a Flexible Management Decision Aid

Landscape management requires estimating impacts of specific actions as well as evaluating plans for achieving a desired future condition. Given that a landscape is a mosaic of private and public ownerships with different land-use goals, evaluating management alternatives or selecting a future condition must include those interests affected by the decisions. Furthermore, the complexity of landscapes and their broad spatial scale prohibit traditional hypothesis testing as a primary tool for landscape-management decision evaluation. An approach that emphasizes input from interested parties, is sufficiently flexible for evaluating land-use options, and provides replication through simulation experiments is called adaptive management (Holling 1978, Lee 1986, Walters 1986). This approach is serving as the design concept for LUCAS.

In adaptive management, a small group of people with experience in integrating information and coordinating resources to solve problems is assembled (Holling 1978). Integration comes from the application of systems analysis techniques like computer modeling. Coordination involves identifying a series of steps needed to evaluate a desired future condition or management action. LUCAS is being designed to address both these tasks. As discussed previously, integration occurs through the transformation of data to a common scale and format and through the links constructed between modules of the KSE. Coordination is accomplished by extracting knowledge about a specific landscape-management issue from land owners and managers, integrating this knowledge into LUCAS for experimentation, and then land owners and managers evaluating the results of the simulations and, hopefully, arriving at a decision.

The success of LUCAS will depend on its functionality and acceptability. The primary function of LUCAS is as a facilitator for making a choice from a set of alternatives. Thus LUCAS must be capable of at least one or more of the following functions during a decision evaluation. First, it must be able to generate estimates for a set of social, economic, and ecological indicators to use in an impact assessment. Second, LUCAS may need to calculate a frequency distribution and perform a sensitivity analysis on these indicators. Third, each management plan or action will need to be evaluated in terms of the behavior of relevant indicators over space and time. Fourth, LUCAS may need to consider tradeoffs between one indicator and another during an analysis session. This may require ranking the indicators in order of importance. Fifth, LUCAS must serve to encourage communication and interaction among research scientists and developers, landscape managers and their staff, and those who must endure management policies or actions. For example, misunderstandings between sponsoring land-management agencies and research institutions about deliverables might be avoided if a standard existed for product specifications. A system like LUCAS can provide this standard.

System acceptability among users is just as important as functionality. For acceptability, LUCAS must maximize "user" involvement so that the local scientists, managers, owners, and other interested parties specify system utility. LUCAS sessions also must be made fully transparent and interactive, typically done through the extensive use of graphics and an interactive computer environment. In other words, direct interactions with databases and models are minimized or made transparent to the user, while system flexibility is emphasized. An effective communication network among scientists, managers, and land owners must be maintained. It is through this network that a foundation of confidence and understanding among system contributors is constructed. These networks are often initiated through workshops that include representatives of various landscape management, research and development, environmental, or ownership groups.

In the southern Appalachians, efforts are being directed toward landscape management. These efforts benefit greatly from advances in remote sensing technology and GIS. Improvements in remotely sensed data have increased our ability to interpret changes in land cover. Geographic information systems have simplified the integration of multidisciplinary spatial information. We have described two approaches—spatially explicit models and the KSE—that can take advantage of improvements in these technologies. We also described an application of these approaches and technologies, LUCAS, that will bring these new developments into practice by providing a flexible and interactive environment that caters to those people interested in managing the landscape.

Acknowledgments

We thank S. Pearson, W. Hargrove, and J. Ranney for their comments on early drafts of this manuscript. This research received funding from the Ecological Research Division, Office of Health and Environmental Research, U.S. Department of Energy, under Contract No. DE-AC05-840R21400 with Martin Marietta Energy Systems, Inc. Funding was also received from the U.S. Man and Biosphere Program, U.S. Department of State; grant No. 1753-000574. This research was also supported in part by an appointment by the U.S. Department of Energy Laboratory Cooperative Post-Graduate Training Program administered by Oak Ridge Associated Universities. Publication No. 4276 of the Environmental Sciences Division, ORNL.

References

Coulson, R.N., M.C. Saunders, D.K. Loh, F.L. Oliveria, D. Drummond, P.J. Barry, and K.M. Swain. 1987. Knowledge system environment for integrated pest management in forest landscapes: The southern pine beetle (*Coleoptera: Scolytidae*). Bulletin of the Entomological Society of America 34:26–33.

Flamm, R.O., R.N. Coulson, J.A. Jordan, M.E. Sterle, H.N. Brodale, R.J. Mayer, F.L. Oliveria, D. Drummond, P.J. Barry, and K.M. Swain. 1991. The integrated southern pine beetle expert system: ISPBEX. Expert Systems with Applications 2:97–105.

Folse, L.J., J.M. Packard, and W.E. Grant. 1989. AI modeling of animal movements in a heterogeneous habitat. Ecological Modeling 46:57–72.

Forman, R.T., T. and M. Godron. Landscape ecology. 1986. New York. John Wiley and Sons.

Holling, C.S. 1978. Adaptive environmental assessment and management. Chichester, England. John Wiley and Sons.

Lee, K. 1986. Adaptive management: Learning from the Columbia River Basin Fish and Wildlife Program. Environmental Law 16:431.

Lee, R.G., R.Flamm, M.G. Turner, C. Bledsoe, P. Chandler, C. DeFerrari, R. Gottfried, R.J. Naiman, N. Schumaker, and D. Wear. 1992. Integrating sustainable development and environmental vitality: A landscape ecology approach. In Watershed management: Balancing sustainability and environmental change, edited by R.J. Naiman, 499–521. New York. Springer-Verlag.

Saarenmaa, H., N.D. Stone, L.J. Folse, J.M. Packard, W.E. Grant, M.E. Makela, and R.N. Coulson. 1988. An artificial intelligence modeling approach to simulating animal/habitat interactions. Ecological Modeling 44:125–141.

Tanimoto, S.L. 1987. The Elements of artificial intelligence. Rockville, MD: Computer Science Press, Inc.

Northern Lake States Late-Successional Forests: Ecological Classification and Change Detection as Guides in Ecosystem Protection and Restoration

Introduction

Thomas R. Crow

The Ecological and Social Setting

The northern Lake States (Michigan, Minnesota, and Wisconsin) comprise one of the most densely forested regions of the nation, with 41 percent of the total area or 50.5 million acres in forest land (Shands and Dawson 1984). About 46 million acres of the total forest land is considered commercial forestland, defined by its ability to grow at least 20 cubic feet of timber per acre and that it has not been withdrawn from timber harvesting by legislative or administrative action. Twenty-four million acres or 52 percent of the commercial forestland is owned by private, nonindustrial owners (including farmers), while forest industry owns 4 million acres or 9 percent of the commercial forestland. Public ownership of forestland is also substantial in the northern Lake States. In total, municipal, county, state, and federal lands account for 17 million of the 46 million acres of commercial forestland (Skok and Buckman 1983, Shands and Dawson 1984).

The Great Lakes region is characterized by its physiographic and climatic diversity that, in turn, create uniquely diverse terrestrial and aquatic ecosystems. Due to its latitude and position on the North American continent, the region is subject to great temperature extremes and moderate amounts of precipitation distributed throughout the year. Strong climatic gradients create a prairie-hardwood-boreal forest transition within the region (Curtis 1959, Mladenoff and Pastor 1993). The hardwood and conifer forests are an integral part of a regional landscape that includes productive agricultural lands providing food for a global population, industrial and manufacturing complexes located in large urban centers, and an abundance of natural resources, including minerals, water, and timber, that are viewed as cornerstones for future economic development.

The Great Lakes, the defining regional feature, lie along the southern edge of the Canadian or Laurentian Shield, a hard mass of Precambrian rock more than 500 million years old. To the south of this ancient rock is younger and softer Paleozoic rock composed of sandstone, shale, limestone, and dolomite 200 to 500 million years old that have been arranged and rearranged by four glacial advances and declines during the Pleistocene. The regional landscapes are relatively young in the geologic sense. Most of the region was glaciated during the most recent glacial period, the Wisconsin advance of the Pleistocene Epoch, with the northern Lake States mostly ice free only 10,000 years. The resulting landscapes are dominated by large glacial outwashes, morainal features, abundance of lakes, and wetland systems embedded in uplands of glacial till.

The current vegetation of the region is a consequence of the physical environment, post-Pleistocene species migration patterns, and human alteration of lands and plant communities. Based on the examination of pollen cores from small lakes in the north central and eastern United States, Davis (1976, 1981) has characterized the migration patterns of tree species following the retreat of the Wisconsin glaciation. Spruce, tamarack, and fir were among the first species to move northward within the region as the glacier retreated. Oak and pine followed spruce, tamarack, and fir, arriving in the lower Great Lakes region about 10,500 to 11,000 years ago. Hemlock and beech were more recent additions to the post-glacial forest in the Lake States. Hemlock arrived at most sites 500 to 1000 years after white pine. Beech reached lower Michigan 7000 years ago and upper Michigan some 2000 years later. It is clear from paleontological records that each species responded individualistically to past environmental changes, creating new combinations of species (i.e., new communities) in the post-glacial invasion.

With European settlement of mid-America came the destruction of the forest (Flader 1983, Stearns 1990). Demands for lumber to build Midwestern cities sent the lumberman in quest of the white pine, first to Michigan in the 1870s, then to Wisconsin in the 1880s, and finally to Minnesota in the 1890s. Initially, harvesting occurred in areas where large blocks of pine were common and where transportation of logs by river driving was available. Following a short but intense period of pine harvesting, the building of railroads allowed harvesting of hardwoods to continue well into the 1920s.

The rate and the extent of change from presettlement conditions are documented for a township in Green County, Wisconsin, along the Wisconsin-Illinois border (Curtis 1956). The vegetation in 1831 before agricultural development began, as derived from records of the original Government Land Survey, was mostly upland deciduous forest, with a small portion of prairie and oak savanna present. Forest cover was reduced to 30 percent of the original by 1882, to 10 percent by 1902, and to 4 percent by 1954. The remaining forest is now small, isolated fragments, widely scattered throughout a landscape dominated by agriculture.

Disturbances such as intensive logging and subsequent fire, conversion of forests and other vegetative communities to agricultural uses, destruction of wetlands, and eventually fire exclusion have drastically altered the composition and structure of the landscape. Early successional forests dominated by aspen and birch, and second-growth forests dominated by sugar maple now occur where pine or mixed stands of pine, northern hardwoods, and hemlock once existed. Agricultural fields have replaced the oak savanna, the prairie, the Big Woods in Minnesota, and many wetlands that were drained to "improve" the productive potential of the land.

Following logging in the Great Lakes regions, many attempts to farm marginal lands failed, resulting in tax forfeiture in the 1920s and 1930s. These defaults produced the abundance of public ownership that characterizes the northern Lake States today. A large share of this public land is in state, county, or municipal ownership.

Major Forest Policy Issues

In addition to the traditional stand-level and species-based approaches to management, many critical resource problems in the northern Lake States need to be addressed at large spatial scales and over long periods of time (Crow 1991, Mladenoff and Pastor 1993). Assessing the cumulative effects of management actions, monitoring the impacts of resource management on ecosystems properties and processes, conserving biological diversity, producing both commodity and noncommodity values all require applying "big picture" approaches that are integrated as well as coordinated across ownerships. The fragmented and mixed ownership pattern in the northern Lake States makes implementing landscape management extremely challenging. With only a few exceptions (e.g., fire suppression), formal mechanisms are not in place to facilitate cooperative management between public and private landowners, or even among different public ownerships.

Another important characteristic of the region is the large population centers located along the shores of the Great Lakes. A few large metropolitan areas represent a large portion of the approximately 50 million people living within the Great Lakes Basin. Rural communities within the Lake States often depend directly on the use of natural resources for their economic development. In contrast, the urban populations place less value on commodity production and tend to place more emphasis on noncommodity values. A major regional trend affecting the management of natural resources is an increased emphasis on noncommodity values, including recreation, scenery, water, fish, and wildlife, that reflects the demands of a growing urban population.

At the same time, the "second forest" is reaching commercial maturity, and there is increased emphasis on the role that these forests can play in regional economic development (Shands and Dawson 1984). While enhancing forest productivity remains a primary focus, there is growing acceptance of more comprehensive ecosystem management strategies that provide a broader spectrum of values and benefits while maintaining forest productivity. The potential roles that the various ownership categories can play in providing this spectrum of values and benefits need to be better defined. It is neither practical nor desirable for more than one ownership to assume the same role.

An ecosystem approach requires much different information for planning, analysis, and decision-making compared to the traditional approach that stressed commercial tree species and a few important game species. Accessibility to extensive spatial and temporal databases and applying the latest computer technologies are essential for ecosystem management. Data sharing and standardized data formats are needed to facilitate exchange of information among land managers. The topics presented in subsequent chapters, ecosystem classification, remote sensing, and GIS, all represent important tools for applying the "big picture" to resource management.

References

Crow, T.R. 1991. Landscape ecology: The big picture approach to resource management, pp. 55–65. In: D.J. Decker, M.E. Krasny, G.R. Goff, C.R. Smith, and D.W. Gross (eds.). Challenges in the conservation of biological resources, A practitioner's guide. Westview Press, Boulder, Colorado.

Curtis, J.T. 1956. The modification of mid-latitude grasslands and forests by man, pp. 721–735. In: W.L. Thomas, Jr. (ed.). Man's role in changing the face of the Earth. University of Chicago Press, Chicago, Illinois.

Curtis, J.T. 1959. The vegetation of Wisconsin. University of Wisconsin Press, Madison. 657 pp.

Davis, M.B. 1976. Pleistocene biogeography of temperate deciduous forests. Geoscience and Man 13:13–26.

Davis, M.B. 1981. Quaternary history and the stability of forest communities, pp. 132–153. In: D.C. West, H.H. Shugart, and D.B. Botkin (eds.). Forest succession: Concepts and applications. Springer-Verlag, New York.

Flader, S.L. (ed.). 1983. The Great Lakes forest, an environmental and social history. University of Minnesota Press, Minneapolis. 336 pp.

Mladenoff, D.J., and J. Pastor. 1993. Sustainable forest ecosystems in the northern hardwood and conifer forest region: Concepts and management, pp. 145–180. In: G.H. Aplet, N. Johnson, J.T. Olson, and V.A. Sample (eds.). Defining sustainable forestry. Island Press, Washington, D.C.

Shands, W.E., and D.H. Dawson. 1984. Policies for the Lake States forests. The Conservation Foundation. Final report of the Lakes States Forest Policy Workshop held in Rhinelander, WI, Feb. 12–14, 1984. 33 pp.

Skok, R.A., and C.B. Buckman. 1983. A status report on the Great Lakes forest: new resource concepts and management issues, pp. 223–242. In: S.L. Flader (ed.). The Great Lakes forests, an environmental and social history. University of Minnesota Press, Minneapolis.

Stearns, F. 1990. Forest history and management in the northern midwest, pp. 107–122. In: J.M. Sweeney (ed.). Management of dynamic ecosystems. The Wildlife Society, West Lafayette, Indiana.

14

Ecological Perspective: Current and Potential Applications of Remote Sensing and GIS to Ecosystem Analysis

David J. Mladenoff and George E. Host

Introduction

Forest management concepts are changing as growing commodity demands confront emerging concerns of global climate change, biodiversity maintenance, and longer-term ecosystem sustainability (Peters and Lovejoy 1992, Aplet et al. 1993). The U.S. Forest Service's recent policy shift to an ecosystem management approach to forest lands is significant evidence of this change in the United States (Overbay 1992). Similar management programs are being started by other federal land management agencies and various states. The most visible manifestation of a pending fundamental change in forest management necessarily affecting a large geographic area was perhaps seen in the Pacific Northwest during the 1980s, with environmental concern over the loss of old-growth forest generally (Johnson et al. 1991), and specifically as habitat for the threatened northern spotted owl (*Strix occidentalis caurina*) (Thomas et al. 1990).

In fact, impetus for these changes has come from several directions that bear on the effects of traditional forest management. These include concern over long-term cumulative watershed impacts and global climate change as well as perceived conflicts over biodiversity maintenance, human recreational use, and increasing rates of forest harvesting (Naiman 1992, Peters and Lovejoy 1992, Aplet et al. 1993, Jaako Pöyry Inc. 1992, Baskerville 1985, 1988). At the same time, knowledge of forest ecosystem functioning has undergone dramatic increases in various regions, showing the need for a more sophisticated management approach, particularly under conditions of intensifying forest commodity demands and environmental uncertainty (Franklin 1993, 1989, Mladenoff and Pastor 1993, Swanson and Franklin 1992, Hansen et al. 1991, Probst and Crow 1991, Franklin et al. 1989).

More recently, the emerging field of landscape ecology (Forman and Godron 1986) has provided a conceptual framework both for research and management of ecosystems at larger scales as the importance of spatial patterns and processes of ecosystems has become appreciated (Mladenoff et al. 1993a, Crow 1991, Franklin and Forman 1987, Turner 1989, Urban et al. 1987, Noss 1983). Clearly, the technological change that has allowed this growth in applications of landscape ecology to resource management and research has been the improvements and cost-effectiveness of geographic information systems (GIS) and remote sensing, and their more effective integration (Lachowski et al. 1992, Green 1992, Trotter 1991, Ehlers et al. 1989, Hall et al. 1988). Although the usefulness of both GIS and remote sensing in resource management has not fulfilled its promise in the past, it now appears that technology, cost, need, and the conceptual framework are converging in a way that will continue to prove practical and effective (Meyer and Werth 1990, Frank et al. 1991).

Applications of these technologies to ecological research and ecosystem management within the northern Great Lakes states have increased dramatically in recent years, as in many other regions (Johnston and Naiman 1990, Pastor and Broschart 1990, Hall et al. 1991, Maclean et al. 1992, Mladenoff et al. 1993a). In this chapter we will concentrate on ecological information needs that may be met with remote sensing and GIS technologies. We will limit our topic to ecological needs that are most immediately relevant to forest ecosystem management. We will not include needs that relate directly to larger-scale global change processes and biosphere functioning (Hobbs and Mooney 1990). However, we do recognize the relevance of these areas as a large part of the overriding context for the smaller scales we will address.

Lake States Regional Characteristics

The area typically called the northern or western Lake States consists of the forested regions of Minnesota, Wisconsin, and Michigan, situated among Lakes Superior, Michigan, and Huron. The predominantly forested portions of these states are largely northern Minnesota, northern Wisconsin, and the upper peninsula and northern lower peninsula of Michigan. This region is transitional between the true boreal forest of Canada to the north and the central deciduous forest to the south (Pastor and Mladenoff 1992, Curtis 1959). It is also transitional to the oak forests and savannas of the prairie-forest border along its western border (Curtis 1959). The region also has tree species that are characteristic and formerly dominant, such as eastern white pine (*Pinus strobus*), eastern hemlock (*Tsuga canadensis*), and yellow birch (*Betula alleghaniensis*). This region is the western portion of the area described by Braun (1950) as the *white pine-northern hardwood and conifer region*.

Physiography, Landforms, and Soils

Landforms of the northern Lake States are glacial landforms, largely dominated by moraines, ice-contact hills, and outwash plains. The present-day landscape is a result of deposition of parent materials and subsequent post-glacial erosion processes. Northern lower Michigan is characterized by thick and extremely sandy glacial deposits; depth to bedrock often exceeds 100 meters. Several large morainal systems are dominant landforms in a matrix of sandy outwash plains and ice-contact hills (Farrand and Eschman 1974). In these sandy landscapes, the presence of sandy clay loam or finer textural bands has been shown to have a significant influence on forest composition and productivity (Cleland et al. 1985, Host et al. 1988). The eastern upper peninsula of Michigan consists primarily of low-elevation, flat glacial lake plain topography, whereas the physiographically distinct western upper peninsula comprises higher-elevation bedrock-controlled morainal and outwash topography (Albert et al. 1986).

In northern Wisconsin, morainal and outwash systems characterize the state. Sands and silty tills form the dominant soils of the region, although these are often covered with loess. The final pulse of ice, occurring about 10,000 years ago, left the water-worked red clays that characterized the Superior lowlands in northern Wisconsin and upper Michigan. In Minnesota, as well as Wisconsin and Michigan, numerous eskers and drumlin fields occur throughout the state. The physiography created by ice movement and disintegration processes, coupled with climate, led to the development of the extensive peatlands that characterize northern Minnesota. Depths to bedrock in Minnesota tend to be much shallower than in Michigan and Wisconsin.

Glacial landforms of the northern Lake States have played an important role in structuring present-day forest composition. Parent materials and physiography, through their influence on microclimate, drainage, and fire frequency (Grimm 1984), impose significant constraints on the codevelopment of soils and plant communities (Rowe 1984). The distribution patterns of both overstory and ground-flora species show strong probabilistic correlations with landscape patterns produced at the scale of landforms (Host and Pregitzer 1992). In the Lake States, these patterns have been further modified by disturbance, including the long-term histories of fire and windthrow, and the more recent history of fire suppression and logging.

Forest Ecosystem Characteristics

Because of their transitional nature, the northern Lake States contain a surprising diversity of tree species. Although we think of the southern Appalachian Mountains as characteristically high in tree species diversity (Braun 1950), the northern Lake States contain approximately 40 tree species, including subcanopy species. More than 300 shrub and herb species commonly occur in the upland forests alone (Maycock and Curtis 1960). Again because of its transitional nature, many of the tree species in this region are at or near the edges of their ranges (Little 1971), both north and south (Table 14.1). The group of early-successional species, aspens, white birch, and pin cherry, are also characteristic of both the boreal forest region and the deciduous region to the south, particularly on mesic sites (Pastor and Mladenoff 1992). There is also not a consistent physiognomic successional pattern that exists in regions such as the Pacific Northwest or Rocky Mountains, where conifers often make up all later-successional stages following the early deciduous species (Franklin and Dyrness 1974, Steele 1984). In this region pines may succeed on sandy or mesic sites, but will be replaced by northern hardwoods such as sugar maple (*Acer saccharum*) on mesic sites (Curtis 1959). Either sugar maple or hemlock may be the most tolerant species on mesic sites. In locations with less favorable soils or microclimates, boreal conifers may dominate in later stages as well (Curtis 1959).

Characterizing species interactions is further complicated by the interaction of two characteristic disturbance regimes as well. In the western portions of the region where boreal ecosystems and pine are more important, fire at short return intervals (<100 years) is the predominant natural disturbance, while in the eastern areas where northern hardwoods predominate, windthrow at longer return intervals (>1000 years) is more important (Heinselman 1973, Curtis 1959). Tree species ecosystem characteristics such as litter quality and nutrient requirements are similarly complex, varying largely between the low-quality, less demanding conifers and higher-quality, more demanding deciduous species. These relationships, however, do not segregate simply along seral stages (Mladenoff 1987, Pastor and Mladenoff 1992). The interaction of these ecosystem properties, disturbance, and site factors produces a fine-grained

TABLE 14.1 Major tree species of the northern Lake States region.

Species with broader southern distributions, at northern range limits	Species with broader northern distributions, at southern range limits
sugar maple (*Acer saccharum*)	balsam fir (*Abies balsamea*)
beech (*Fagus grandifolia*)	white spruce (*Picea glauca*)
basswood (*Tilia americana*)	jack pine (*Pinus banksiana*)
Characteristic northern hardwood and conifer region species	**Common early successional species**
yellow birth (*Betula alleghaniensis*)	aspen (*Populus tremuloides/grandidentata*)
eastern hemlock (*Tsuga canadensis*)	white birth (*Betula papyrifera*)
white pine (*Pinus strobus*)	pin cherry (*Prunus pensylvanica*)

Source: Modified from Mladenoff and Pastor (1993)

complexity in the landscape, particularly in the disturbed landscapes of today (Mladenoff et al. 1993a, Mladenoff and Pastor 1993).

Management History

Forest change in the Lakes States has been described recently from both ecological perspectives (Mladenoff and Pastor 1993, Stearns 1990) and within social and environmental contexts (Cronon 1991, Williams 1989, Flader 1983). The largely old-growth forests of the region underwent destructive logging and repeated fires during the late 1800s and early 1900s. Major changes included the elimination of conifer-dominated forests, predominantly white pine and hemlock. Along with the typical conifer ecosystem characteristics described above, these species in particular were both long-lived and large-structured, and dominated sandy and mesic soils, respectively. Historical cutting patterns have eliminated these species from a dominant position. Hemlock, for example, is now present at only 0.5 percent of its former extent (Eckstein 1980), and reestablishes poorly for numerous reasons (Mladenoff and Stearns 1993). The regional landscape is now dominated by aspen-birch and second growth northern hardwoods (Mladenoff and Pastor 1993, Pastor and Mladenoff 1994).

The result of these profound changes is a forest landscape with radically altered ecosystem processes, landscape pattern, and habitat composition and structure (Mladenoff and Pastor 1993, Pastor and Mladenoff 1992). Recent, traditional forest management has sought to maintain particular forest states, primarily early-successional forests, rather than ecological processes. This approach has simplified the ecosystems and also increased forest fragmentation. Forest fragmentation is more complex in this largely contiguously forested region than occurs where forest islands are isolated in agricultural landscapes. Fragmentation in northern forests is expressed as a large reduction in patch sizes, and a great increase in the number and types of successional patches at the expense of late-successional ecosystems (Mladenoff et al. 1993a). Edge effects and reduction in forest habitat occur, but in a lower contrast context than in an agricultural matrix or regions with large areas of remaining old-growth forest.

Regional Applications

A focus on managing ecological processes rather than a concentration on particular forest states is required for an ecosystem approach to management. This approach also requires that processes be managed and monitored across scales, with goals that incorporate a greater temporal context as well. This suggests a broader approach that might be described as *Dynamic Landscape Heterogeneity* (Mladenoff and Pastor 1993), which embodies an emphasis on spatial and temporal landscape change, driven by process management across scales. Implicit in this approach is the notion that over larger temporal and spatial scales, states that appear constant and processes that appear linear actually vary. Greater attention must be paid to these scaled, linked processes and the positive and negative feedbacks between them that cause state variability over time (Pastor and Mladenoff 1992, Wein and El-Bayoumi 1982). This variation must be considered in management, particularly under the combination of increased demands on ecosystems and greater ecological uncertainty (Mladenoff and Pastor 1993, Regier and Baskerville 1986, Baskerville 1988).

Current Examples

Current ecological applications within the northern Lake States are concentrated in several areas, with overlapping purposes including (1) forest and land-cover mapping, (2) ecological land classification, (3) regional change detection, and (4) detection and modeling of ecosystem properties and processes. These efforts are driven by traditional inventory and monitoring needs—the desire to expand the spatial and temporal scale of information considered in management decisions. Furthermore, there are growing needs to detect and monitor changes in ecosystem processes that may vary with management regimes, land ownership and long-term ecosystem change across different ecoregions. Many of these applications can also be used to address other, more specific objectives such as biodiversity maintenance and restoration of old-growth or other ecosystems greatly reduced from their former extent.

Forest and Land-Cover Mapping

Forest classification is a key area that bears on most of the other topics we will discuss regarding ecosystem management. Many past efforts using remote sensing technology for forest mapping in the Lake States have suffered from technological limitations when applied to regional conditions. As described in the introduction, the region has certain characteristics that have made forest classification difficult. These include abundant tree species, the lack of strong elevational or other environmental gradients, and a history of widespread regional forest disturbance. The fine-grained environmental variation of these glacial landscapes and multi-species combinations has challenged the spatial and spectral resolution of available sensors and classification methods to provide useful forest type differentiation. Forest classifications done from Landsat satellite multi-spectral scanner (MSS) data, available since 1972, have not proved adequate for many ecological or research applications. Such classifications have often discriminated only to Anderson Level I (forest/nonforest) (Anderson et al. 1976), with 90 percent or better accuracies (Williams and Nelson 1986). Classifications of Level II (conifer/mixed/deciduous) or better have generally attained only 70–80 percent accuracy (Williams and Nelson 1986, Hall et al. 1991). Similarly, the spatial resolution of MSS data (80 meters) provides only a coarse representation of the fine-grained northern Lake States landscapes.

These accuracies have not usually been considered adequate for many research or management purposes except for very general questions at large spatial scales. Newer Landsat thematic mapper (TM) data have improved spectral and spatial (30-meter) resolution and have been available since 1982. Classifications in the Lake States based on TM data have increased forest species group discrimination and overall accuracy (Hopkins et al. 1988).

However, significant needs remain to be met in this key area for both research and management. Deciduous tree species groups still pose problems, with poor discrimination of oak (*Quercus* spp.) dominated hardwoods, which predominate on sandier soils, from sugar maple-dominated northern hardwoods, which eventually dominate more mesic sites. Separability of successional aspen-paper birch forests from late-successional maple-dominated stands is also not reliable. As a result, there is serious confusion across both successional and land-type gradients, reducing the usefulness of these classifications for fine-grained analysis or decision-making.

Further improvement in TM classification has been obtained by integrating GIS with remote sensing. This has been done using an expert system approach, and using ancillary data layers such as soils and topography to guide the classification (Bolstad and Lillisand 1992). This method successfully improves classification accuracy, but some drawbacks still remain. An advantage of Landsat data is its synoptic nature and frequency of coverage. Incorporating ancillary data layers can only be done where such information exists, and ideally in digital form. This eliminates most of the northern Lake States region from using this approach without often cost-prohibitive additional work. In many cases where such GIS data layers exist they may not have the spatial accuracy or resolution of the Landsat imagery. One result is that the number of distinguishable classes may be increased but with a sacrifice in spatial resolution of the final map.

We have applied another approach to the forest classification problem by using multitemporal satellite imagery. The method requires analysis of satellite images from various seasons of the year to use differential tree species phenology and physiognomy to separate a greater variety of forest type classes (Wolter et al. 1993). Several major tree species that are not distinguishable on peak summer imagery have characteristics that make them separable at other times of the year (Table 14.2). These include obvious differences such as early fall coloring of maples, late retention of oak leaves, early spring leaf-out of aspen, and detection of understory conifer layers during winter.

An initial classification using this seasonality approach to species discrimination has been promising, with mapping possible down to a finer classification than is possible with other techniques, often to the species level where there is strong single-species dominance (Wolter et al. 1993). However, this approach too has room for improvement. Currently, TM data are very costly ($4400 per scene). In this test effort we used an early summer base scene, with less expensive ($200–$1000 per scene) MSS imagery from the other seasons. As a result of our preliminary success, the classification could be immediately improved by taking advantage of the increased spatial and temporal resolution of TM data from the same dates. Finding dates with clear, cloud-free imagery is always problematic, and the need for usable multiseasonal dates compounds this problem. Also, many of the species phenological characteristics vary annually in the date and magnitude of their occurrence, and are very ephemeral.

TABLE 14.2 **Major tree species with distinguishing phenological or physiognomic characteristics useful for multitemporal (seasonal) forest classification with satellite imagery**

		Major spectral characteristics		
Species	Species characteristics	Near infrared	Visible	Season
Maples (*Acer rubrum*, *A. saccharum*)	Peak scenesence	high	high	early fall
Oaks (*Quercus rubra*, *Q. ellipsoidalis*)	Peak scenescence	high	high	mid-late fall
Black ash (*Fraxinus nigra*)	lowland, uniform leaf drop	low	high	late summer/early fall
Aspen (*Populus tremuloides*, *P. grandidentata*)	early leaf flush	high	low	mid-spring
Tamarack (*Larix laricina*)	lowland, deciduous conifer	low	high	winter
Conifer understory	evergreen	low	high	winter

Obtaining imagery from the same year is desirable for this approach, but we have not found it to be possible. As a result, as the imagery dates become farther spaced over several years, information is lost due to land-cover changes and eventually successional changes. Sun-angle changes over the seasons is another problem, but we found this not to be a major factor in our region. However, using this technique in areas of greater topographic relief may be more difficult.

Forest Structural Attributes

We include under forest structure physical characteristics such as canopy cover, tree size classes, tree mass and form, and aspects of physiognomy. Lake States forest ecosystems have also provided a challenge in detecting and mapping structural features. Tree species factors similar to those that are problematic for species discrimination also contribute to problems with distinguishing structure. A large number of species have overlapping characteristics, such as small canopy diameter and bole and branch mass in relation to sensor spatial resolution. In the Pacific Northwest, for example, larger canopy features of this type allow discrimination of the structural attributes of mature conifer forests with Landsat TM and SPOT imagery (Cohen and Spies 1992). Mass of large, dead branch wood was an important characteristic in the analysis of Cohen and Spies (1992). The Lake States lack such structural characteristics to the degree necessary for the resolution of current sensors.

Canopy closure is also a less reliable characteristic for discriminating late-successional forest. Unlike more arid regions, canopy closure occurs early in stand regeneration, often under 10 years. The small range of tree sizes and early canopy closure also affects the ability to discriminate size classes, although this can be done to some extent. Those Lake States tree species that can reach a stature great enough to contribute to good discrimination of canopy structural attributes (white pine and hemlock) have been largely eliminated from the landscape (Mladenoff and Pastor 1993, Tyrrell and Crow 1993, Stai et al. 1993).

Forest structural variability associated with a broad range of tree-age classes and physiognomic features of conifers are known to be very important habitat characteristics, particularly for forest birds (Jaako Pöyry Inc. 1992). These features and characteristics such as old-growth treefall gaps and large woody debris have been largely eliminated from forests in the region (Mladenoff and Pastor 1993, Mladenoff 1990, Stai et al. 1993, Tyrrell and Crow 1993). Developing effective methods for detecting and mapping these attributes for research and management purposes remains a high priority.

Ecological Land Unit Mapping

Managing forests from an ecosystem perspective requires the identification of spatially delineated ecological units on the land that account for variability in ecosystem composition, structure, and functional processes. In the Lake States, researchers have been developing and using ecological classification systems (ECS), in which multiple factors are used to identify ecological land units at hierarchically nested spatial scales (Barnes et al. 1982). Each level in the spatial hierarchy is defined in terms of the dominant controlling factors (Damman 1979).

At the broadest scale (1:2,000,000), the northern Lake States have been divided into ecoregions defined by macroclimate and regional physiography (Bailey 1980, Albert et al. 1986). Climate and physiography are also the main factors in delineating land units at scales of 1:100,000 to 1:250,000. The spatial analysis capabilities of geographic information systems are important tools for integrating these different data types. In a regional analysis of northwestern Wisconsin, for example, we used a raster-format climatic database in combination with a vector-based map of Pleistocene geology to identify the predominant ecoregions in a 1:250,000 quadrangle (Host et al. 1993). Delineation of these large-scale land units is more than an academic exercise: numerous important ecological processes, such as the spread of insects or disease or the movement patterns of moose, wolves, and other mammals, occur at these large spatial scales. Moreover, these ecoregional processes invariably cross land-ownership boundaries, requiring that management strategies be coordinated across multiple ownerships. GIS has proven to be an important tool for assessing the distribution of land ownership within ecoregions, which in turn identifies the key public agencies and private ownership groups involved in making management policy decisions (White et al. 1993).

At finer spatial scales, topographic position, drainage, soil, and vegetation are key factors in delineating ecological land units (Cleland et al. 1992). Research has shown that ecological land units delineated at scales of 1:10,000–1:24,000 differ in fundamental ecological processes, such as productivity (Host et al. 1988), succession (Host et al. 1987), and nutrient cycling (Zak et al. 1986, Zak and Pregitzer 1990). The maps produced from an ecological classification are therefore an integration of ecosystem processes. As efforts are directed toward management of ecological processes rather than forest species compositional states, ecological land classification becomes the language through which the potentials of the forested landscape are defined.

The mapping of ecological land units in the Lake States currently is done almost exclusively, on the National Forests and state lands, by contract. At rates ranging from $0.40–$1.00 per acre, field mappers use a legend or field guide to traverse the landscape and delineate ecological boundaries as expressed by variation in soils or vegetation. The resulting map lines, if placed on a georeferenced map base, can be readily placed into a GIS framework. Testing of current remote sensing capability to efficiently accomplish this task is in the planning stages.

As cartographic entities, ecological land units are readily amenable to storage, manipulation, and analysis with geographic information systems. Numerous forest management applications require the types of information stored in ECS databases. Managing for wildlife species such as black bear, for example, requires information on the availability and abundance of important forage species over the course of a year, the availability of habitat features such as seeps, springs, rock outcrops, and den and escape trees, and the distance and arrangement of resource patches relative to a bear's home range. In a habitat suitability analysis, we used a vector-based GIS to query ECS map layers and associated databases to assess black bear habitat on the Ottawa National Forest (Johnson et al. 1991). The ECS floristic database provided information on the distribution and abundance of the predominant forage species for each season. An associate soil/landform database located seeps, rock outcrops, and other required habitat features. Ancillary databases (e.g., U.S. Geological Survey [USGS] Digital Line Graph files) were used to quantify factors such as road and house densities, which may have adverse impacts on bear populations. The end product was a series of maps of habitat quality for black bear, summarizing habitat quality by resource (forage, structure) and season, as well as an integrated map of overall habitat quality. Development of this type of decision support system is applicable not only to wildlife habitat analysis, but also silvicultural prescription development, old-growth management, and biodiversity assessment.

Change Detection

Landsat satellite imagery, which has been archived since its beginning in 1972, now provides a 20-year record of regional change, although some early data have not been maintained. Multitemporal analysis of Landsat data has been done for few locations in the Lakes States. Transition analysis of successional, disturbance, and land use changes can provide a powerful tool in understanding regional landscape dynamics, particularly as the data archive lengthens. Hall et al. (1991) used two dates of an area of northeastern Minnesota to examine forest change from 1973–1983. Use of these early dates requires the use of MSS data, with its inherent limitations as noted previously in *Forest and Land-Cover Mapping*. The classes Hall et al. (1991) were able to discriminate reliably approximate only Anderson Level II, with the addition of a regenerating class. They attempted to show differences in forest landscape dynamics between the managed portions of the Superior National Forest and the Boundary Waters Canoe Area Wilderness (BWCAW). While the temporal transitions within the same area are real, comparisons between the two areas are partly confounded because of differing landforms and soil regions. This is a problem with all large-scale analyses, which nearly always cross environmental boundaries or ecoregions. Part of this is an information problem as described in *Ecological Land Unit Mapping* above.

We are conducting a multitemporal analysis of a large region in northwestern Wisconsin, using three dates from 1974–1987 (Mladenoff et al. 1993b). This analysis is also based on MSS data because of the years covered. Change vectors across the three dates for unique classes are based on a single combined classification. We also found the data source and associated problems limited the classification to classes only somewhat better than Anderson Level II. However, the information content of the classification can be increased by stratifying the regional landscape by ecoregions (Host et al. 1993) and land ownership categories during interpretation (White et al. 1993). Other types of historical data can be used to supplement the classified dates, or to include dates prior to the Landsat archive in the analysis (White and Mladenoff 1994, Goldman 1990, Morrison and Ribansky 1989). However, when these approaches are taken, the comparison is limited by the lowest common denominator (coarsest data source).

For example, White and Mladenoff (1994) used three dates based on current aerial photography, a 1930s land inventory, and a mapped reconstruction of the presettlement forest from original government survey notes. GIS was then used in a transition analysis over the past 100 years. Because of the coarse early data sources, spatial change in general forest types could be analyzed, but change in explicit landscape pattern could not. In a related study, Mladenoff et al. (1993a) used aerial photography as the base for an analysis of change in forest landscape pattern from presettlement to present. This was done by comparing an unlogged old-growth landscape to a managed second growth area. Significant differences were found in forest patch size, complexity, and adjacency relationships between the two areas on similar landforms. Care must be taken that transitions across dates indicate actual change, whether successional or land use, and not noncomparable classifications. Significant problems remain to be solved to make fuller use of this powerful technique.

Ecosystem Properties and Process Modeling

The direct application of remote sensing to measuring important ecosystem parameters such as photosynthesis, production, carbon allocation, and soil processes such as organic matter turnover or nutrient mineralization is perhaps the least developed area of research. To some extent this has been due to inherent technological limitations (Waring et al. 1986) and also a lack of an ecological emphasis to remote sensing programs and sensor designs (Wickland 1991). While direct measurement remains a goal to which some progress has been made (Aber et al. 1990, Wessman 1990), current sensors have been used to derive such ecosystem parameters indirectly by linking remotely sensed data with large-scale ecosystem process models in the Rocky Mountain west (Running 1990). This has been done largely by using remote measurement of leaf area index (LAI) coupled with ground calibration measurements and intensive climatic data (Running 1990).

Applications of this approach in the northern Lake States, directed at characterizing forest productivity across a range of forest types by differing habitats, may suffer from the same factors that limit remote sensing applications of general forest and ecosystem mapping. These factors include the inherent high number of tree species, small tree mass, and fine-scale landscape heterogeneity, coupled with the spatial, spectral, and radiometric limitations of current sensors. Similarly, such approaches that require ancillary GIS data layers would benefit from finer spatial resolution as well as greater general availability. The attractiveness of this approach, as with many remote sensing applications, is the potential for integrating measurement of relatively fine-grained processes across wide geographic areas. This would be particularly useful in assessing regional forest productivity and in monitoring change.

Forest Succession and Land-Cover Modeling

Remotely sensed input to simulation models of landscape change and forest succession provides another application where the combination of fine resolution and wide extent can be useful to understanding change in landscape patterns as it relates to global change and various management and natural disturbance regimes. Once again, the same combination of technological limitations and regional forest ecosystem characteristics constrain these efforts. Forest maps derived from remotely sensed imagery must be created at a consistent classification level to be useful in a model. For example, all forest types must be identified to the same classification level by species and age or structural classes if the model operates on those attributes.

There are then several limitations on this application. (1) In general there are the tradeoffs between computer model complexity and computational load and efficiency, with the related scale tradeoff of model resolution and spatial extent. Because of this limitation, the actual intended applications and need must be carefully considered in creating any model. (2) Because of the consistency required in the classified input map, the lowest common denominator prevails as the usable level. Because of the limitations of (2), the computational limitations may not be reached in actual application, which only comes into play with availability of ideal input data.

We are continuing to refine a forest landscape model designed to simulate forest landscape change due to management scenarios and natural disturbance (LANDIS) (Mladenoff et al. 1994). The model is generally derived from the LANDSIM model of Roberts (1993), which is a vector-based fire and succession model for the central Rocky Mountains that operates on principles of species vital attributes (Noble and Slatyer 1980). Our model deviates from the approach of LANDSIM in several respects. LANDIS is object-oriented, raster-based, and designed to operate at different scales in the northern Lake States. In some operations we have sacrificed part of the computational ease of LANDSIM for greater detail of dynamic processes. To model forest landscape dynamics in the northern Lake States, we have designed disturbance routines that include windthrow as well as fire, and their interactions. The greatest opportunity for widespread application of such an approach is the ability to incorporate satellite-derived forest maps. Simulation of attributes at finer resolution than the lowest common denominator classification level require their assignment using some ancillary data source or objective process. For example, age classes within a mapped forest type could be assigned randomly based on a known general distribution of classes and a range of stand sizes. However, while very useful for gaining understanding of ecological processes, such an approach lacks the spatial accuracy desirable for management purposes. For smaller landscapes or stand-level applications, more detailed maps that contain this level of detail can be manually digitized into a GIS and used in the model. For large areas (>10,000 hectares), however, this becomes very costly.

Current hardware and GIS software limitations have not been severe when the model is based on efficient computer code and algorithms (Mladenoff et al. 1993c). Improvement in forest and land-cover classification accuracy derived from satellite data would do most to broaden the usefulness of this approach.

Artificial Intelligence Approaches to GIS and Spatial Modeling

The ability to model spatial components of forest ecosystem states and processes has been enhanced through linkages between GIS and artificial intelligence (AI)-based simulation tools. Among these tools are hierarchical data structures and rule-based or event-based modeling approaches. The hierarchical data structures characteristic of several AI and object-oriented programming languages allow for the relatively simple development or quantification of heterogeneous and complex environments (Saarenmaa et al. 1988, Mladenoff et al. 1993c). In a hierarchical data structure, fine-scale environments such as forest patches can inherit the properties of higher-level classes, such as the soil texture of the parent landform, or weather patterns characteristic of the regional climate. The LANDIS model described previously uses a hierarchical object-oriented strategy to structure both tree species and ecological land units (Mladenoff et al. 1994).

Hierarchical data structures are not only computationally efficient, but also interface well with rule-based approaches to modeling. To date, rule-based AI models have focused largely on simulating the behavior and movements of wide-ranging wildlife species within a habitat matrix. Folse et al. (1989), for example, used a rule-based approach to predict the goal-oriented behavior of the white-tail deer within a spatially heterogeneous array of habitat patches. Rules for deer behavior related to the animal's needs (food, shelter) coupled with its perception of its immediate environment. A related model for moose incorporated some of the cognitive abilities of assessing food quality and time (Roese et al. 1991). The ability to model individuals responding to spatially and temporally unique resource patches has provided a significant improvement over models of mean population responses to average conditions (Roese et al. 1991).

Expert systems, computer programs that capture the knowledge of human experts for decision support or problem solving, are beginning to play an important role in fundamental GIS operations, such as geographic feature extraction and decision support (Robinson and Frank 1987), as well as in improving the predictive ability of tree and forest simulation models (Rauscher and Isebrands 1990). GIS operations such as identification of watershed boundaries or specific geologic features (valleys, drumlins) can be accomplished by linking an expert system with a GIS database. In a more complex application, the Forestry Expert System (Goldberg et al. 1984) has been used to detect temporal changes in land cover using multitemporal Landsat data. Expert systems such as AS-

PENEX (Morse 1987) and FMS-Aspen (Rauscher et al. 1990) link silvicultural rule bases with forest stand data to develop spatially explicit management recommendations. Aspen is currently the most extensive cover type in Minnesota, Michigan, and Wisconsin (Jones and Markstrom 1973), and occurs across a wide range of site conditions. Associated with this broad distribution pattern is a wide range of variation in productivity, disease susceptibility, and subsequent successional pathways. Expert systems such as FMS-Aspen encode knowledge of relationships between site, growth, and associated ecosystem processes into an AI rule base, and use this information to direct the behavior of simulation models. Given the heterogeneous and relatively fine-grained landscapes characteristic of the northern Lake States, this AI control of simulation models within a spatially explicit GIS framework improves the ability to predict relatively subtle ecosystem responses resulting from geographic variation (Host et al. 1992).

Conclusions

Ecological applications of remote sensing and GIS to ecosystem management are in an early stage. This reflects several factors, such as past failures of these technologies to meet their stated potential, the lack of training and familiarity with GIS and remote sensing on the part of ecologists, and the limitations of costs (Roughgarden et al. 1991). Current technologies have not been fully exploited, although some desired applications will have to await technological improvements.

The ecological characteristics, physiography, and management history of the northern Lake States region have created a current landscape that presents certain challenges. Mapping and measuring conditions and processes at a very fine-grained level stretch the capabilities of individual expertise, technology, and available funds. Also, the Lake States are unlike many regions of the western United States that still have larger areas of primary forest ecosystems for study and comparison. In contrast, the cutting history and subsequent widespread disturbances in the Lake States forests require interpretation of a predominantly mid-successional second-growth landscape. Because of this history, in the Lake States proportionately greater effort will be directed toward ecosystem restoration than in some other regions. These regional characteristics create unique data needs and analytical requirements that must be met if ecosystem management is to address the problems at varied scales that need solving.

Acknowledgments

Several of our research project associates contributed to the various studies reviewed here. These include Joel Boeder, Phil Polzer, Mark White, and Peter Wolter. A portion of this work has been funded through a USDA Forest Service Cooperative Agreement. Cooperators: D.J. Mladenoff and G.E. Host, Natural Resources Research Institute, University of Minnesota, Duluth; and Thomas R. Crow, U.S. Forest Service North Central Experiment Station, Landscape Ecology Program. We also appreciate the comments of two anonymous reviewers. This is Contribution Number 128 of the Center for Water and the Environment, Natural Resources Research Institute, and Number 29 of the Natural Resources GIS Laboratory.

References

Anderson, J.R., E.E. Hardy, J.T. Roach, and R.E. Witmer. 1976. A land use and land cover classification system for use with remote sensor data. U.S. Geological Survey Professional Paper 964, 28 pp.

Aber, J.D., C.A. Wessman, D.L. Peterson, J.M. Melillo, and J.H. Fownes. 1990. Remote sensing of litter and soil organic matter decomposition in forest ecosystems. pp. 87–103. In: R.J. Hobbs and H.A. Mooney (eds.). Remote sensing of biosphere functioning. Springer-Verlag, New York, NY.

Albert, D.A., S.A. Denton, and B.V. Barnes. 1986. Regional landscape ecosystems of Michigan. School of Natural Resources, University of Michigan, Ann Arbor. 32 pp.

Aplet, G.H., J.T. Olson, N. Johnson, and V.A. Sample (eds.). 1993. Defining sustainable forestry. Island Press. Washington, DC.

Bailey, R.G. 1980. Description of the ecoregions of the United States. USDA Forest Service Miscellaneous Publication No. 1391. Washington, DC. 77 pp.

Barnes, B.V., K.S. Pregitzer, T.A. Spies, and V.H. Spooner. 1982. Ecological forest site classification. Journal of Forestry 80:493–498.

Baskerville, G. 1985. Adaptive management: Wood availability and habitat availability. The Forestry Chronicle 61:171–175.

Baskerville, G.L. 1988. Redevelopment of a degrading forest system. Ambio 17:314–322.

Bolstad, P.V., and T.M. Lillesand. 1992. Improved classification of forest vegetation in northern Wisconsin through rule-based combination of soils, terrain, and Landsat thematic mapper data. Forest Science 38:5–20.

Braun, E.L. 1950. Deciduous forests of eastern North America. Blakiston Co., Philadelphia, PA.

Cleland, D.T., J.B. Hart, K.S. Pregitzer, and C.W. Ramm. 1985. Classifying oak ecosystems for management. pp 120–134. In: J.E. Johnson (ed.). Proceedings of Challenges in Oak Management and Utilization. University of Wisconsin-Madison, WI.

Cleland, D.T., T.R. Crow, P.E. Avers, and J.R. Probst. 1992. Principles of land stratification for delineating ecosystems. pp. 40–50. In: Proceedings, National Workshop: Taking an ecological approach to management. USDA Forest Service WO-WSA-3. Washington, D.C.

Cohen, W.B., and T.A. Spies. 1992. Estimating structural attributes of Douglas-fir/western hemlock forest stands from Landsat and SPOT imagery. Remote Sensing of the Environment 41:1–17.

Cronon, William. 1991. Nature's metropolis: Chicago and the great West. W.W. Norton. New York.

Crow, T.R. 1991. Landscape ecology: The big picture approach to resource management. In: D.J. Decker, M.E. Krasny, G.R. Goff, and others (comps., eds.). Challenges in the conservation of biological resources: A practitioner's guide. Westview Press, Boulder, CO. pp. 55–65.

Curtis, J.T. 1959. The Vegetation of Wisconsin. University of Wisconsin Press, Madison, WI. 657 pp.

Damman, A.W.H. 1979. The role of vegetation analysis in land classification. Forestry Chronicle. 55:175–182.

Eckstein, R.G. 1980. Eastern hemlock (*Tsuga canadensis*) in north central Wisconsin. Wisconsin Department of Natural Resources Research Report 104. Madison, WI. 20 pp.

Ehlers, M., G. Edwards, and Y. Bédard. 1989. Integration of remote sensing with geographic information systems: A necessary evolution. Photogrammetric Engineering and Remote Sensing 55:1619–1627.

Flader, S.L. (ed.). 1983. The Great Lakes forest: An environmental and social history. University of Minnesota Press, Minneapolis, MN.

Folse, L.J, J.M. Packard, and W.E. Grant. 1989. AI modeling of animal movements in a heterogeneous habitat. Ecological Modeling 46:57–72.

Forman, R.T.T., and M. Godron. 1986. Landscape ecology. John Wiley. New York.

Frank, A.U., M.J. Egenhofer, and W. Kuhn. 1991. A perspective on GIS technology in the nineties. Photogrammetric Engineering and Remote Sensing 57:1431–1436.

Franklin, J.F., and C.T. Dynress. 1973. Natural vegetation of Oregon and Washington. Oregon State University Press, Corvallis.

Franklin, J.F. 1993. The fundamentals of ecosystem management with applications in the Pacific Northwest. In: G.H. Aplet, J.T. Olson, N. Johnson, and V.A. Sample (eds.). Defining sustainable forestry. Island Press, Washington, DC.

Franklin, J.F., and R.T.T. Formann. 1987. Creating landscape patterns by forest cutting: Ecological consequences and principles. Landscape Ecology 1:5–18.

Franklin, J. 1989. Toward a new forestry. American Forests, November/December, pp. 37–44.

Franklin, J.F., D.A. Perry, T.D. Schowalter, M.E. Harmon, A. McKee, and T.A. Spies. 1989. Importance of ecological diversity in maintaining long-term site productivity. pp. 82–97. In: D.A. Perry, R. Meurisse, B. Thomas, R. Miller, J. Boyle, J. Means, C.R. Perry, and R.F. Powers (eds.). Maintaining the long-term productivity of Pacific Northwest forest ecosystems. Timber Press, Inc., Portland, OR.

Goldberg, M., M. Alvo, and G. Karam. 1984. The analysis of Landsat imagery using an expert system: Forestry applications. Proceedings AutoCarto 6:493–503.

Goldmann, R.A. 1990. Using multiresolution remote sensing data and GIS technology to update forest stand compartment maps under Lake States conditions. M.S. thesis, Univ. of Wisconsin, Madison.

Green, K. 1992. Spatial imagery and GIS. Journal of Forestry 90:32–36.

Grimm, E.C. 1984. Fire and other factors controlling the big woods vegetation of Minnesota in the mid-nineteenth century. Ecological Monographs 54:291–311.

Hall, F.G., D.E. Strebel, and P.J. Sellers. 1988. Linking knowledge among spatial and temporal scales: Vegetation, atmosphere, climate, and remote sensing. Landscape Ecology 2:3–22.

Hall, F.G., D.B. Botkin, D.E. Strebel, K.D. Woods, and S.J. Goetz. 1991. Large-scale patterns of forest succession as determined by remote sensing. Ecology 72:628–640.

Hansen, A.J., T.A. Spies, F.J. Swanson, and J.L. Ohmann. 1991. Conserving biodiversity in managed forests. BioScience 41:382–392.

Heinselman, M.L. 1973. Fire in the virgin forests of the Boundary Waters Canoe Area, Minnesota. Quaternary Research 3:329–382.

Hopkins, P.F., A.L. Maclean, and T.M. Lillesand. 1988. Assessment of thematic mapper imagery for forestry applications under Lake States conditions. Photogrammetric Engineering and Remote Sensing 54:61–68.

Host, G.E., K.S. Pregitzer, C.W. Ramm, D.P. Lusch, and D.T. Cleland. 1988. Variation in overstory biomass among glacial landforms and ecological land units in northwestern Lower Michigan. Canadian Journal of Forestry Research 18:659–668.

Host, G.E., H.M. Rauscher, and D. Schmoldt. 1992. SYLVATICA: an integrated framework for forest landscape simulation. Landscape and Urban Planning 21:281–284.

Host, G.E., K.S. Pregitzer, C.W. Ramm, J.B. Hart, and D.T. Cleland. 1987. Landform-mediated differences in successional pathways among upland forest ecosystems in northwestern Lower Michigan. Forest Science 33:445–457.

Host, G.E., and K.S. Pregitzer. 1992. Geomorphic influences on ground-flora and overstory composition in upland forests of northwestern Lower Michigan. Canadian Journal of Forest Research. 22:1547–1555.

Host, G.E., D.J. Mladenoff, P. Polzer, M.A. White, and T.R. Crow. 1993. A climatic and physiographic classification of regional landscape ecosystems in northwestern Wisconsin (Abs.). p. 59. In: Proceedings, U.S.-International Association for Landscape Ecology annual meeting. Oak Ridge, TN.

Jaakko Pöyry Consulting, Inc. 1992. Generic environmental impact statement on timber harvesting and forest management in Minnesota: Wildlife technical paper. Prepared for Minnesota Environmental Quality Board, St. Paul, MN.

Johnson, K.N., J.F. Franklin, J.W. Thomas, and J. Gordon. 1991. Alternatives for management of late-successional forests of the Pacific Northwest. A report to the U.S. House of Representatives by The Scientific Panel on Late-Successional Forest Ecosystems. 59 pp.

Johnson, L.B., G.E. Host, J.K. Jordan, and L.L. Rogers. 1991. Use of GIS for landscape design in natural resources management: Habitat assessment and management for the female black bear. pp. 507–517. In: Proceedings, GIS/LIS '91, Atlanta, GA.

Johnston, C.A., and R.J. Naiman. 1990. In: R.J. Naiman (ed.). Watershed management: Balancing sustainability and environmental change. Springer-Verlag. New York.

Jones, J.R., and D.C. Markstrom. 1973. Aspen—an American wood. USDA Forest Service, FS-217. 8 pp.

Keane, R.E., II. 1987. Forest succession in western Montana—A computer model designed for resource managers. Research. Note INT-376. USDA Forest Service, Intermountain Research Station. Ogden, UT.

Kercher, J.R., and M.C. Axelrod. 1984. Analysis of SILVA: A model for forecasting the effects of SO_2 pollution and fire on western coniferous forests. Ecological Modeling 23:165–184.

Kessell, S.R., and M. Potter. 1980. A quantitative model of nine Montana forest communities. Journal of Environmental Management 4:227–240.

Lachowski, H., P. Maus, and B. Platt. 1992. Integrating remote sensing with GIS. Journal of Forestry 90:16–21.

Little, E.L. 1971. Atlas of United States trees. USDA Forest Service. Miscellaneous Publication No. 1146. Washington, DC.

Lloyd, D. 1991. A phenological classification of terrestrial vegetation using shortwave vegetation index imagery. International Journal of Remote Sensing 11:2269–2279.

Maclean, A.L., D.D. Reed, G.D. Mroz, G.W. Lyon, and T. Edison. 1992. Using GIS to estimate forest resource changes. Journal of Forestry 90:22–25.

Maycock, P.F., and J.T. Curtis. 1960. The phytosociology of boreal conifer-hardwood forests of the Great Lakes Region. Ecological Monographs 30:1–35.

Meyer, M., and L. Werth. 1990. Satellite data: Management panacea or potential problem? Journal of Forestry 88:10–13.

Mladenoff, D.J. 1987. Dynamics of nitrogen mineralization and nitrification in hemlock and hardwood treefall gaps. Ecology 68:1171–1180.

Mladenoff, D.J. 1990. The relationship of the soil seed bank and understory vegetation in old-growth northern hardwood-hemlock treefall gaps. Canadian Journal of Botany 68:2714–2721.

Mladenoff, D.J., G.E. Host, J. Boeder, and T.R. Crow. 1994. LANDIS: A spatial model of forest landscape disturbance and succession at multiple scales. In: Proceedings, Second International Conference on Integrating GIS and Environmental Modeling. National Center for Geographic Information and Analysis (NCGIA). Santa Barbara, CA.

Mladenoff, D.J., and J. Pastor. 1993. Sustainable forest ecosystems in the northern hardwood and conifer region: Concepts and management. In: G.H. Aplet, J.T. Olson, N. Johnson, and V.A. Sample (eds.). Defining sustainable forestry. Island Press, Washington, DC, pp. 145–180.

Mladenoff, D.J., M.A. White, J. Pastor, and T. Crow. 1993a. Comparing spatial pattern in unaltered old-growth and disturbed forest landscapes. Ecological Applications 3:293–305

Mladenoff, D.J., P. Polzer, P. Wolter, G.E. Host, and T.R. Crow. 1993b. Multitemporal analysis of regional forest landscape change across ownership and landtype categories (Abs.). Bulletin Ecological Society of America 74:363.

Mladenoff, D.J., and F. Stearns. 1993. Eastern hemlock regeneration and deer browsing in the northern Great Lakes Region: A reexamination and model simulation. Conservation Biology: 889–900.

Morrison, L., and S. Ribansky (eds.). 1989. The development and demonstration of a combined remote sensing/geographic information system for the North Temperate Lakes Long-Term Ecological Research Site. Institute for Environmental Studies, University of Wisconsin-Madison, WI.

Morse, B. 1987. Expert interface to a geographic information system. Proceedings AutoCarto 8:535–541.

Naiman, R.J. (ed.). 1992. Watershed management. Springer-Verlag, New York. 542 pp.

Noble, I.R., and R.O. Slatyer. 1980. The use of vital attributes to predict successional changes in plant communities subject to recurrent disturbances. Vegetatio 43:5–21.

Noss, R.F. 1983. A regional landscape approach to maintain diversity. BioScience 33:700–706.

Overbay, J.C. 1992. Ecosystem management. In: Proceedings, National Workshop, Taking an Ecological Approach to Management. USDA Forest Service, Washington, DC. pp. 3–15.

Pastor, J., and D.J. Mladenoff. 1994. Modeling the effects of timber management on population dynamics, diversity, and ecosystem processes. In: D.C. LeMaster (ed.). Modeling sustainable forest ecosystems. American Forests, Washington, DC. In press.

Pastor, J., and D.J. Mladenoff. 1992. The southern boreal-northern hardwood border. In: H.H. Shugart, R. Leemans, and G.B. Bonan (eds.). A systems analysis of the global boreal forest. Cambridge University Press, Cambridge, UK. pp. 216–240.

Peters, R.L., and T.E. Lovejoy. 1992. Global warming and biological diversity. Yale University Press. New Haven, CT.

Probst, J., and T.R. Crow. 1991. Integrating biological diversity and resource management. Journal of Forestry 89:12–17.

Rauscher, H.M., and J.G. Isebrands. 1990. Using expert systems to model tree development. In: L. Wensel and G. Biging (eds.). Forest simulation systems: Proceedings 1988 IUFRO Conference. Berkeley, CA pp. 129–138.

Rauscher, H.M., D.A. Perala, and G.E. Host. 1990. An aspen forest management advisory system. In: R.D. Adams (ed.). Proceedings, Aspen Symposium '89. USDA Forest Service North Central Forest Experiment Station. General Technical Report NC-140. St. Paul, MN.

Regier, H.A., and G.L. Baskerville. 1986. Sustainable development of regional ecosystems degraded by exploitive development. Chapter 3. In: W.C. Clark and R.E. Munn (eds.). Sustainable development of the biosphere. International Institute for Applied Systems Analysis, Laxenburg, Austria. Cambridge University Press, Cambridge, UK.

Roberts, D.W. 1987. VITAL: A forest succession model based on vital attributes theory. Department of Forest Resources, Utah State University. Logan, UT. Unpublished report.

Roberts, D.W. 1993. Landscape vegetation modeling with vital attributes and fuzzy systems theory. Ecological Modeling: In press.

Robinson, V.B., and A.U. Frank. 1987. Expert systems for geographic information systems. Photogrammetric Engineering and Remote Sensing 53:1435–1441.

Roese, J.L., K.L. Risenhoover, and L.J. Folse. 1991. Habitat heterogeneity and foraging efficiency: An individual-based model. Ecological Modeling 57:133–143.

Roughgarden, J., S.W. Running, and P.A. Matson. 1991. What does remote sensing do for ecology? Ecology 72:1918–1922.

Rowe, J.S. 1984. Forestland classification: Limitations of the use of vegetation. In: J.G. Bockheim (ed.). Proceedings, Symposium for forest land classification: Experience, problems, perspectives. Department of Soil Science, University of Wisconsin- Madison, WI pp. 132–147.

Running, S.W. 1990. Estimating terrestrial primary productivity by combining remote sensing and ecosystem simulation. In: R.J. Hobbs and H.A. Mooney (eds.). Remote sensing of biosphere functioning. Springer-Verlag, New York. pp. 65–86.

Saarenmaa, H., N.D. Stone, L.J. Folse, J.M. Packard, W.E. Grant, M.E. Makela, and R.N. Coulson. 1988. An artificial intelligence modeling approach to simulating animal/habitat interactions. Ecological Modeling 44:125–141.

Stai, S.A., D.J. Mladenoff, and K. Rusterholz. 1993. Structural features of old-growth pine and northern hardwood forests in Minnesota (Abs.). Bulletin Ecological Society of America. In press.

Stearns, F. 1990. Forest history and management in the northern midwest. In: J.M. Sweeney (ed.). Management of dynamic ecosystems. North Central Section, The Wildlife Society, West Lafayette, IN pp. 107–122.

Steele, R. 1984. An approach to classifying seral vegetation within habitat types. Northwest Science 58:29–39.

Swanson, F.J., and J.F. Franklin. 1992. New forestry principles from ecosystem analysis of Pacific Northwest forests. Ecological Applications 2:262–274.

Thomas, J.W., E.D. Forsman, J.B. Lint, E.C. Meslow, B.R. Noon, and J. Verner. 1990. A conservation strategy for the northern spotted owl. USDA Forest Service, USDI Bureau of Land Management, U.S. Fish and Wildlife Service, and USDI National Park Service, Portland, OR. 427 pp.

Trotter, C.M. 1991. Remotely sensed data as an information source for geographical information systems in natural resource management: A review. International Journal of Geographical Information Systems 5:225–239.

Turner, M.G. 1989. Landscape ecology: The effect of pattern on process. Annual Review of Ecology and Systematics 20:171–197.

Tyrrell, L.E., and T.R. Crow. 1993. Analysis of structural characteristics of old-growth hemlock-hardwood forests along a temporal gradient. In: J.S. Fralish, R.P. McIntosh, and O.L. Loucks (eds.). John T. Curtis: Fifty years of Wisconsin Plant Ecology. Wisconsin Academy Press. Pp. 237–246.

Urban, D.L., R.V. O'Neill, and H.H. Shugart. 1987. Landscape ecology. Bioscience 37:119–127.

Waring, R.H., J.D. Aber, J.M. Melillo, and B. Moore III. 1986. Precursors of change in terrestrial ecosystems. Bioscience 36:433–438.

Wein, R.W., and M.A. El-Bayoumi. 1983. Limitations to predictability of plant succession in northern ecosystems. In: R.W. Wein, R.R. Riewe, and I.R. Methven (eds.). Resources and dynamics of the boreal zone. Association of Canadian Universities for Northern Studies. Ottawa. pp. 214–225.

Wessman, C.A. 1990. Evaluation of canopy biochemistry. In: R.J. Hobbs and H.A. Mooney (eds.). Remote sensing of biosphere functioning. Springer-Verlag, New York. pp. 135–156.

White, M.A., and D.J. Mladenoff. 1994. Old-growth forest landscape transitions in northern Wisconsin, USA, from presettlement to present. Landscape Ecology: in press.

White, M.A., D.J. Mladenoff, G.E. Host, P. Wolter, and T.R. Crow. 1993. Analyzing regional forest landscape structure across ownership categories and ecological land units (Abs.). In: Proceedings, U.S.-International Association for Landscape Ecology, Oak Ridge, TN p. 106.

Wickland, D.E. 1991. Mission to planet earth: The ecological perspective. Ecology 72:1923–1933.

Williams, M. 1989. Americans and their forests: A historical geography. Cambridge University Press, Cambridge, UK.

Wolter, P., D.J. Mladenoff, P. Polzer, G.E. Host, and T.R. Crow. 1993. Use of multitemporal LANDSAT imagery linked to tree species phenology in a classification of the Chequamegon National Forest region of Wisconsin (Abs.). Bulletin Ecological Society of America 74:492–493.

Zak, D.R., K.S. Pregitzer, and G.E. Host. 1986. Landscape variation in nitrogen mineralization and nitrification. Canadian Journal of Forest Research 16:1258–1263.

Zak, D.R., and K.S. Pregitzer. 1990. Spatial and temporal variability of nitrogen cycling in northern Lower Michigan. Forest Science 36:367–380.

Zhu, Z., and D.L. Evans. Mapping mid-south forest distributions. Journal of Forestry 90:27–30.

15

Resource Management Perspective: Remote Sensing and GIS Support for Defining, Mapping, and Managing Forest Ecosystems

David T. Cleland, Thomas R. Crow, James B. Hart, and Eunice A. Padley

Introduction

On June 4, 1992 the USDA Forest Service adopted a policy of ecosystem management for 191 million acres of national forests and grasslands across the United States. Since then, the agency has formed teams and partnerships at all levels of the organization to begin implementing this direction. Questions remain both within and outside the Forest Service, however, about what the terms "ecosystem" and "ecosystem management" actually mean. Some wonder if ecosystem management is even possible given our limited knowledge of complex ecological systems.

In 1935, Tansley introduced the term "ecosystem," and the explicit idea of ecological systems defined by abiotic and biotic factors of climate, physiography, soil, water, plants, and animals was formally expressed in our language (Major 1969). The ecosystem concept brings the biological and physical worlds together into a holistic framework within which ecological systems can be described, evaluated, and managed (Rowe 1992). Ecosystems are places where life forms and environment interact; they are three-dimensional or volumetric segments of the earth (Rowe 1980). Ecosystems are defined by associations of multiple biotic and abiotic factors (Figure 15.1), and are distinguished from one another by dissimilarities in their structure and function.

According to Webster's Third New International Dictionary (1981), the term "manage" means to render submissive, to dominate, to achieve objectives, to use with judgment, and to guide by careful or delicate treatment. To the Forest Service, management now means to carefully achieve objectives using judgment and scientifically based methods. The

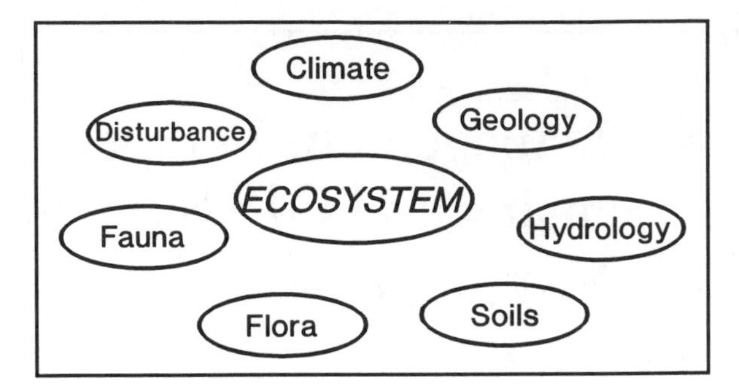

Figure 15.1 Multiple factors comprise ecosystems.

objectives of ecosystem management are to meet people's needs while ensuring that National Forests and Grasslands represent diverse, healthy, productive, and sustainable ecosystems (Robertson 1992).

To effectively manage ecosystems, we need to define what ecosystems are, identify where they occur, and understand how they function. Admittedly, these will not be easy tasks. We may never fully comprehend the myriad elements and interactions that constitute ecosystems. In lieu of comprehensive information, we will have to use heuristic models and a great deal of common sense to manage ecosystems. We will need to adjust our thinking and efforts as we acquire new information and develop new technologies. The argument that ecosystems are too complex to understand or manage, however, has become moot. Environmental concerns and changing public expectations compel us to no longer ask "can this be done," but rather "how well" and "at what expense."

In this chapter, we present several concepts that are useful for defining, mapping, and managing ecosystems, and offer a brief prospectus of the use of remote sensing and GIS technology in support of these activities. The topics presented here include (1) multiple factors, hierarchical structures, and ecosystems, (2) spatial and temporal variability, (3) ecosystem classification and mapping, (4) effects of scale on attributes of change, and (5) ecosystem management.

Multiple Factors, Hierarchical Structures, and Ecosystems

Biotic distributions and ecological processes are largely regulated along energy, moisture, nutrient, and disturbance gradients. These gradients in turn are affected by climate, physiography, soils, hydrology, flora, and fauna (Barnes et al. 1982, Cleland et al. 1985). These factors vary at different spatial scales. Thus it is useful to conceive of ecosystems as occurring in a nested geographic arrangement, with smaller ecosystems contained within larger ones (Allen and Starr 1982, O'Neill et al. 1986, Albert et al. 1986). Conditions and processes occurring in larger ecosystems modify and often override those of smaller, embedded ecosystems, and properties of smaller ecosystems emerge in the context of larger systems (Bailey 1985). The integration of multiple ecological factors at their respective scales of change provides the basis for defining and delineating ecosystems (Spies and Barnes 1985).

While the association of multiple factors is all-important in understanding ecosystems, all factors are not equally important in defining ecosystems at all spatial scales. The important factors at coarse scales are largely abiotic or physical variables (e.g., macroclimate, gross physiography), while both biotic and abiotic factors are important at finer scales (e.g., microclimate, soils, vegetation).

Macroclimate dominates ecosystems at all spatial scales. At macro scales, ecosystem patterns correspond with climatic regions, which change mainly due to latitudinal, orographic, and maritime influences (Bailey 1987, Spurr and Barnes 1980). Within climatic regions, physiography or landforms modify the intensity and flux of solar energy and moisture. Varying elevations, slopes, aspects, and geologic parent materials cause large variations in temperature and moisture regimes within climatic regions (Bailey 1988a, Rowe 1980). Landforms affect the movement of organisms, the orientation of watersheds, and the frequency and spatial pattern of disturbances such as wind and fire (Swanson et al. 1988).

Within physiographic regions, topography causes variations in microclimate and water drainage patterns, and hence in vegetation (Bailey 1987, Omernik 1987). Soils develop in parent materials on the mantle of landforms. Hence landforms often exhibit patterns in soil characteristics as well as patterns in vegetation (Rowe 1980, Host et al. 1987). Within plant rooting zones, soil physics, chemistry, and microbial populations govern moisture and nutrient availability. Thus soils exert a strong influence on vegetation, and indirectly affect wildlife populations.

Flora and fauna are integral components of ecosystems (McNaughton 1983, Pregitzer and Barnes 1982). Photoperiod, temperature, and precipitation exert primary control over the structure, composition, and genetic differentiation of plant populations (Denton and Barnes 1988). In turn, flora mediate in situ levels of light, temperature, and moisture and affect soil development processes, nutrient cycling, and carbon storage (Waring and Schlesinger 1985). Fauna rely on vegetation for food and shelter, and influence ecosystem development through seed predation or dispersal, selective herbivory, and a host of other mechanisms (Marquis and Brenneman 1981, Gysel and Stearns 1968). Microflora and microfauna are critical biogeochemical processors that all life forms depend on; they produce oxygen, fix nitrogen, and reduce carbon from organic to inorganic forms.

Disturbance regimes alter species distributions, community structure, and landscape patterns. Wildfires, blowdowns, insects and disease, and other forms of disturbance modify ecosystems at a variety of spatial and temporal scales (Heinselman 1973, Shugart and West 1981, Runkle 1982, Grimm 1984, Canham and Loucks 1984, Knight 1987, Swanson et al. 1990). Human activities have long influenced ecosystems throughout the world, but increased population pressures and advanced technologies are accelerating the rate of human-caused change as never before. Humans both influence and depend on ecosystems, and are therefore fundamental components of ecosystems in today's world.

Spatial and Temporal Variability

The structure and function of ecosystems change through space and time. Therefore we need to consider both spatial and temporal variability while evaluating, mapping, or managing ecosystems (Delcourt et al. 1983, Forman and Godron 1986). Figure 15.2 illustrates spatial and temporal variations that affect ecosystems. This figure shows spatial variations measured at local and regional scales, and temporal variations measured in years and centuries.

Within a local area, particular locations are wetter or drier, or more or less fertile than other locations because of differences in soil properties or hydrology. Each of these conditions supports certain assemblages of plants and animals. In midwestern forests, this environmental-biotic continuum could include oak savannas, xeric jack pine, dry-mesic pine-oak, mesic northern hardwood, and hydric hardwood or conifer communities.

At broader spatial scales, temperature and moisture gradients vary with latitude, elevation, and proximity to major bodies of water. For example, in the midwest, northern hardwoods are replaced by central hardwoods, and black oak by post oak, along a north-to-south axis because of differences in macroclimate. These changes at macro scales represent

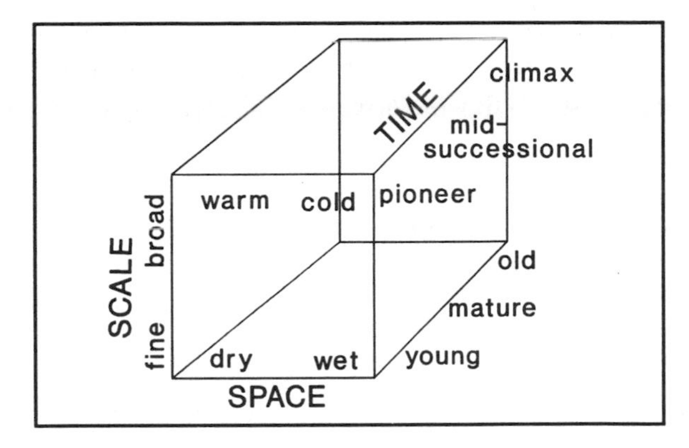

Figure 15.2 Spatial and temporal sources of ecosystem variability

spatial sources of environmental variability that affect regional, landscape, and local ecosystem structure and function.

Figure 15.2 also shows changes occurring through time. At temporal scales measured in years, a given ecosystem may be supporting vegetation that is young, mature, or old growth. Each of these conditions benefits certain plant and animal species and assemblages. For example, old-growth forests provide for species such as the spotted owl and pileated woodpecker. Mature forests with substantial amounts of coarse, woody debris benefit tree species that establish on rotting logs, such as eastern hemlock and yellow birch. Young, regenerating forests favor shade-intolerant vegetation and animal species such as white-tailed deer and ruffed grouse. These changes at finer time scales represent temporal variations that affect local ecosystem structure and function.

At temporal scales measured in centuries, ecosystems undergo change such as succession. Successional developments affect the nature and complexity of food webs, ratios of net primary production to respiration, and rates of nutrient and carbon cycling (Odum 1969). From a landscape perspective, these changes form a shifting mosaic of local ecosystems at different successional stages (Borman and Likens 1979). Changes occurring at these time scales represent temporal variations affecting landscape and local ecosystem structure and function.

Three examples illustrate the effects of spatial and temporal variability on ecosystems with regard to animal species. These examples involve the local and landscape ecosystems of the Kirtland's warbler (*Dendroica kirtlandii*), the Karner blue butterfly (*Lycaeides melissa samuelis*), and the black bear (*Ursus americanus*).

The Kirtland's warbler is a federally listed endangered species that is endemic to jack pine ecosystems in the northeastern part of lower northern Michigan. This area has a relatively cold macroclimate (short growing seasons, low winter temperatures), and is composed primarily of glacial outwash plains containing xeric sandy soils. This cold macroclimate constrains species composition and succession within outwash sands to conifer-dominated communities. Variations in regional climate, landscape-level landforms, and local-level soil conditions, then, determine the Kirtland's warbler spatial niche.

The Kirtland's warbler inhabits stands of jack pine that are generally between the ages of six and 23 years old and five to 16 feet high (Probst 1991). As these stands grow older, or change temporally, they become unsuitable habitat for this endangered species, favoring other species under changed conditions. Thus, both spatial and temporal variations govern the distribution of this species' habitat and, consequently, its population viability. The Kirtland's warbler has a very narrow spatial and temporal niche, or a narrow ecological amplitude in space and time.

The Karner blue butterfly is also a federally listed endangered species. This species inhabits oak savannas in outwash plains in southwestern lower Michigan, an area that has a relatively warm macroclimate when compared to the northeast. This warm macroclimate allows succession on xeric outwash sands to proceed to pin and black oak communities, as opposed to the conifer-dominated communities common in northeastern lower Michigan. The Karner blue's spatial niche, then, differs from the Kirtland's warbler due to regional climate, although both species occupy habitats within very similar landforms and soils.

The Kirtland's warbler and Karner blue butterfly are both fire-dependent species that occupy xeric soils in outwash landforms in Michigan. These species' habitats differ, however, due to environmental variability occurring at relatively coarse spatial scales. The narrow niches of the Kirtland's warbler and the Karner blue butterfly need to be available in both space and time if the dependent species are to survive. In today's society, these niches will be actively restored or maintained through ecosystem management. If left to chance, populations of these species will probably decline further given unmanaged rural development, fire suppression, and other human activities.

A third example, the black bear, is a state-listed sensitive species in Michigan. The black bear uses xeric jack pine communities to forage on huckleberry, blueberry, and bearberry; dry-mesic pine-oak communities to forage on juneberry, hawthorn, and maple leaf viburnum; mesic northern hardwood communities to forage on gooseberry, jack-in-the-pulpit, and wild leeks; and wetlands to forage on several species as well as to escape human predation (Rogers and Allen 1987). The black bear uses regenerating forests to forage on light-loving plants including blackberry and raspberry, mature forests to forage on aforementioned species, and over-mature and old-growth forests to forage on insects inhabiting dead and dying trees.

The black bear uses many spatial and temporal niches throughout its range, and has an extremely wide ecological amplitude when compared to the Kirtland's warbler or the Karner blue butterfly. Resource managers can enhance the viability of bear populations, however, by evaluating its requirements in the context of spatial and temporal sources of ecosystem variability, in terms of the seasonality of forage, and so forth.

Ecosystem Classification and Mapping

Ecosystems are places where life forms and environment interact, and are distinguished from one another by differences in their structural and functional characteristics. In forested ecosystems, structure and function vary according to the age and composition of existing biota, as well as environmental factors. Therefore, to classify and map ecosystems we need to address both biotic and abiotic conditions. Furthermore, an understanding of ecological processes provides useful information for defining and delineating ecosystems. Thus biotic and environmental conditions, and the interactions thereof, are all considered in classifying and mapping ecosystems.

In practice, we need to combine two types of inventories to classify and map ecosystems. These are inventories of existing conditions that change readily through time, and inventories of potential conditions that are relatively stable. Existing conditions are inventoried as current vegetation, wildlife, water quality, and so forth. Potential conditions are inventoried as natural associations of ecological factors at their respective scales of occurrence. When these inventories are combined, biotic distributions and ecological processes can be evaluated, and results of such evaluations can then be extrapolated to similar ecosystems (Figure 15.3). GIS will provide a means of combining these separate themes of information to define and map ecosystems.

To develop classification units for mapping areas of uniform ecological potentials, scientists filter spatial sources of variability from temporal

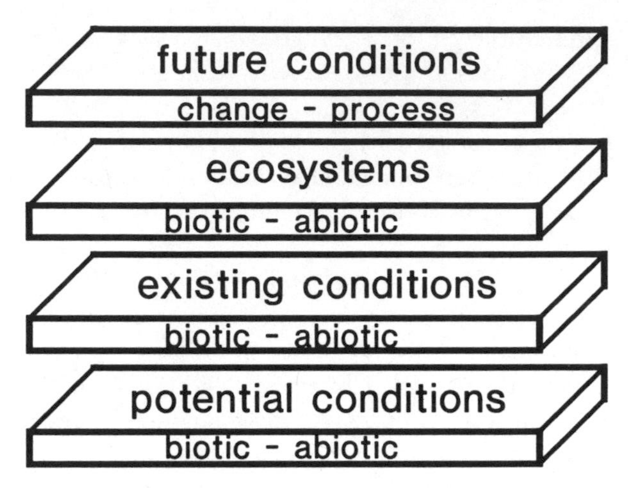

Figure 15.3 Templates for ecosystem mapping and management

sources by examining biotic-abiotic relationships within a particular time reference. In forested ecosystems, mature, climax, or late-successional ecosystems are often initially evaluated, with follow-up studies conducted to ascertain structural and functional changes due to time. In this classification and mapping process, plants are used as indicators of environmental conditions, and environmental conditions are evaluated and mapped using both this indirect expression as well as direct gradient analyses.

The Forest Service is developing a national hierarchical framework for inventorying land and water units of differing ecological potentials (Avers and Schlatterer 1991). This framework is entitled the Ecological Classification and Inventory System. This system integrates multiple biotic and abiotic factors to classify and map ecological units at national, regional, landscape, and local scales. The Ecological Classification and Inventory System builds upon concepts and systems developed by numerous researchers and National Forests (Cajander 1926, Koppen 1931, Fenneman 1938, 1950, Hills 1952, Whitaker 1960, Kuchler 1964, Daubenmire and Daubenmire 1968, Wertz and Arnold 1972, Corliss 1974, Cowardin et al. 1979, Rowe 1980, Eyre 1980, Jordan 1982, Barnes et al. 1982, Bailey 1983, Jones et al. 1983, Driscoll et al. 1984, Smalley 1986, McNab 1987, Omernik 1987, Host et al. 1988, Hix 1988, Jensen et al. 1991, Cleland et al. 1992). When combined with information on existing conditions, this system can be used to evaluate the spatial and temporal relationships of ecosystems, and results can be used in ecosystem management.

Effects of Scale on Attributes of Change

Ecosystems are dynamic and subject to constant change. How, then, can we define, map, or manage ecosystems? What may initially appear to be an overwhelming problem becomes more feasible when one considers that the attributes of change through space and time vary according to scale of observation. Understanding how differences in spatial and temporal scales affect our perceptions of the world around us is paramount to understanding ecosystems.

There is a general relationship between space and time. Changes over large spatial areas generally require long time periods in which to occur, whereas changes over small areas often occur within short periods (Figure 15.4). Although change is intrinsic to ecosystems, the rates of change of most phenomena of concern to ecosystem managers are not so great as to make them incomprehensible or unquantifiable. Indeed, the evolution of management into an ecological approach was fostered by human concerns regarding the nature of change, particularly change we are responsible for or affected by. Concepts of change like homeostasis, resiliency, and succession become more meaningful when viewed in the context of natural associations of space and time.

Figure 15.4 shows examples of environmental and biological changes that have occurred at commensurate spatial and temporal scales (Forman and Godron 1986). Plate tectonics has altered evolution at a global scale over periods of millions of years. Climate change has altered species distributions at a continental scale over periods of thousands of years. Disturbance through fire has altered landscapes over periods of centuries. Drought has altered growth and mortality of organisms and local ecosystems over periods of decades. Seasons have altered the phenology and physiology of plants and animals over periods of months. Diurnal cycles have altered individual organism's habits daily (e.g., photosynthesis, sleep cycles). Within organisms, molecular reactions such as metabolism have occurred within seconds. Within molecules, atomic reactions have occurred within nanoseconds. There are many implications of this natural space-time association. The first implication is that ecosystem managers should consider rates of change with respect to the particular phenomena of interest or concern.

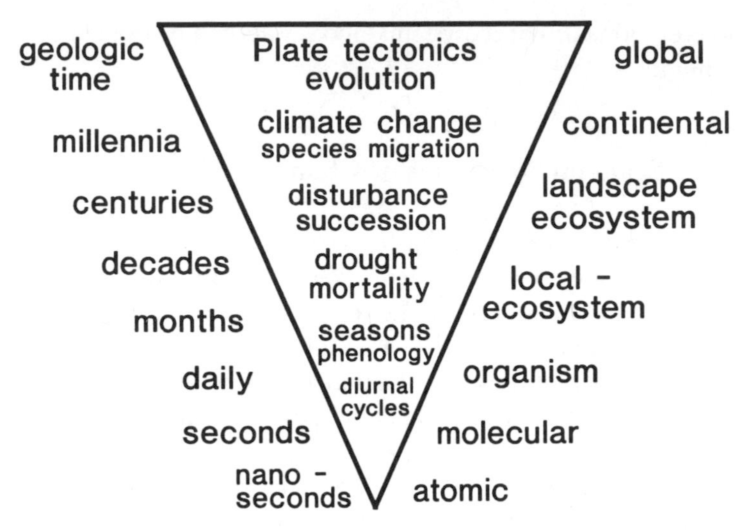

Figure 15.4 Examples of change at commensurate spatial and temporal scales

A second implication is that we may want to evaluate the effects of human activities on complex ecological systems based on our knowledge of natural space-time associations. At a global scale, for example, there is concern that climate is warming at an unprecedented rate. If this is true, it would be an example of a human-induced alteration of a natural space-time relationship. If climate changes within periods several orders of magnitude less than natural rates of change, the ability of species and ecological systems to adapt and compensate through migration or other mechanisms at commensurate spatial scales may be exceeded.

At a landscape scale, our suppression of wildfires has extended intervals between major fire events. These efforts have resulted in events such as the infamous Yellowstone National Park fire. This fire not only had a different character than past natural fires, but it also burned a far larger area. In a reductionistic sense, human intervention through fire suppression caused increased fuel loadings and formation of fuel ladders. These atypical fuel conditions led to extreme crown fires that burned extensive areas. In a holistic sense, however, it may have been the extension of time between wildfires that led to a more extensive spatial effect.

At local scales, ecosystems develop under or co-develop with different types of disturbance regimes (Urban et al. 1987). Xeric and dry-mesic pine-oak ecosystems, for example, tend to occupy droughty soils, accumulate litter, and produce volatile foliar substances, thereby increasing their susceptibility to burning. In contrast, mesic northern hardwood ecosystems occupy moist soils and produce litter that decomposes rapidly, thereby decreasing their susceptibility to burning. Pine-oak ecosystems in the midwest experienced broad-scale, high-intensity disturbance through fire over intervals generally ranging from one to several centuries (Whitney 1986). Mesic northern hardwood ecosystems, conversely, experienced frequent fine-scale, low-intensity disturbance through tree-fall and infrequent, widespread blowdowns (Runkle 1982, Canham and Loucks 1984).

These two ecosystems have markedly different reproductive strategies that are adaptations to these different types of perturbations. Pine-oak ecosystems are resilient to catastrophic wildfires because the thick bark of pine monarchs allow them to withstand fire, thereby providing a seed source for the next generation; and although the stems of oak species usually die, the root systems often survive and regenerate through stump and seedling sprouts. In contrast, mesic northern hardwood ecosystems are not adapted to fire, and suffer high mortality in all structural layers when burned. Mesic ecosystems are resilient to low-intensity, fine-scale disturbance caused by tree-fall, however, because the understory is composed of shade-tolerant seedlings that persist in high densities beneath closed forest canopies. This understory responds to increased light levels following fine-scale disturbance by filling gaps through rapid growth. Furthermore, the decaying logs left lying along the forest floor of mesic ecosystems provide germination and establishment niches for species such as eastern hemlock and yellow birch, setting the stage for these species to compete for canopy positions following the next fine-scale disturbance.

These ecosystems may also have developed different nutrient conservation strategies that are responsive to these different types of disturbance. For example, Zak et al. (1989) found that the end product of nitrogen mineralization in xeric and dry-mesic pine-oak systems is ammonium, whereas nitrate forms are also produced in mesic hardwood systems. Ammonium is a cation that is readily held onto cation exchange sites in soil carbon and clay fractions, whereas nitrates are anions that leach through soils if not rapidly sequestered into living vegetation. Pine-oak systems conserve nitrogen released as ammonium following high-intensity disturbance through fire because of soil properties. Mesic hardwood systems conserve nitrogen released as nitrates and ammonium following low-intensity disturbance through rapid growth of understory vegetation. These ecosystems, then, have different nitrogen-cycling mechanisms that effectively conserve this nutrient following the type of natural disturbance with which each system co-developed.

The reproductive and nutrient conservation mechanisms of these ecosystems can be related to natural space-time relationships. Pine-oak ecosystems are adapted to high-intensity disturbance that historically occurred at broad spatial and temporal scales, whereas northern hardwood ecosystems are adapted to low-intensity disturbance that occurred at fine scales.

Ecosystem Management

Ecosystem management represents a change in the way we view the interrelationships between ecological and economic systems. Under this evolving paradigm, management will continue to follow multiple-use principles, but with a difference. Elements that we have traditionally emphasized, such as popular game or commercial tree species, will be placed into a broader context in which they are considered along with the intrinsic and spiritual values that exist for all ecosystems (e.g., their history, complexity, beauty, and cultural significance). This broader context includes considerations of stability and change at relevant spatial and temporal scales. This more comprehensive approach to management does not mean that resource utilization to meet human needs is being diminished, however. On the contrary, understanding and protecting ecosystems is essential to providing a lasting supply of the materials and experiences that people require. Sustaining both ecological and economic systems is an imperative of ecosystem management, since these systems are inextricably linked and mutually dependent.

Ecosystem management depends on our ability to define and locate ecosystems, and to reasonably predict how existing conditions will change because of human actions and natural processes. Our knowledge of existing conditions, underlying potentials, and ecological processes will enable us to evaluate the potential of the physical environment to support existing or introduced biota, and the interactions and processes that contribute to the ecosystems development or alteration.

Part of ecosystem management is based on a realization that we can improve management and analyze the effects thereof by examining conditions and processes at multiple scales. In other words, ecosystem management considers effects at, above, and below the level under consideration (Rowe 1992). For example, the effects of timber harvesting, road building, and rural development are manifest not only at a local level, but also at microsite and landscape levels; these effects may be deleterious or beneficial (Noss 1983, Andren and Angelstam 1988). The use of hierarchical concepts therefore has practical implications for ecosystem management. Applying these concepts in operational planning and management will enable us to better integrate management objectives, evaluate outputs and tradeoffs, and estimate the indirect and cumulative effects of management actions at regional to local scales.

Ecosystem management is concerned with the effects of both natural and human induced change or disturbance on the structure and function of ecosystems. Species and ecosystems are already being managed, deliberately or inadvertently, by the introduction or exclusion of different types of disturbance (Swanson et al. 1990). The exclusion of fire in jack pine and oak savanna ecosystems is a disturbance that has adversely affected these ecosystems and their dependent populations, including the Kirtland's warbler and Karner blue butterfly. The harvesting of timber, use of prescribed burning, and planting of jack pine seedlings are also forms of disturbance that have advantageously affected populations of the Kirtland's warbler.

Ecosystem management seeks to maintain the natural diversity and processes intrinsic to ecosystems. There is a growing appreciation that much of this diversity and many of these processes were maintained in the past by different forms of natural disturbance (Urban et al. 1987, Swanson et al. 1990). Information on natural disturbance regimes, including the nature (e.g., fire, wind, flooding), spatial and temporal scale, and pattern of disturbance that occurred in pre-European settlement time will provide useful models for ecosystem management. These models can be tested, verified, adjusted, or refuted as appropriate. The emulation of natural disturbance regimes will require long-term research and monitoring, however, to determine when and where such practices are effective, and to develop better techniques. Research partnerships and adaptive management strategies will be critical to the overall success of these efforts.

Human use in forested ecosystems often includes various vegetative management practices that represent a disturbance intensity continuum. We may be able to mimic natural disturbance regimes while managing forested ecosystems using various silvicultural systems or modifications thereof. These practices could include the establishment of artificial plantations; even-aged, multi-aged, or uneven-aged vegetative management; or the absence of active manipulation. Each system imposes a different form and intensity of disturbance that creates conditions benefitting certain assemblages of species while discriminating against others (Thomas 1979, Hunter 1987). Each method has advantages and disadvantages that depend on management objectives and local conditions (Probst and Crow 1991).

Ecosystem management will strive to avoid or reduce conflicting land uses by analyzing ecosystems at several spatial scales and placing management emphasis areas in suitable locations (Harris 1984, Lord and Norton 1990, Crow 1990). In terms of vegetative management, intensively managed systems such as conifer plantations could be established or maintained in highly productive areas that have already incurred substantial investments in road systems or existing plantations. Even-aged management could be applied in landscapes that historically experienced catastrophic disturbance. Uneven-aged management could be applied in ecosystems that seldom experienced catastrophic disturbance. Old-growth forests could be maintained or allowed to mature in ecosystems where the life span of the dominant trees exceeds the time period between catastrophic disturbance events. Of course, departures from these strategies would also be acceptable across the different land holdings of agencies and private landowners with different management objectives and mandates. Moreover, considerations including water quality, wildlife habitat, cultural resources, human benefits and safety, and so forth would be weighted in actual ecosystem analysis, planning, and management.

A major challenge of ecosystem management is to distinguish associations of the factors that comprise ecosystems at different spatial scales, and to understand interactions or processes occurring within and across these different levels of organization. Our ability to recognize natural associations among life forms, environment, space, time, and scale, and to emulate natural processes as we learn, may have practical implications for the welfare of humankind as well as other species. For example, research indicates that the Mack Lake ecosystem, the primary habitat of the Kirtland's warbler, burned in areas greater than 10,000 acres every 35 years on average in presettlement times. During the period of settlement and early suppression, between 1850 and 1929, fires of this size occurred every 27 years. Now, in an era of rigorous fire control, fires of 10,000 acres or larger have occurred every 25 years (Simard and Blank 1982).

The frequency of modern era wildfires is not significantly different in the Mack Lake ecosystem than pre-European settlement times. It appears that human efforts to alter the space-time relationship of large wildfires in the Mack Lake area have failed. Recognizing this natural space-time relationship, and taking appropriate measures to manage this ecosystem by creating extensive fuel breaks through timber harvest and/or prescribed fire in temporal and spatial patterns that mimic natural disturbance regimes might be warranted. This would not be an effort to simply imitate nature for the sake of it, but more to follow nature's lead as we learn to better provide for the well-being of species and ecosystems, including humankind.

Application of Remote Sensing and GIS in Ecosystem Management

Remote sensing and GIS have great potential to advance the science and art of ecosystem management. Remote sensing will assist managers and scientists in identifying conditions of climate, physiography, vegetation, and soils from which we will be able to identify associations through careful analysis and synthesis. Weather satellite imagery, for example, will provide information that was previously only available from weather stations. In many parts of the country, there are too few weather stations to detect climatic trends, or their locations may be more related to cultural trends in settlement as opposed to locations allowing meaningful observation or extrapolation. Weather satellites will also show spatial extent of patterns formerly observed on a point-by-point basis.

Remote sensing will also be useful in measuring structural characteristics of ecosystems at different scales, such as identifying land use patterns at landscape scales or leaf area indices and forest gaps at local scales. Measurements of ecosystem function expressed by surface temperatures, evapotranspiration, plant-water stress, and energy budgets will be facilitated through remote sensing. Information needed for conducting spatial and temporal analyses and detecting changes over time will be made available through remote sensing. We may be able to eventually forecast future conditions by relating existing conditions to ecological potentials and our knowledge of process. Nowcasting, or detecting changes that defy prediction due to randomness or nonlinearities such as climatic (e.g., tornadoes) and wild fire events, or insect and disease outbreaks, will be made possible through remote sensing.

GIS will be useful in evaluating land capabilities, portraying existing and desired future conditions, and evaluating alternative landscape management designs for both commodity and conservation objectives. GIS will help us in determining the effects of altering structure and patterns at regional, landscape, and local scales. By providing a means of conducting multiscaled analyses, GIS will help us to better describe and understand processes and interactions occurring among adjacent ecosystems, as well as interactions occurring across hierarchical levels. Multiscaled phenomena such as the metapopulation dynamics of neotropical birds or wide-ranging mammals will be more readily investigated when GIS technology and attribute databases are widely available to natural resource managers and scientists.

Although remote sensing and GIS technology will be helpful in the evolution of thought and action needed for ecosystem management, these technologies are only tools. The actual integration of ecological factors, comprehension of processes, and formulation of research and management strategies ultimately lie with the scientists and managers using these tools (Bailey 1988b).

Summary

Ecosystems are volumetric segments of the earth that are distinguished from one another by differences in their structural and functional characteristics. Ecosystems are composed of, and defined by, interactions of multiple biotic and abiotic factors. The structure and function of ecosystems change through space and time, and these changes occur at different and often commensurate spatial and temporal scales.

To analyze, plan for, and manage ecosystems, we need information on their nature and distribution, information on the elements that compose ecosystems, and information on the various processes that maintain or change ecosystems. Inventories of existing biotic conditions and inventories of environmental factors that regulate ecological processes and potentials are used together to identify and manage ecosystems.

Ecosystem management is concerned with the nature of change that is inherent to ecosystems. Ecosystems are altered through biotic processes including reproduction, aging, mortality, and succession and ecological processes such as disturbance. These changes are rendered by both natural and human causes. As we learn about the nature and effects of change with respect to species and ecosystems, we can follow nature's lead by emulating the natural disturbance regimes that maintained or altered ecosystems throughout time.

By improving our comprehension of the associations and interactions of organisms and their environment that change through space and time at different scales, we can begin to understand how conditions and processes influence the stability and resiliency of ecosystems. Given this understanding, we can develop management strategies that satisfy objectives ranging from commodity production to preservation, and identify the assemblages of species that benefit along this continuum through monitoring, research, and adaptive management. By doing so, we will be able to better provide for the diversity, health, productivity, and sustainability of ecosystems and dependent populations, including humankind, through ecosystem management.

Conclusion

In this chapter, several concepts and examples have been presented to describe what the terms "ecosystem" and "ecosystem management" mean. Although these concepts have not been widely applied to date, they are not new. In 1966, the prominent scientist Eugene P. Odum presented a presidential address at the Ecological Society of America that he later published (Odum 1969). Odum stated "Until recently mankind has more or less taken for granted the gas exchange, water purification, nutrient cycling, and other protective functions of self-maintaining ecosystems, chiefly because neither his numbers nor his environmental manipulations have been great enough to affect regional and global balances. Now, of course, it is painfully evident that such balances are being affected, often detrimentally. The 'one problem, one solution' approach is no longer adequate and must be replaced by some form of ecosystem analysis that considers man as part of, not apart from the environment."

As a management recommendation, Odum proposed that "We can compromise so as to provide moderate quality and moderate quantity on all the landscape, or we can deliberately plan to compartmentalize the landscape so as to simultaneously maintain highly productive and predominantly protective types as separate units subject to different strategies (strategies ranging, for example, from intensive cropping on the one hand to wilderness management on the other). If ecosystem development theory is valid and applicable to planning, then the so-called multiple-use strategy, about which we hear so much, will work only through one or both of these approaches, because in most cases, the projected multiple uses conflict with one another."

These statements are more relevant today than they were when made 27 years ago. Public opinion and expectations have caught up with this thinking, and the policy of ecosystem management was born as a consequence. Although ecosystems are extremely complex, we can formulate meaningful hypotheses, conduct research, and improve management by taking an ecological approach. It is time to marry ecological concepts with natural resource management and research to ensure that natural systems and the well-being of species that are dependent on them, including humankind, are sustained.

References

Albert, D.A., S.R. Denton, and B.V. Barnes. 1986. Regional landscape ecosystems of Michigan. School of Natural Resources, Univ. of Michigan, Ann Arbor. 32 pp.

Allen, T.H.F., and T.B. Starr. 1982. Hierarchy: Perspectives for ecological complexity. Univ. of Chicago Press, Chicago. 310 pp.

Andren, H.P., and P. Angelstam. 1988. Elevated predation rates as an edge effect in habitat islands: Experimental evidence. Ecology 69:544–47.

Avers, P.E., and E.F. Schlatterer. 1991. Ecosystem classification and management on national forests. In: Proceedings of the 1991 Symposium on Systems Analysis in Forest Resources.

Bailey, R.G. 1983. Delineation of ecosystem regions. Environ. Manage. 7:365–373.

Bailey, R.G. 1985. The factor of scale in ecosystem mapping. Environ. Manage. 9:271–276.

Bailey, R.G. 1987. Suggested hierarchy of criteria for multiscale ecosystem mapping. Landscape and Urban Planning 14:313–319.

Bailey, R.G. 1988a. Ecogeographic analysis: a guide to the ecological division of land for resource management. USDA Forest Service Misc. Pub. 1465. 18 pp.

Bailey, R.G. 1988b. Problems with using overlay mapping for planning and their implications for geographic information systems. Environ. Manage. 12:11–17.

Barnes, B.V., K.S. Pregitzer, T.A. Spies, and V.H. Spooner. 1982. Ecological forest site classification. J. Forestry 80:493–498.

Barnes, B.V. 1984. Forest ecosystem classification and mapping in Baden-Wurttemberg, West Germany. In: Forestland classification: Experience, problems, perspectives. Proceedings of the Symposium, Madison, WI, March 18–20. pp. 49–65.

Borman, F.H., and G.E. Likens. 1979. Pattern and process in a forested ecosystem. Springer-Verlag, New York. 253 pp.

Cajander, A.K. 1926. The theory of forest types. Society for the Finnish Literary, Acta For. Fenn 29:1–108.

Canham, C.D., and O.L. Loucks. 1994. Catastrophic windthrow in the presettlement forests of Wisconsin. Ecology 65(3):803–809.

Cleland, D.T., J.B. Hart, K.S. Pregitzer, and C.W. Ramm. 1985. Classifying oak ecosystems for management. In: J.E. Johnson (ed.). Proceedings of Challenges for Oak Management and Utilization, March 28–29, University of Wisconsin, Madison. pp. 120–134.

Cleland, D.T., T.R. Crow, P.E. Avers, and J.R. Probst. 1992. Principles of land stratification for delineating ecosystems. In: Proceedings of Taking an Ecological Approach to Management national workshop. April 27–30: Salt Lake City, Utah. pp. 40–50.

Corliss, J.C. 1974. ECOCLASS—A method for classifying ecosystems. In: Foresters in Land-Use Planning. Proc. 1973 Nat. Convention Soc. Am. For., Washington, D.C. pp. 264–271.

Cowardin, L.M., V.Carter, F.C. Golet, and E.T. LaRoe. 1979. Classification of wetlands and deep-water habitats of the United States. U.S. Fish and Wildlife Service. Washington, D.C. 131 pp.

Crow, T.R. 1990. Biological diversity and silvicultural system. In: Proceedings of the National Silvicultural Workshop. July 10–13: Petersburg, AK. Washington D.C. U.S. Department of Agriculture, Forest Service. pp. 180–185.

Daubenmire, R., and J.B. Daubenmire. 1968. Forest vegetation of eastern Washington and northern Idaho. Washington State Tech. Bull, 60, Washington Agric. Exp. Stn. 104 pp.

Delcourt, H.R., T.A. Delcourt, and T. Webb. 1983. Dynamic plant ecology: The spectrum of vegetative change in space and time. Quant. Sci. Rev. 1:153–75.

Denton, D.R., and B.V. Barnes. 1988. An ecological climatic classification of Michigan: A quantitative approach. For. Sci. 34:119–138.

Driscoll, R.S., and others. 1984. An ecological land classification framework for the United States. USDA Forest Service. Miscel. Pub. 1439. Washington, D.C. 56 pp.

Eyre, F.H. 1980. Forest cover types of the United States and Canada. Society of American Foresters. Washington, DC. 148 pp.

Fenneman, N.M. 1938. Physiography of eastern United States. McGraw-Hill. 714 p.

Forman, R.T.T., and M. Godron. 1986. Landscape Ecology. New York: John Wiley and Sons.

Grimm, E.C. 1984. Fire and other factors controlling the big woods vegetation of Minnesota in the mid-nineteenth century. Ecol. Monog. 54:2913–311.

Gysel, L.W., and F. Stearns. 1968. Deer browse production of oak stands in central lower Michigan. USDA Forest Service Res. Note NC-48.

Harris, L.D. 1984. The Fragmented Forest. Univ. Chicago Press, Chicago.

Heinselman, M.L. 1973. Fire in the virgin forests of the Boundary Waters Canoe Area. Quat. Res. 3:275–289.

Hills, G.A. 1952. The classification and evaluation of site for forestry. Ontario Dep. Lands and Forests. Res. Div. Rep. 24.

Hix, D.M. 1988. Multifactor classification of upland hardwood forest ecosystems of the Kickapoo River watershed, southwestern Wisconsin. Can. J. For. Res. 18:1405–1415.

Host, G.E., K.S. Pregitzer, C.W. Ramm, J.B. Hart, and D.T. Cleland. 1987. Landform mediated differences in successional pathways among upland forest ecosystems in northwestern Lower Michigan. For. Sci. 33:445–457.

Host, G.E., K.S. Pregitzer, C.W. Ramm, D.T. Lusch, and D.T. Cleland. 1988. Variations in overstory biomass among glacial landforms and ecological land units in northwestern Lower Michigan. Can. J. For. Res. 18:659–668.

Hunter, M.L. 1987. Managing forests for spatial heterogeneity to maintain biological diversity. Trans. N. Amer. Wildl. and Nat. Res. Conf. 53:61–69.

Jensen, M.C., C. McMicoll, and M. Prather. 1991. Application of ecological classification to environmental effects analysis. J. Env. Qual. 20:24–30.

Jones, R.K., and others. 1983. Field guide to forest ecosystem classification for the clay belt, site region 3e. Ministry of Natural Resources. Ontario, Canada. 123 pp.

Jordan, J.K. 1982. Application of an integrated land classification. In: Proceedings, Artificial Regeneration of Conifers in the Upper Lakes Region. October 26–28, 1982. Green Bay, WI. pp. 65–82.

Knight, D.H. 1987. Parasites, lightning, and the vegetation mosaic in wilderness landscapes. In: M. Turner (ed.). Landscape heterogeneity and disturbance. Springer-Verlag, NY. pp. 59–83.

Koppen, J.M. 1931. Grundriss der Klimakunde. Walter de Grayter, Berlin. 388 pp.

Kuchler, A.W. 1964. Potential natural vegetation of the conterminous United States. Am. Geog. Soc. Spec. Pub. 36. 116 pp.

Lord, J.M., and D.A. Norton. 1990. Scale and the spatial concept of fragmentation. Conserv. Biol. 4:197–202.

Major, J. 1969. Historical development of the ecosystem concept. In: G.M. Van Dyne (ed.). The ecosystem concept in natural resource management. Academic Press, New York. pp. 9–22.

Marquis, D.A., and R. Brenneman. 1981. The impact of deer on forest vegetation in Pennsylvania. USDA FS Gen. Tech. Rep. NE-65. pp. 1–7.

McNab, W.H. 1987. Rationale for a multifactor forest site classification system for the southern Appalachians. In: Proceedings of 6th Central Hardwood Forest Conference. Feb. 24–26. Knoxville, TN. pp. 283–294.

McNaughton, S.J. 1983. Serengeti grassland ecology: The role of composite environmental factors and contingency in community organization. Ecol. Monog. 53:291–320.

Meentemeyer, V., and E.O. Box. 1987. Scale effects in landscape studies. In: M.G. Turner (ed.). Landscape Heterogeneity and Disturbance. Springer-Verlag, New York. pp. 15–34.

Noss, R.F. 1983. A regional landscape approach to maintain diversity. BioScience 33:700–706.

O'Neill, R.V., D.L. DeAngelis, J.B. Waide, and T.F.H. Allen. 1986. A hierarchical concept of ecosystems. Princeton University Press, Princeton, N.J.

Odum, E.P. 1969. The strategy of ecosystem development. Science 154:262–270.

Omernik, J.M. 1987. Ecoregions of the conterminous United States. Annals of the Association of American Geographers 77:118–125.

Pregitzer, K.S., and B.V. Barnes. 1984. Classification and comparison of upland hardwood and conifer ecosystems of the Cyrus H. McCormick Experimental Forest, upper Michigan. Can. J. For. Res. 14:362–375.

Probst, J.R., and T.R. Crow. 1991. Integrating biological diversity and resource management. J. For. 89:12–17.

Probst, J.R. 1991. Kirtland's Warbler. In: R. Brewer, G.A. McPeek, and R.J. Adams Jr. (eds.). The atlas of breeding birds of Michigan. Mich. State Univ. Press, East Lansing. pp. 414–417.

Robertson, F.D. 1992. June 4th USDA Forest Service-wide memorandum on ecosystem management.

Rogers, L.L., and A.W. Allen. 1987. Habitat suitability index models: Black bear, upper Great Lakes region. Biological Report 82 (10.144). USDI, Fish and Wildlife Service. 54 pp.

Rowe, J.S. 1980. The common denominator in land classification in Canada: An ecological approach to mapping. For. Chron. 56:19–20.

Rowe, J.S. 1984. Forestland classification: Limitations on the use of vegetation. In: Forest Land Classification: Experience, Problems, Perspectives. Proceedings of the symposium, Madison, WI, March 18–20. pp. 132–147.

Rowe, J.S. 1992. The ecosystem approach to forestland management. For. Chron. 68:222–224.

Runkle, J.R. 1982. Patterns of disturbance in some old-growth mesic forests of eastern North America. Ecology 63:1533–1546.

Russell, W.E., and J.K. Jordan. 1991. Ecological classification system for classifying land capability in midwestern and northeastern U.S. national forests. In: Proceedings of the symposium, Ecological Land Classification: Applications to Identify the Productive Potential of Southern Forests, Charlotte, NC, January 7–9. USDA Forest Service, General Technical Report SE-68.

Shugart, H.H., Jr., and D.C. West. 1981. Long-term dynamics of forest ecosystems. Am. Sci. 69:647–652.

Simard, A.J., and R.W. Blank. 1982. Fire history of a Michigan jack pine forest. Michigan Academy 1982:59–71.

Smalley, G.W. 1986. Site classification and evaluation for the Interior Uplands. USDA Forest Service. Technical Publication R8-TP9. Southern Region, Atlanta, GA.

Spies, T.A., and B.V. Barnes. 1985. A multifactor ecological classification of the northern hardwood and conifer ecosystems of Sylvania Recreation Area, Upper Peninsula, Michigan. Can. J. For. Res. 15:949–960.

Spurr, S.H., and B.V. Barnes. 1980. Forest ecology. Third edition. John Wiley and Sons, New York. 687 pp.

Swanson, F.J., T.K. Kratz, N. Caine, and R.G. Woodmansee. 1988. Landform effects on ecosystem patterns and processes. BioScience 38:92–98.

Swanson, F.J., J. Franklin, and J. Sedell. 1990. Landscape patterns, disturbance and management in the Pacific Northwest, USA. In: Changing landscapes: An ecological perspective. Springer-Verlag, New York. 286 pp.

Thomas, J.W. (ed.). 1979. Wildlife habitats in managed forests: The Blue Mountains of Oregon and Washington. USDA Forest Service, Ag. Handbook 553. 512 pp.

Urban, D.L., D. O'Neil, and H.H. Shugart. 1987. Landscape ecology. BioScience 37:119–127.

Waring, R.W., and W.H. Schlesinger. 1985. Forest ecosystems concepts and management. Academic Press, Inc. 340 pp.

Wertz, W.A., and J.A. Arnold. 1972. Land systems inventory. USDA Forest Service, Intermountain Region, Ogden, UT.

Whitaker, R.H. 1960. The vegetation of the Siskiyou Mountains of Oregon and California. Ecol. Monogr. 30:279–338.

Zak, D.R., G.E. Host, and K.S. Pregitzer. 1989. Regional variability in nitrogen mineralization, nitrification, and overstory biomass in northern lower Michigan. Can. J. For. Res. 19:1521–1526.

16

GIS Applications Perspective: Remote Sensing and GIS as a Bridge Between Ecologists and Resource Managers in the Northern Lake States

Mark D. MacKenzie

Introduction

Chapter 14 presented a discussion of ecological land classification and the application of remote sensing and GIS in the northern Lake Sates. The discussion mentioned four general application needs: (1) forest and land-cover mapping, (2) ecological land classification, (3) regional change detection, and (4) detection and modeling of ecosystem properties and processes. Many examples were cited of the use of remote sensing and GIS to address these application areas. This chapter will provide specific examples of the use of remote sensing and GIS to address three of these application areas in the Northern Highland Lakes District (NHLD) of north central Wisconsin. This chapter will address issues of sensor spatial resolution and its effect on interpretation and sensor utility. It will also illustrate the integration of remote sensing and GIS to address forest ecosystem management issues. Other examples of such integration can be found throughout this book and in Dobson (1993). The chapter concludes with a brief discussion of the importance of communication among remote sensing and GIS practitioners and methods to help in this communication.

Study Area

The majority of the research described in this chapter has been done in conjunction or collaboration with the North Temperate Lakes Long-Term Ecological Research site (NTL-LTER), which is located in the NHLD of Wisconsin. The Long-Term Ecological Research (LTER) program is supported by the National Science Foundation (NSF) and is composed of 18 sites distributed throughout the United States, Puerto Rico, and Antarctica. The sites are ecologically diverse and range from lakes to forest, prairie, desert, tundra, and salt marsh (Van Cleve and Martin 1991). Research at the sites deals with time scales of years to decades to a century, and with spatial scales of meters to kilometers, to cross-continent intersite comparisons in an attempt to study the spatial and temporal dynamics of ecological processes (Franklin et al. 1990, Magnuson 1990, Swanson and Sparks 1990). The LTER network of sites provides a unique scientific infrastructure for performing long-term and large-scale ecological research on a coordinated basis. The LTER network has had a strong commitment to remote sensing and GIS due, in part, to an NSF Advisory Committee report (Shugart et al. 1988), which stated that "Scientists in the LTER network need to be facile on the use of remotely sensed data to test theories of ecological patterns and processes," and "acquiring a geographic information system (GIS) capability was the single technological addition that would most strongly advance the state of ecosystem science across the entire LTER network." While the focus of the NTL-LTER has dealt with long-term trends in physical, chemical, and biological properties of lake systems (Magnuson and Frost 1982), questions dealing with terrestrial and aquatic interactions have driven various studies of the forest ecosystems in the region. The focal point of research at the NTL-LTER is the Trout Lake Station located in Vilas County, Wisconsin. NTL-LTER remote sensing efforts in the region have been concentrated in Forest, Oneida, Price, and Vilas Counties, Wisconsin (Figure 16.1).

The descriptions of the northern Lake States presented in the introduction to this section and in Chapter 14 provide a general description of the NTL-LTER study area and the NHLD. To be more specific, the climate of the NTL-LTER is cold, continental with January and July temperatures averaging −10°C and 19°C, respectively. Annual precipitation averages 760 millimeters and snowfall ranges from 1270 to 1520 millimeters (Van Cleve and Martin 1991). Geologic landform, and consequently soil type, varies greatly within the NHLD and can be grouped into five classes: (1) outwash plains, (2) pitted outwash plains, (3) ground moraines, (4) end moraines, and (5) glacial lacustrine deposits (Attig 1985, Simpkins et al. 1987). Vegetation also varies greatly in the NHLD but appears to be highly correlated to geologic landform and soil type (Kotar et al. 1988). The vegetation is also strongly affected by natural and anthropogenic disturbance events. Superimposed on the predominantly forested landscape is one of the highest densities of lakes in the world (Magnuson and Frost 1982). As a result, the landscape in the NHLD, as is typical of the

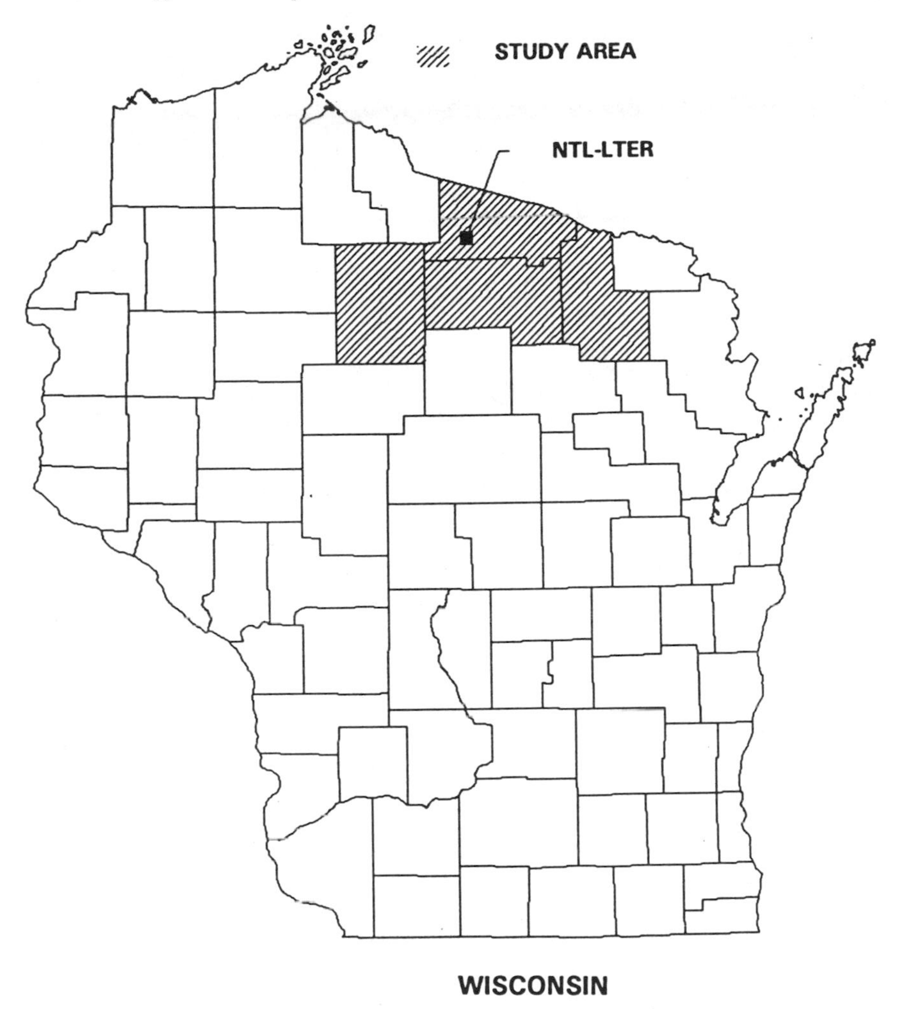

Figure 16.1 Location of the NHLD study area

northern Lake States region (Mladenoff and Host, Chapter 14), is very heterogeneous.

Sensor Resolution

Hoffer (Chapter 3) and Lillesand and Kiefer (1994) present the characteristics of various sensors that are being used to study forest ecosystems. This section will concentrate on three characteristics of satellite sensors and how these characteristics affect the ability of satellite sensors to resolve forest attributes in northern Lake States forests. One can compare characteristics of satellite sensors in terms of spatial, spectral, and temporal resolution. For example, the Advanced Very High Resolution Radiometer (AVHRR) has a coarse spatial resolution (1.1 kilometers), moderate spectral resolution (5 bands), and very fine temporal resolution (0.5 days). The SPOT panchromatic sensor has a fine spatial resolution (10 meters), coarse spectral resolution (1 band), and moderate temporal resolution (4 days). Many remote sensing practitioners would say that the "perfect" sensor would be one that combines fine spatial, spectral, and temporal resolutions. However, there would be a negative side to this "perfect" sensor due to the massive quantities of data that it would collect. It is not clear that we currently have the appropriate technology to analyze such quantities of data. Using today's sensor technology, the remote sensing practitioner has to make tradeoffs in choosing which sensor(s) to use. The competent remote sensing practitioner will choose a sensor that best answers the questions being asked by the research and/or management activities being performed.

The following provides an example of the importance of spatial resolution relative to land-cover classification. The AVHRR sensor is being used fairly extensively in studies pertaining to regional and global issues (Roller and Colwell 1986, Loveland et al. 1991, Townshend et al. 1991, Burgan and Hartford 1993). It is the ability of AVHRR to acquire a moderate amount of spectral information of the earth's surface on a daily basis that has made it a valuable tool. The fine temporal resolution yields a data set that can be used to study spectral characteristics of the vegetation of the conterminous United States on a biweekly basis (Eidenshink 1992). This temporally fine resolution is achieved by sacrificing spatial resolution. The spectral information has been used to prepare land-cover maps of the conterminous United States (Loveland et al. 1991) and, in time, will do the same for the world (Townshend et al. 1991). The land-cover map for the conterminous United States has been prepared using the Anderson et al. (1976) classification system (Table 16.1). Townshend et al. (1976) suggest that this map is accurate to Level II of the Anderson system but do not provide an accuracy assessment to support such detail.

Benson and Mackenzie (1994) have shown that the AVHRR sensor has limitations in detecting spatially small landscape features. As mentioned earlier, lakes represent a significant portion of the surface area within the NTL-LTER. Based on analysis of USGS 7.5-minute topographic maps for the NTL-LTER area, lakes represent 18 percent of the surface area. Benson and Mackenzie (1994) have shown that there is great variation in satellite sensors and their ability to measure landscape param-

TABLE 16.1 The Anderson land-use and land-cover classification system for use with remote sensor data

Level I	Level II
1 Urban or built-up land	11 Residential
	12 Commercial and service
	13 Industrial
	14 Transportation, communications, and utilities
	15 Industrial and commercial complexes
	16 Mixed urban or built-up land
	17 Other urban or built-up land
2 Agricultural land	21 Cropland and pasture
	22 Orchards, groves, vineyards, nurseries, and ornamental horticultural areas
	23 Confined feeding operations
	24 Other agricultural land
3 Rangeland	31 Herbaceous rangeland
	32 Shrub and brush rangeland
	33 Mixed rangeland
4 Forestland	41 Deciduous forestland
	42 Evergreen forestland
	43 Mixed forestland
5 Water	51 Streams and canals
	52 Lakes
	53 Reservoirs
	54 Bays and estuaries
6 Wetland	61 Forested wetland
	62 Nonforested wetland
7 Barren land	71 Dry salt flats
	72 Beaches
	73 Sandy areas other than beaches
	74 Bare exposed rock
	75 Strip mines, quarries, and gravel pits
	76 Transitional areas
	77 Mixed barren land
8 Tundra	81 Shrub and brush tundra
	82 Herbaceous tundra
	83 Bare-ground tundra
	84 Wet tundra
	85 Mixed tundra
9 Perennial snow or ice	91 Perennial snowfields
	92 Glaciers

Source: After Anderson et al. 1976

TABLE 16.2 Comparison of landscape parameters from the SPOT multispectral scanner (SPOT), Landsat Thematic Mapper (TM), and the Advanced Very High Resolution Radiometer (AVHRR)

Parameter	SPOT	TM	AVHRR
Nominal spatial resolution (m)	20	30	1100
Percent water	11.9	10.9	6.7
Number of lakes	3428	2829	63
Average lake area (ha)	11.7	7.8	360.0
Average lake perimeter (m)	1324	1123	8600

Source: After Benson and MacKenzie 1994

eters (Table 16.2). For example, within the study area, the SPOT multispectral sensor was able to discriminate 3428 lakes with an average lake area of 11.7 hectares. The AVHRR sensor was only able to discriminate 63 lakes with an average lake area of 360 hectares. The difference between the measures of landscape parameters was expected and can be attributed to the difference in sensor spatial resolution.

This example is not intended to suggest that the AVHRR should not be used. On the contrary, it is meant to illustrate some of the limitations of the AVHRR sensor and to enforce the statement that sensor selection should be based on research and/or management objectives.

Land-Cover Classification

Prior to 1982, attempts at land-cover classification within the northern Lake States using satellite sensors were hindered by the spatial and spectral resolution of the Landsat multispectral scanner (MSS), which was the only sensor available at the time. These early classifications (Mead and Meyer 1977, Bryant et al. 1980, Roller and Visser 1980) used MSS to produce classifications to Level I of the Anderson system. The deployment of the Landsat thematic mapper (TM) sensor in 1982, with its improved spatial, spectral, and temporal resolution, greatly enhanced land-cover classification in the northern Lake States. Hopkins et al. (1988) showed that TM data could be used to classify northern Lakes States forests to Anderson Level II with species-level identification of jack pine (*Pinus banksiana* Lamb.) and red pine (*P. resinosa* Ait.) with an overall classification accuracy of 93 percent.

TM data have successfully been used for land-cover classification at the

TABLE 16.3 Land cover classes used in the classification of the NTL-LTER

Class
Northern hardwoods (aspen, birch, maple, oaks)
Hardwood-conifer mix
Red or white pine
Jack pine
White pine, Norway spruce
Upland brush (inluding young northern hardwoods)
Lowland conifer
Mixed lowland vegetation (mixture of two previous types)
Herbaceous vegetation (recently disturbed land)
Grass, sedge, pasture
Aquatic vegetation (rooted aquatic macrophyte)
Urban-bare soil
Water

Source: After Morrison and Ribanszky 1989

NTL-LTER site (Morrison and Ribanszky 1989, Lillesand et al. 1989, Bolstad and Lillesand 1992a,b). Classifications were performed to Anderson Level II types (Table 16.3) with some discrimination beyond Level II including a jack pine type and a red pine/white pine (*Pinus strobus* L.) mixed class. Classification based solely on single-date TM imagery yielded an overall classification accuracy of 83 percent. Bolstad and Lillesand (1992a,b) have shown that classification accuracy can be improved by adding ancillary thematic information into the classification methodology through the use of rule-based classification models. The rule-based methodology is described in detail elsewhere (Bolstad 1990, Bolstad and Lillesand 1992b) but can be summarized as a method of combining non-image spatial data with image spatial data in the classification methodology. In their rule-based analysis, soil texture and topographic position information were integrated into the classification through the use of a GIS (Bolstad and Lillesand 1992a,b). The rule-based approach increased the overall classification accuracy of the NTL-LTER study area from 83 percent to 94 percent. This specific example helps to illustrate that the integration of remote sensing and GIS is not always unidirectional (remote sensing output being used as input to a GIS) but can also be bidirectional (GIS output being used as input in remote sensing functions).

Mladenoff and Host (Chapter 14) suggest that the disadvantage of the rule-based approach to classification is the general lack of appropriate digital data sets. This may be the case in some instances but is rapidly changing. There are a number of federal and state programs (see Communication below) that have as their mission the development of appropriate data sets for enhanced classification. A number of statewide data sets already exist for Wisconsin. A subset of these data sets include (1) digital elevation models (DEMs) at the 1:100,000 scale for the state (FDGC 1993) and 1:24,000 scale for select areas within the state; (2) digital line graph (DLG) data at the 1:100,000 scale including transportation, hydrography, and political boundaries (FGDC 1993); and (3) soil map unit information (STATSGO and SSURGO) prepared by the Soil Conservation Service (FDGC 1993). The problem facing the remote sensing and GIS practitioner is locating these data sets (see Communication below).

While the addition of ancillary thematic data has increased the overall accuracy of classification it still has not resolved a fundamental concern of forest managers and researchers in the northern Lakes States—the ability to discriminate the individual deciduous species that comprise the northern hardwood forest type (see Mladenoff and Host). Wolter et al. (1994) have shown the utility of incorporating imagery acquired at various dates throughout the growing season. The analysis of this multitemporal imagery uses phenological differences in species leaf-out, leaf production, and leaf senescence to aid in species-level discrimination. Analysis of multitemporal imagery in conjunction with the addition of ancillary thematic information will provide the mechanism necessary to classify northern hardwoods to the species level. Due to the spatial heterogeneity of the northern Lakes States landscape, in terms of both species composition and disturbance regimes, it is truly questionable whether we will be able to use currently available satellite remote sensing techniques to measure such structural attributes as canopy closure and dead organic matter, as described by Cohen (Chapter 7) for western coniferous forests.

Regional Change Detection

Within the context of the NTL-LTER, reconstructing historic vegetation patterns and documenting changes in vegetation over time is critical to interpreting the spatial and temporal variation in numerous site characteristics (e.g., forest ecosystem structure and function, watershed hydrologic processes, lake water chemistry). A historical vegetation change analysis of the NTL-LTER was performed through compilation and interpretation of field notes associated with the original Public Land Survey for

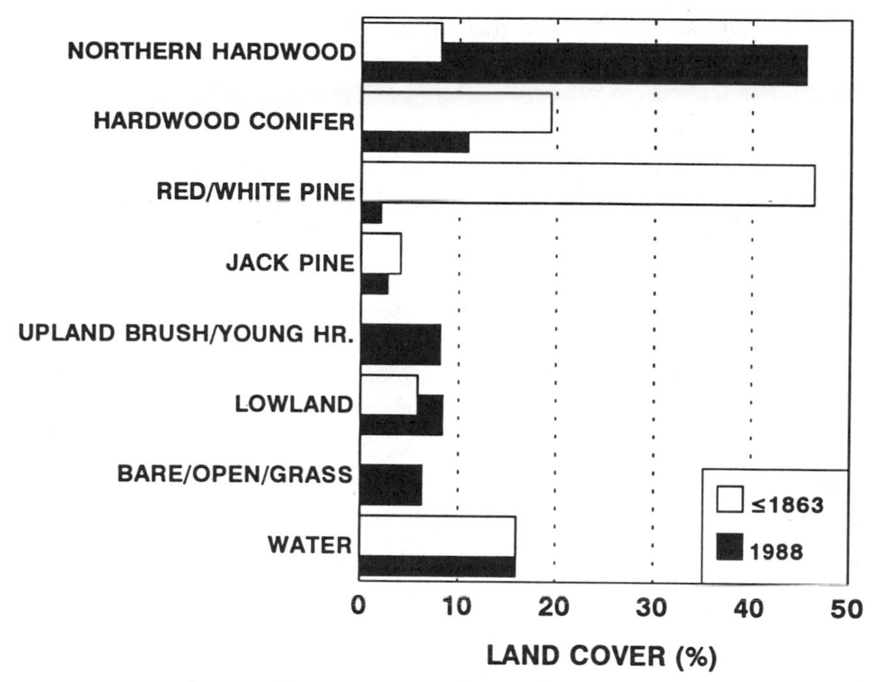

Figure 16.2 Land-cover change from presettlement (≤1863) to current (1988) (After Morrison and Ribanszky 1989)

Wisconsin (Finely 1951) and the current (1988) TM-derived land cover discussed above (Morrison and Ribanszky 1989, Lillesand et al. 1989). During the time of the original analysis, presettlement vegetation information was digitized from original source material (Morrison and Ribanszky 1989). Since that time, presettlement vegetation information for the entire state has been digitized and is available through the Wisconsin Department of Natural Resources (Laedlein 1993).

Comparison of presettlement vegetation and current land cover (Figure 16.2) reveals a dramatic conversion of presettlement pine forests to the current predominance of northern hardwood species. Presettlement pine forests accounted for 44 percent of the total land cover and hardwood forests accounted for 8 percent. In 1988, pine and northern hardwood species accounted for 5 percent and 46 percent of the total land cover, respectively. It has been hypothesized that this dramatic conversion of pine to northern hardwoods has had an effect not only on the structure and function of the terrestrial system but also on hydrologic balance and lake water chemistry (see below). This research is another example of the benefits that can be achieved through the integration of remote sensing and GIS.

Detection and Modeling of Ecosystem Properties and Processes

Research is currently being conducted in the NHLD in an attempt to use remote sensing to measure leaf area (Fassnacht et al. unpublished data). Leaf area is an important factor controlling CO_2, water, and energy exchange between the terrestrial biosphere and the atmosphere. Leaf-area index (LAI) has been recognized as one of the most important structural characteristics for quantifying energy and mass exchange of terrestrial ecosystems by satellite (Wittwer 1983). Major factors influencing LAI include climatic conditions such as temperature, and water and nutrient availability (Grier and Running 1977, Golhs 1982, Gower et al. 1992, Snowden and Benson 1992). Several studies have shown that the LAI of evergreen conifers in the Pacific Northwest can be remotely sensed (Running et al. 1986, Peterson et al. 1987). Preliminary results in the NHLD suggest the LAI of northern Lake States forest ecosystems can also be remotely sensed (Figure 16.3). Direct measurements indicate that LAI may differ by threefold among the major forest types in the northern Great Lakes with jack pine stands supporting an LAI of 1.6 and sugar maple (*Acer saccharum* Marsh) dominated northern hardwood stands supporting an LAI of 5.9 (Gower, unpublished data).

Researchers at the NTL-LTER are currently using remotely sensed measures of land cover and LAI in combination with digital soils information and meteorological information for the NHLD to drive a model of forest biogeochemical cycling (FOREST-BGC) (Running and Coughlan 1988, Running and Gower 1991) in an effort to understand the nutrient dynamics, carbon allocation, and water status of northern Lakes States forests (MacKenzie and Gower 1992). This is part of an overall effort at the NTL-LTER to develop a spatially explicit hydrologic model for the region. It is anticipated that this model will be used to investigate the effects of conversion of presettlement pine forest to current-day northern hardwood forests and effects of anticipated climate change on the hydrologic process of the NTL-LTER. The overall hydrologic model is made up of three components: (1) a terrestrial component, (2) a groundwater component, and (3) a lake chemistry component. All the components of the model are "loosely linked" using the GIS. In other words, input and output from each component of the model are being stored and transferred through the GIS (Figure 16.4). Within the context of the overall

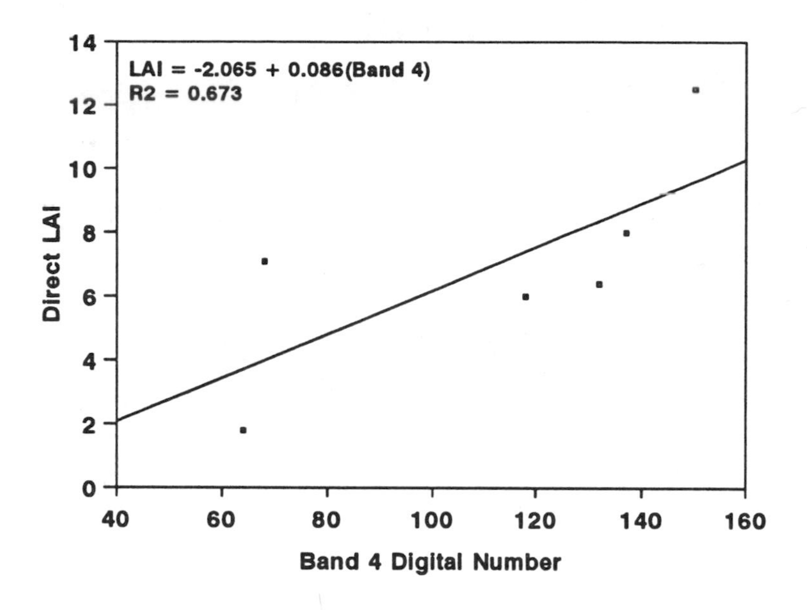

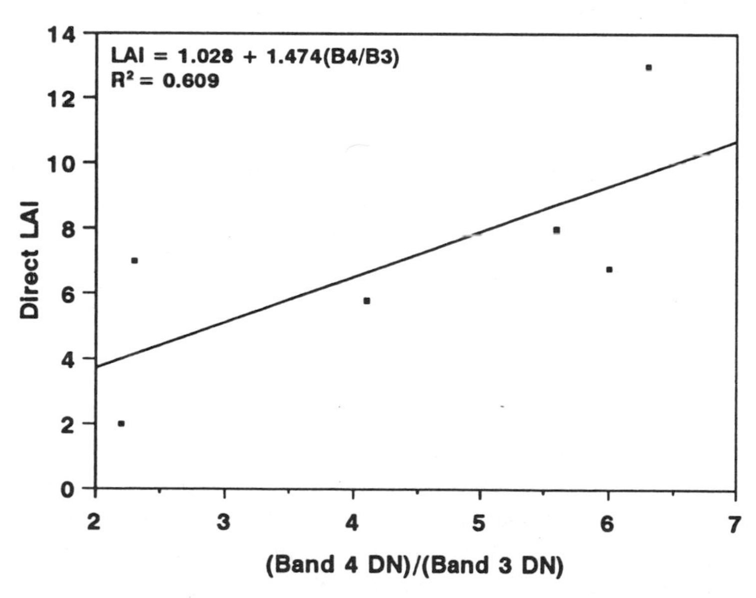

Figure 16.3 Preliminary data illustrating the relationship between select Landsat TM bands and direct estimates of leaf area index (LAI) for northern hardwood and conifer forests in the NHLD (Fassnacht, Gower, and MacKenzie, unpublished data)

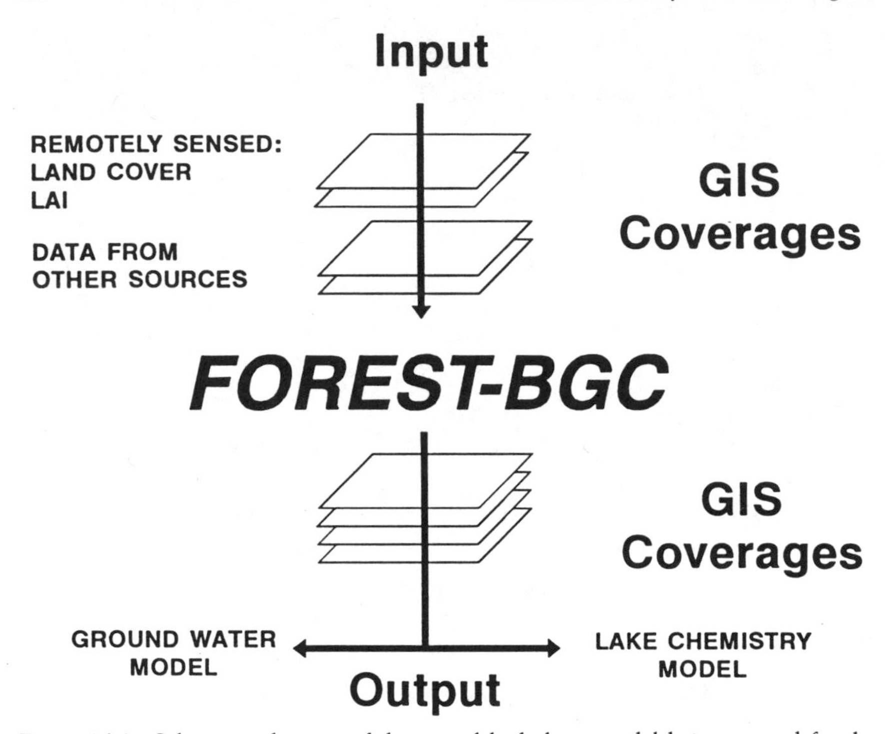

Figure 16.4 Schematic diagram of the spatial hydrologic model being created for the NTL-LTER showing the use of both remote sensing and GIS. The various components of the model are linked through the GIS.

spatial hydrologic model, the GIS has been an important tool in integrating spatial data from a variety of sources and scales while remote sensing has been shown to be useful in estimating structural properties at the regional scale.

Communication

The state of Wisconsin is fortunate to have a number of federal, state, local, and private entities involved in the compilation, manipulation, and analysis of spatial data. Due in part to this diversity of interests, the state and various collaborators have developed the Wisconsin Land Information Program (WLIP) (Ventura et al. 1993). The WLIA serves as a review board and a clearinghouse for spatial data collection, documentation, and distribution. The state, under the direction of the State Cartographer's Office and the Department of Natural Resources, and a number of collaborators are actively involved in the Wisconsin initiative for statewide cooperation on land-cover analysis and data (WISCLAND) (Gurda 1993). The WISCLAND initiative was stimulated by a report on the need for automation of statewide land-cover mapping using remote sensing techniques (Lillesand 1992). This report highlighted the need for land-cover information at all levels of government by a variety of land management and planning agencies. It also stated that due to uncoordinated data collection at various levels, individual agencies are forced to collect land-cover data with a narrowly focused programmatic and/or geographic perspective (Lillesand 1992). WLIA and WISCLAND (of which WLIA is a participant) represent two major, state-supported efforts for the exchange and standardization of spatial data. Even with this infrastructure in place, it is sometimes difficult for the remote sensing and GIS practitioner to acquire necessary digital data sets.

The challenge to the remote sensing and GIS practitioner is to obtain requisite digital data sets in the appropriate form and with appropriate documentation. A number of federal (FGDC 1992) and state (see above) agencies are currently involved in developing data transfer and documentation standards. Practitioners should be prepared to follow the standards when they become established. The establishment of standards will help in the transfer of data but the problem of actually getting the data in hand still exists. Where does one go to get digital data? The first place to look is the federal government. Many federal agencies are preparing spatial data sets in digital format. Some of these data sets are described in the *Manual of Federal Data Products* (FDGC 1993). At the state level, it is possible that there are a number of agencies working with digital spatial data. In Wisconsin, sources of digital data sets include, but are not limited to, (1) the WLIA, (2) the State Cartographer's Office, (3) the Department of Natural Resources, (4) the Department of Transportation, (5) the Wisconsin Geologic and Natural History Survey, and (5) various departments within the University of Wisconsin system. The Wisconsin Department of Natural Resources has even published a *GIS Database User's Guide* (Laedlein 1993). There are also a number of federal agencies with state offices that have been cooperative in data exchange. These include, but are not limited to, (1) the Fish and Wildlife Service, (2) the National Park Service, (3) the U.S. Geologic Survey, and (4) the Soil Conservation Service. Some of these agencies make it easy to track down data sets. With others it is a little more difficult. Carter (1992) provides an interesting discussion of the need for and problems associated with data sharing.

The best way to obtain data sets is to be prepared to offer data sets in return. This means that when forced to digitize a map, be willing to make

TABLE 16.4 Internet listservers, as of December 1993, that deal with remote sensing and GIS applications to forest ecosystem management

List Name	Listserver Address	Description
IMAGRS-L	CSEARN.BITNET	Image processing and remote sensing
GIS-L	UBVM.BITNET	GIS
ECOLOG-L	UMDD.UMD.EDU	Ecology
CANSPACE-L	UNB.CA	GPS information, etc.

To subscribe to a particular list, send the following EMail message to the appropriate *Listserver Address:*

SUB *List_Name Your_Name*

where *List_Name* is the List Name from the above table and *Your_Name* is your name. For example, if I wanted to subscribe to the GIS-L list, I would send the following EMail to UBVM.BITNET:

SUB GIS-L Mark MacKenzie

If you have subscribed correctly, you will receive a message from the listserver with information about the list and other advice.

Source: See Krol (1992) for more information.

the digital product available to others. Spend the time preparing the appropriate documentation (metadata) and keep track of all the processing steps involved in creating the data set. Meeting other remote sensing and GIS practitioners in the region is also a good way to find out what is going on relative to digital data and who has or wants what. In this age of the "information highway" none of us should be living in a vacuum. For those with access to the Internet (Krol 1992) there are a number of lists (i.e., bulletin boards) that might be worth subscribing to (Table 16.4). The Internet is generally available to universities, federal agencies, and some state agencies. Internet access is also available through various commercial vendors. These lists represent a mechanism for exchange of ideas and requests for help, various information, and data. Before jumping in and creating new digital data sets, it is worth trying to determine if someone else will provide you with the data in digital form.

Summary

Mladenoff and Host (Chapter 14) describe four general application areas for remote sensing and GIS in the Lake States regions: (1) forest and land-cover mapping, (2) ecological land classification, (3) regional change detection, and (4) detection and modeling of ecosystem properties and processes. This chapter presents examples of the use of remote sensing and

GIS in three of the four application areas within the context of the NTL-LTER. Two important themes throughout this chapter have been (1) the importance of sensor resolution (spatial, spectral, and temporal) and its effect on interpretation and sensor utility and (2) the value of integrating remote sensing and GIS to address these application areas. It is important for remote sensing and GIS practitioners to communicate among themselves and minimize the difficulty in acquiring and exchanging information and data.

Acknowledgments

Portions of the research described in this chapter have been supported by the National Science Foundation's Long-Term Ecological Research Program grant BSR8514330 and through McIntire-Stennis funding to the University of Wisconsin-Madison. I acknowledge the many contributions of Dr. Tom Gower to the research described in this chapter.

References

Anderson, J.R., E.E. Hardy, J.T. Roach, and R.E. Witmer. 1976. A land use and land cover classification system for use with remote sensor data. U.S. Geological Survey Professional Paper 964, U.S. Government Printing Office, Washington, D.C.

Attig, J.W. 1985. Pleistocene geology of Vilas County, Wisconsin. Wisconsin Geological and Natural History Survey Information Circular 50, WGNHS, Madison, WI.

Benson, B.J., and M.D. MacKenzie. 1994. Effects of sensor resolution on landscape structure parameters. Landscape Ecology, in press.

Bolstad, P.V. 1990. The integration of remote sensing, geographic information systems, and expert system technologies for landcover classification. Ph.D. dissertation, University of Wisconsin-Madison.

Bolstad, P.V., and T.M. Lillesand. 1992a. Improved classification of forest vegetation in northern Wisconsin through a rule-based combination of soils, terrain, and Landsat thematic mapper data. Forest Science 38(1):5–20.

Bolstad, P.V., and T.M. Lillesand. 1992b. Rule-based classification models: Flexible integration of satellite imagery and thematic spatial data. Photogrammetric Engineering and Remote Sensing 58(7):965–971.

Bryant, E., A.G. Dodge, and S.D. Warren. 1980. Landsat for practical forest type mapping: a test case. Photogrammetric Engineering and Remote Sensing 46(12):1575–1584.

Burgan, R.E., and R.A. Hartford. 1993. Monitoring vegetation greenness with satellite data. General Technical Report INT-297. U.S. Department of Agriculture, Forest Service, Intermountain Research Station, Ogden, UT.

Carter, J.R. 1992. Perspectives on sharing data in geographic information systems. Photogrammetric Engineering and Remote Sensing 58(11):1557–1560.

Dobson, J.E. 1993. Commentary: A conceptual framework for integrating remote sensing, GIS, and geography. Photogrammetric Engineering and Remote Sensing 59(10):1491–1496.

Eidenshink, J.C. 1992. The 1990 conterminous U.S. AVHRR data set. Photogrammetric Engineering and Remote Sensing 58(6):809–813.

FDGC. 1992. Content standards for spatial metadata: Federal Geographic Data Committee (Draft). FDGC Secretariat, U.S. Geological Survey, Reston, VA.

FDGC. 1993. Federal Geographic Data Committee: manual of federal geographic data products. FDGC Secretariat, U.S. Geological Survey, Reston, VA.

Finley, R.W. 1951. The original vegetation cover of Wisconsin. Ph.D. dissertation, University of Wisconsin-Madison.

Franklin, J.F., C.S. Bledsoe, and J.T. Callahan. 1990. Contributions of the long-term ecological research program. BioScience 40:509–523.

Gholz, H.L. 1982. Environmental limits to aboveground net primary production, leaf area and biomass in vegetation zones of the Pacific Northwest. Ecology 63:469–481.

Gower, S.T., K.A. Vogt, and C.C. Grier. 1992. Carbon dynamics of Rocky Mountain Douglas-fir: Influence of water and nutrient availability. Ecological Monographs 62:43–65.

Grier, C.C., and S.W. Running. 1992. Leaf area of mature northwestern coniferous forests: relation to site water balance. Ecology 58:893–899.

Gurda, B. 1993. WISCLAND: Wisconsin initiative for statewide cooperation on land cover analysis and data. Available from the Wisconsin State Cartographer's Office, Madison, WI.

Hopkins, P.F., A.L. Maclean, and T.M. Lillesand. 1988. Assessment of Thematic Mapper imagery for forestry applications under Lake States conditions. Photogrammetric Engineering and Remote Sensing 54(1):61–68.

Kotar, J., J.A. Kovach, and C.T. Locey. 1989. Field guide to the forest habitat types of northern Wisconsin. Department of Forestry, University of Wisconsin-Madison.

Krol, E. 1992. The whole Internet: User's guide & catalog. O'Reilly & Associates, Sebastopol, CA.

Laedlein, J. 1993. Wisconsin DNR GIS database user's guide. Wisconsin Department of Natural Resources, Bureau of Information Management, Geographic Services Section, Madison, WI.

Lillesand, T.M. 1992. Toward automation of statewide land cover mapping using remote sensing techniques: Final report for cooperative river basin study. Environmental Remote Sensing Center, University of Wisconsin-Madison.

Lillesand, T.M. and R. W. Kiefer. 1994. Remote sensing and image interpretation (3rd ed.). John Wiley and Sons, New York.

Lillesand, T.M., M.D. MacKenzie, J.R. Vande Castle, and J.J. Magnuson. 1989.

Incorporating remote sensing and GIS technology in long-term and large-scale ecological research. Proceedings, GIS/LIS '89, Volume 1, pp. 228–242.

Loveland, T.R., J.W. Merchant, D.O. Ohlen, and J.F. Brown. 1991. Development of a land-cover characteristics database for the conterminous U.S. Photogrammetric Engineering and Remote Sensing 57(11):1453–1463.

MacKenzie, M.D., and S.T. Gower. 1992. The use of remote sensing and GIS to examine terrestrial ecosystem properties that influence regional hydrology of NTL-LTER. Bulletin of the Ecological Society of America 73(2) (Abstract).

Magnuson, J.J. 1990. Long-term ecological research and the invisible present. BioScience 40:495–501.

Magnuson, J.J., and T.M. Frost. 1982. Trout Lake Station: A center for North Temperate Lakes studies. Bulletin of the Ecological Society of America 63:223–225.

Mead, R.A., and M.P. Meyer. 1977. Landsat digital data application to forest vegetation and land use classification in Minnesota. Proceedings, Symposium on Machine Processing of Remotely Sensed Data, LARS, Purdue Univ. West Lafayette, IN. pp. 270–279.

Morrison, L., and S. Ribanszky (eds.). 1989. The development and demonstration of a combined remote sensing/geographic information system for the North Temperate Lakes Long-Term Ecological Research site. Institute for Environmental Studies, University of Wisconsin.

Peterson, D.L., M.A. Spanner, S.W. Running, and K.B. Tueber. 1987. Relationship of Thematic Mapper Simulator data to leaf area of temperate coniferous forests. Remote Sensing of Environment 22:223–241.

Roller, N.E.G., and J.E. Colwell. 1986. Coarse-resolution satellite data for ecological surveys. BioScience 36(7):468–475.

Roller, N.E.G., and L. Visser. 1980. Accuracy of Landsat forest cover type mapping in the Lake States region of the U.S. Proceedings, 14th International Symposium on Remote Sensing of Environment, ERIM, Michigan, pp. 1511–1520.

Running, S.W., and J.C. Coughlan. 1988. A general model of forest ecosystem processes for regional applications. I. Hydrologic balance, canopy gas exchange and primary production processes. Ecological Modelling 42:125–154.

Running, S.W., and S.T. Gower. 1991. FOREST-BGC: A general model of forest ecosystem processes for regional application. II. Dynamic carbon allocation and nitrogen budgets. Tree Physiology 9:147–160.

Running, S.W., D.L. Peterson, M.A. Spanner, and K.B. Tueber. 1986. Remote sensing of coniferous forest leaf area. Ecology 67:273–276.

Shugart, H.H., W.J. Parton, G.R. Shaver, and S.G. Stafford. 1988. Report of the NSF Advisory Committee on Scientific and Technological Planning for long-term ecological research projects. Long-term Ecological Network Office, University of Washington.

Simpkins, W.W., M.C. McCartney, and D.M. Mickelson. Pleistocene geology of

Forest County, Wisconsin. Wisconsin Geological and Natural History Survey Information Circular 61, WGNHS, Madison, WI.

Snowdon, P., and M.L. Benson. 1992. Effects of combination of irrigation and fertilization on the growth and above-ground biomass production of Pinus radiata. Forest Ecology and Management 52:87–116.

Swanson, F.J., and R.E. Sparks. 1990. Long-term ecological research and the invisible place. BioScience 40:502–508.

Townshend, J., C. Justice, W. Li, C. Gurney, and J. McManus. 1991. Global land cover classification by remote sensing: Present capabilities and future possibilities. Remote Sensing of Environment 35:243–255.

Van Cleve, K., and S. Martin. 1991. Long-term ecological research in the United States: a network of research sites 1991 (6th ed.). Long-Term Ecological Research Network Office, University of Washington, College of Forest Resources, AR-10, Seattle, WA.

Ventura, S., P. Kishor, B. Niemann, Jr., K. Kuhlman, E. Epstein, and W. Holland. 1993. Laws that drive change: GIS/LIS development by local governments, and the Wisconsin Land Information Program. Proceedings, GIS/LIS '93, Volume 2, pp. 681–690.

Wittwer, S. (chair). 1983. Land related global habitability science issues. NASA Technical memorandum 85841. NASA, Washington, D.C.

Wolter, P.T., D.J. Mladenoff, G.E. Host, and T.R. Crow. 1994. Improved forest classification in the northern Lake States using multitemporal Landsat imagery. Photogrammetric Engineering and Remote Sensing (in review).

Part III

Potential Use of Military and Space Remote Sensing Technologies to Address Ecological Information Needs

Introduction

V. Alaric Sample

Since the dawn of the Space Age, the National Aeronautics and Space Administration (NASA) has been at the forefront in developing the science and technology of space-based and high-altitude aerial remote sensing. The reconnaissance and intelligence-gathering responsibilities of the Department of Defense have led it to develop specialized technologies, darkly rumored to be capable of reading license plates on the Zil limousines at the Kremlin from 100 miles up. Whether or not reality can quite measure up to such speculation, it is clear that remote sensing technologies developed for military and space exploration purposes, some of which might prove useful in meeting ecological information needs, are not yet available to or being fully utilized by the civilian and commercial sectors.

Two different trends suggest that this situation could be changing soon. First is the rapidly expanding need for new kinds of remotely sensed ecological information, discussed at length in the foregoing chapters. NASA has begun to respond to this need in a systematic fashion with its Mission to Planet Earth program, described in part by Forrest Hall in the following chapter. Second, with the easing of Cold War tensions, defense-related agencies and organizations are recognizing that environmental protection and the sustainable use of natural resources will be increasingly important factors in international political stability and national security (Myers 1993). As signified by the recent creation of a new office of the Undersecretary for Environmental Security within the Department of Defense, understanding and sustaining ecological systems in this nation and around the world promises to become a growing part of the U.S. defense mission.

The communications gap between ecologists, resource managers, and remote sensing/GIS specialists is perhaps widest at this point, due to the minimal interaction in the past between the conservation community and space/defense-related reconnaissance experts. Although reading license plates from 100 miles up might be stretching it a little, no doubt there are remote sensing capabilities within the defense sector that ecologists and resource managers haven't yet dreamed of. Remote sensing experts in the defense sector are eager to apply military technology to meet ecological information needs and have shown a willingness to declassify data and processes where the need has been clearly demonstrated. But the doors have not been thrown wide open, nor should they be. Rather than lay out all the things they are technically capable of doing and then leave it to the ecologists and resource managers to determine how these capabilities can be applied, space/defense remote sensing specialists have invited potential users to identify and articulate their information needs. The available information can then be declassified and even adapted to specifically suit those needs. In some cases, agencies have expressed a willingness to place additional sensors on satellites yet to be launched in order to provide information that is not currently gathered but that is within their technical capability to gather.

In Part III, remote sensing/GIS experts from NASA's Goddard Space Flight Center and the U.S. Army's Topographic Engineering Center discuss their agencies' efforts to understand and respond to specific needs for information useful in environmental protection and ecosystem analysis. They also describe a process whereby ecologists and resource managers can articulate emerging information needs so that the agencies can determine whether and how they might be of assistance. While an unchecked imagination may be somewhat disappointed in the type of remote sensing information these agencies are actually capable of providing, this can be the start of a productive and mutually beneficial cooperation between two communities that, by combining their efforts, can advance the cause of peace by helping to ensure the ecological health and productivity of this nation and of others.

Reference

Myers, Norman. 1993. Ultimate Security: The Environmental Basis of Political Stability. New York: W.W. Norton. 308 pp.

17

Adaptation of NASA Remote Sensing Technology for Regional-Level Analysis of Forested Ecosystems

Forrest G. Hall

Introduction

The first part of this chapter briefly summarizes the major thrust of and rationale for NASA's ongoing global change program and the associated use of satellite remote sensing, and describes how this global earth-systems science focus relates to NASA's historically regional approach. The second part of the chapter summarizes the major parameters for regional forest surveys and describes components of the data analysis and display systems required for extraction of these parameters from remotely sensed and other data. The remainder of the chapter discusses the use of satellite remote sensing technology for regional-level analysis of forested ecosystems and the specific analysis steps required. Of particular importance are the physics of spectral signatures and their relationship to forest information classes and how to use this knowledge to label the information classes in the satellite image. The chapter concludes with a discussion on how NASA's global change program can benefit regional application issues in the future and suggestions for some improvements in data analysis and management systems.

NASA's Global Change Programs

Fossil-fuel combustion and deforestation are almost certainly the cause of rapidly increasing concentrations of atmospheric CO_2, observed originally in the 1950s by Dave Keeling at Mauna Loa. His series of measurements, when extrapolated forward, show that the earth's atmospheric CO_2 concentration will double in just 50 years. General circulation models (GCMs) of the atmosphere predict that this doubling will increase the earth's average surface temperature somewhere between 1.5°C and 4.5°C, and will produce wetter winters and dryer growing seasons than at present; indeed, because the geographic distribution of the projected changes are not uniform, temperature increases for continental interiors, at higher latitudes, are projected to be as much as 10°C. Clearly, if such warming and drying were to occur, it would alter the structure and function of many terrestrial ecosystems.

There is, however, a great deal of uncertainty surrounding these projections. The atmospheric GCMs on which these predictions are based contain overly simple mathematical and physical representations of surface-atmosphere radiation, energy, and mass exchange. For example, GCM models of boundary-layer convective processes largely ignore the effects of "cloud feedback" as the atmosphere warms and moistens. Increased stratospheric cloud generation would act to retard global warming by reflecting more sunlight, while increased lower-level convective cloud cover would enhance warming by trapping radiation. As another example, GCM representations of the effects of surface vegetative stomatal control on the evaporative component of energy balance are overly simple; in many GCMs the land surface at a scale of 200 kilometers is considered to be horizontally homogeneous and biologically inert.

Improved GCMs are not the only advance required to better understand and manage the biosphere. Uncertainty in the flux of carbon dioxide among the atmosphere, the oceans, and the land also limits our ability to predict the exact magnitude of future global warming. We know currently that the land and the oceans are absorbing somewhere between 3 and 5 gigatons per year of the 6 to 8 gigatons we dump annually into the atmosphere from fossil-fuel emissions and global deforestation. Would the strength of this combined land-ocean sink increase or decrease with global warming? Better ecosystem and biogeochemical models are needed to answer this question.

In the 1970s a number of agencies stimulated efforts to assess the state of our knowledge of the earth as a system, and to define approaches reducing our uncertainty about the processes underlying global change. How does the sun's radiative forcings of the atmosphere depend on land-cover composition and distribution? How will the atmospheric state in the near term and climate change in the long term modify the land cover? How do the fluxes of biogenic, radiatively active gases between the atmosphere, the biosphere, and the oceans depend on ecosystem state and climate?

Better answers to these and other questions depend on better process models and better data inputs to the models. The data are required globally and include radiation and meteorological fields, global land cover and associated characteristics (albedo, leaf area, phenology, etc.), soils physical properties, soil moisture fields, topographic data, and so on. The spatial and temporal resolutions needed differ from one parameter to the next, ranging from 1 to 10^4 km^2 and from daily to annually. The only practical means for acquiring many of these data sets, in a consistent and timely fashion on a global scale, is satellite remote sensing.

In the early 1980s a number of satellites were already in orbit that could satisfy at least some of these data needs, but a review of the status of this technology quickly revealed that better satellite sensors and algorithms for reducing the sensor data to parameter fields were needed. In response to these needs, NASA in the early 1980s assembled teams of scientists to define a satellite-based earth-observing system that could satisfy them. This system, christened the "Earth Observing System" (EOS), has gone through an evolutionary series of design alternatives since that time and is currently planned for launch in the late 1990s. EOS will include a suite of land, atmosphere, and ocean sensors designed to satisfy global parameter needs.

At the same time, it was recognized that improved remote sensing algorithms would also be needed to produce the desired parameter sets on a global scale. A program of fundamental research in remote sensing science was already in place within NASA, but to develop and test such algorithms, field experiments in different ecosystems were also needed. Such field experiments would also be vital to developing and testing the various earth science process models. As a result, a series of field experiments was planned and executed, including the First International Satellite Land Surface Climatology Field Experiment, (the First ISLSCP Field Experiment, FIFE), conducted over the Konza Prairie in central Kansas during 1987 and 1989 (see the Journal of Geophysical Research FIFE special issue 1992), to be followed in 1994 by a similar experiment, BOREAS, over a Boreal Forest Biome in Canada (Hall et al. 1993). The objectives of these experiments are to acquire measurements at the variety of spatial scales necessary to develop and test both remote sensing algorithms and process models.

FIFE and BOREAS are relatively large, interdisciplinary field experiments involving atmospheric boundary layer physicists, micrometeorologists, biologists, ecologists, hydrologists, atmospheric chemists, and remote sensing specialists, and form the major thrust of NASA's research and development program in biospheric sciences. In addition however, there are a number of smaller-scale efforts, some interdisciplinary, some single-discipline studies, to look at a broad number of issues involving remote sensing science and surface processes.

NASA's Regional Application Programs

NASA's large earth-resource survey applications programs of the 1970s, the Large Area Crop Inventory Experiment (MacDonald and Hall 1980) and AgRISTARS (Remote Sensing of Environment, AgRISTARS Special Issue 1979), have been followed by more modest individual applications investigations. The Earth Observations Commercial Applications Program (EOCAP) has funded a number of investigations in this vein. In addition to applications development efforts at universities, EOCAP funds industry-government partnerships to implement and integrate existing technology into industry in known market areas. Much of the remote sensing technology developed as a part of NASA's global change program will be directly applicable to the regional-scale studies that constitute the EOCAP program.

In addition to EOCAP, needs among a number of agencies for improved ecosystem management information at regional scales have stimulated a large number of regional applications of remote sensing outside of NASA sponsorship. Many of these efforts rely on digital remotely sensed images from aircraft and satellite and geographic information systems to produce map overlays of land-cover classes, ownership data, topography, roads, towns, and other resources. Satellite images from Landsat and SPOT, in spite of their current high cost per scene, are proving cost effective in comparison to other modes of obtaining such resource information over large areas.

Many of the agencies actively engaged in employing digital imagery and GIS use commercially available technology and workstations for the GIS and remote sensing analysis. In some cases, the agencies contract directly with firms who digitize the ancillary data provided by the agency, process the satellite data for specific land-cover classes, and produce the needed forest resource information.

Remote Sensing and Data Systems Technology, Current Status

For many regional forest applications, major thrusts are inventory and mapping. Mapping parameters most often consist of their landscape distribution and change over time, including (1) community composition and structure, (2) successional stage, (3) landscape structure (patch size, shape, connectivity, etc.), (4) stand age, (5) crown closure, and (6) biomass and height. In the future, as ecosystem process models (e.g., biogeochemical cycling models, forest succession models, hydrological models) become more mature and ready for use in forest ecosystem management, additional process parameters such as photosynthetic capacity, energy balance, temperature, leaf area index, and so on will need to be mapped. In the material to follow, we will refer to the above parameters simply as "forest information classes."

The forest information classes are usually needed for each "homogeneous" patch within the ecosystem. Forest managers also need the landscape areal proportions occupied by each forest information class. They need to know how the landscape is structured at any one time, to assess how landscape-level parameters such as patch size distribution and landscape connectivity relate to habitat suitability for various species. Maps of these forest information classes are frequently overlayed in a GIS with other maps, such as topography, streams, fires, and roads, to further understand and display their implications for forest resource and management decisions. Changes in these maps over time are also critical to analyze temporal trends and the impacts of various management and mitigation strategies.

Status of Current Data Analysis and Display Systems for Information Extraction

Perhaps as much as any other development, the emergence of geographic information systems (GIS) in the last decade has stimulated widespread use of remotely sensed data by the regional applications community. The use of GIS has not only greatly facilitated the analysis and integration of disparate types of landscape-level resource data, it has also provided a powerful tool for displaying and understanding the results of such analyses. Resource managers are increasingly turning to GIS to create data displays to convince themselves and their constituents of the wisdom and costs of certain mitigation and management strategies.

While the GIS is a central component of an analysis system, it must communicate easily and rapidly with other data analysis functions, as shown in Figure 17.1. In addition to the map data on a GIS, the analyst often must be able to access, store, and manage data collected at isolated points on the landscape: complex and disparate data such as meteorological measurements, nutrient cycling measurements, stream flow data, biomass measurements, and so on. Query and search of large amounts of this

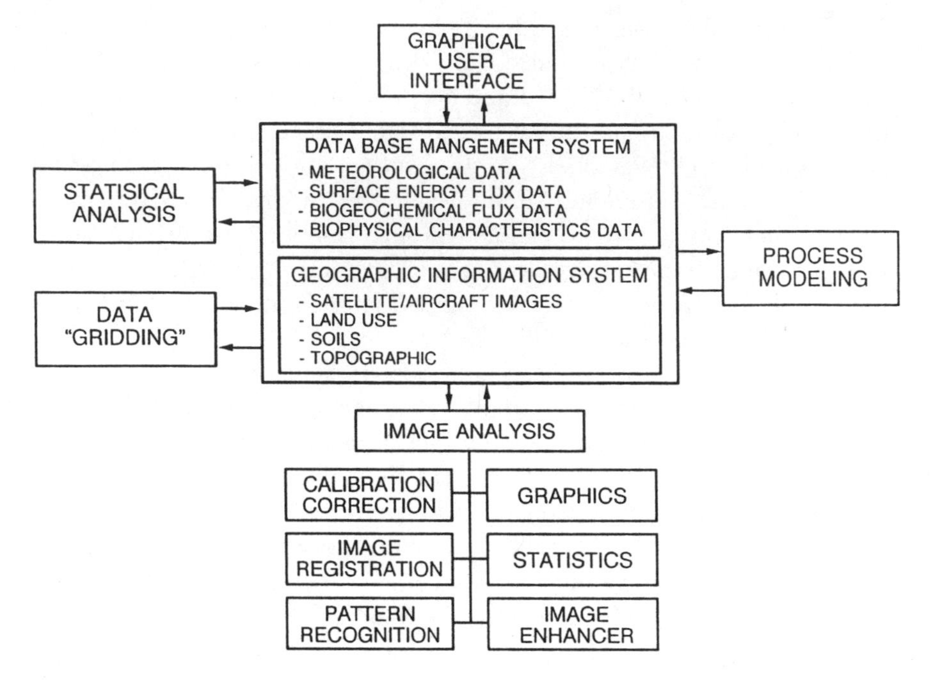

Figure 17.1 A hypothetical data analysis and display system designed to permit the analyst to conveniently access, query, analyze, and display point and area data. The Database Management System should permit the analyst to search and query large numbers of "point" data and areal data, acquired at specific locations, to find relevant subsets of data (times, locations, data types, etc.). The analyst should be able to conveniently interpolate these point data sets to spatial grids, using data gridding software, then overlay them into a geographic information system. The GIS data should then be accessible to analysis software including image processing, proces models, and statistical analysis subroutines. The derived results of the models and analysis should be easily storable back into the GIS and DBMS, and available to a graphical user interface for three-dimensional color displays of plots, geographic overlays, etc. Components of this ideal data analysis and display system exist, but are currently not integrated into a single "shell."

data is greatly aided by a database management system (DBMS) that permits the analyst to view menus of available data—specific dates, times, and places—and finally, to combine these disparate data tables into edited tables, customized for a particular application.

Another critical subsystem necessary to support the data analysis and display system is a statistical analysis package that is integrated into the system and can easily access both the GIS map data and the point data tables. The statistical analysis packages should be not only capable of the standard kinds of regression analyses, but also of more elaborate statistical testing procedures, factor analysis, and so on as well as spatial statistical analysis routines such as Kriging, semivariagram analysis, and others.

The analyst must also be able to access the data stored in the GIS and DBMS with an image-processing system. The image processing system (IPS) itself has many submodules including image registration, image correction and calibration, graphics, statistics, display, and perhaps even expert systems modules. The IPS must be able not only to retrieve data from the DBMS and GIS, but also to insert processed remotely sensed image information back into them.

In addition, if the ecosystem manager is using ecosystem models, the modeling software must be able to communicate with the GIS and DBMS for data input and output. This requires, first, computer algorithms to overlay the map-based GIS data with point data such as biomass measurements, meteorological station data, and so on. This step can involve, for example, Kriging algorithms or two-dimensional relaxation algorithms to extrapolate from landscape points to the landscape itself. Second, these "gridded" data sets, that is, overlays of GIS and contoured point data, must be accessible as input to the ecosystem models. Executing the models can, in turn, produce "derived data," such as ecosystem productivity estimates, which need to be overlayed back on the GIS data planes for further display and assessment.

Overlaying and displaying the source and derived data can be greatly facilitated by a good graphical user interface (GUI). This permits the analyst to generate displays of the results of data analysis that can be easily understood and manipulated; the GUI should be able to generate and display three-dimensional color displays of data-plane overlays: topographic data, roads, trails, cover type, modeling output, and so on, as well as graphical results plotting various relationships important to understanding ecosystem structure and function.

All of the subcomponents of the data analysis and display system described above currently exist. There are several good commercial sources for each. Some systems have begun the integration of some of the algorithms (e.g., image processing and GIS packages) into linked data analysis systems. However, so far there is no integrated data analysis and display system that can accomplish all of the above functions easily and painlessly. Passing data from one subsystem to another requires reformatting and is often tedious and time consuming. Future developments integrating the data analysis and display subsystems will facilitate remote sensing applications as much as the GIS developments did initially. Efforts are needed to integrate existing commercial packages into a common "shell" that permits the exchange of data between modules and a user-

Table 17.1 EOS Data Information System Sponsored Data Centers

Center	Initial system	Discipline(s)	Key spacecraft/instruments
UAF-University of Alaska-Fairbanks	ASF System	Sea ice, polar processes from synthetic aperture radars	SAR-ERS-1, JERS-1, ERS-2, Radarsat, and ongoing role as ground station
EDC-EROS Data Center	GLIS, Landsat Processing	Land process imagery	ASTER, MODIS (Level 2 and above land), SAR (land), AVHRR (1 km), SIRC/XSAR, Landsat (USGS providing access to TM/MSS)
GSFC-Goddard Space Flight Center	NCDS, PLDS, CDDIS	Upper atmosphere, atmospheric dynamics, global biosphere, geophysics	MODIS, AIRS, MHS, AMSU, SeaWiFS (I+II), GLAS, HIRDLS, TOMS, VIRS, Atlas, SOLSTICE (I+II), SAFIRE or MLS, AVHRR Pathfinder (land, atmos.), TOVS Pathfinder, UARS (all)
JPL-Jet Propulsion Laboratory	NODS	Ocean circulation and air-sea interaction	TOPEX/Poseidon, NSCAT, DORIS, SSA, TMR, AVHRR Pathfinder (SST)

LaRC-Langley Research Center	ERBE Procesing	Radiation budget, aerosols, tropospheric chemistry	CERES, ERBE, ACRIM, MOPITT, MISR, EOSP, SAGE (I+II), and TES
MSFC-Marshall Space Flight Center	WetNet	Hydrology	MIMR, TMI, TRMM PR, LIS AND SSM/I
UC-National Snow and Ice Data Center (NSIDC), University of Colorado	CDMS	Cryosphere	MODIS (snow & ice), SMMR, SSM/I, OLS, GLAS, MIMR
ORNL-Oak Ridge National Laboratory	TGDDIS	Biogeochemical dynamics	Ground-based data relating to biogeochemical dynamics
CIESIN-Consortium for International Earth Science Information Network	—	Human dimensions of global change and policymaking applications	Socioeconomic data

friendly, mouse-driven interface that permits easy operation of and communication between individual modules.

One final capability beginning to emerge within the applications community that will greatly speed applications development is data sharing. The electronic capability to engage in data sharing is currently emerging somewhat informally through the development of high-speed communications infrastructure with networks like the Internet, Telenet, ARPAnet, SPAN, and others. As users around the country begin to develop electronic databases for their particular regions and put these databases on line, sharing of these data among networks will permit users to be quickly informed of techniques and approaches developed by their colleagues, as well as algorithm software and data exchange. A few such databases already exist, such as the FIFE Information System and the National Science Foundation LTER site database. As the applications communities become better organized, funding should be devoted at a minimum to developing electronic bulletin boards where data and algorithm holdings can be posted, or perhaps to developing a more organized database management system where such data can be integrated and organized for general access. For example, NASA is currently experimenting with such a system of data analysis and archive centers (DAACs). The location and envisioned function of these DAACs are shown in Table 17.1.

Use of Remote Sensing Algorithms for Land-Cover Mapping

Using remotely sensed images and image-processing algorithms to map forest information classes is part art and part science. It involves matching satellite "spectral classes" to landscape "forest information classes." When we view a three-band color image of a forest scene, we see a mosaic of colors. Each of the distinct colors is a different "spectral class." Two landscape patches give rise to different colors because they reflect radiation differently in at least one of the three spectral bands generating the image. If we then "overlay" the image on a forest cover map, most often we will find that the different forest information classes correspond to different color patches in the remotely sensed image; but not always—in some cases two distinct forest information classes will fall into the same spectral class—in other cases, the same forest information class will be represented by two or more spectral classes.

The process of finding the matches and the differences between the spectral classes and the forest information classes is both art and science. The science involves understanding how different forest information classes reflect radiation differently. Such understanding can guide the matching process, can define which forest information classes are likely to be spectrally distinct, and can help define the remote sensing limitations. This matching process can also be aided considerably by a cleverly designed GUI module within the data analysis and display system. For example, we would want to be able to display just the patches that belong to a particular spectral class, then compare these with independently derived maps of the forest information classes. The art (i.e., subjectivity) in the process is the tradeoffs involved in deciding which information classes correspond to which spectral classes, and the procedure by which the available spectral and ancillary data are used to identify the information classes.

In the next two sections, I will review (1) the most commonly used processes for matching the spectral and information classes, (2) the physics of canopy-radiation interaction that can help guide this matching process, and (3) how a well-designed data analysis and display system can facilitate forest information extraction and display.

Supervised and Unsupervised Classification

A remote sensing imager recording radiation reflected from a forest scene in a number of spectral bands will generate spectral classes that can be thought of as clouds of data points, concentrated around cluster centers in an *n*-dimensional, Cartesian "spectral space." Two- or three-band sensor data for example can be visualized with two- or three-dimensional "scatter plots" using graphic displays, where the band values form the axes of the scatter plots (see Figures 17.5 and 17.7, later).

The forest information classes are defined by the user. As discussed earlier, these classes can involve stand composition, age classes, crown closure classes, roads, clearings, trails, and so on. To the degree that the forest information classes reflect radiation differently, the spectral data points recorded by the sensor from each information class will occupy different locations in spectral space.

Matching the forest information classes to the spectral space classes can be done with either (1) supervised classification techniques or (2) unsupervised classification techniques. Both supervised and unsupervised techniques can involve the use of ground-mapped stands (training data) to determine empirically which spectral values are associated with which forest information classes. With supervised algorithms, the ground-mapped stands are used to "train" the algorithm to assign each satellite pixel in the mapping region to an information class. Unsupervised techniques involve using clustering algorithms to group the spectral values of the image into distinct spectral or cluster classes, producing a cluster map of the image. Once completed, groups of pixels (i.e., spectral clusters) are then labeled according to which information classes they best represent. The labeling process can involve matching a cluster image to a forest information class map, as in the supervised approach.

For this process to be a *cost-effective* use of satellite data, the cost of ground mapping the stands used for training data must be small in comparison with the total costs of mapping the region by satellite; otherwise the satellite processing resources could be spent on ground visits and produce the regional map without the use of the satellite data. To reduce reliance on ground visits, the training of both supervised and unsupervised algorithms can be aided by radiative transfer models of how forest information classes reflect radiation. Such models are based on understanding of how the structural and optical properties of forest information classes affect the amount of reflected radiation in different spectral bands; this understanding has progressed rapidly in the last few years, aided by the development of radiative transfer models and extensive field measurement programs. These models have progressed to the point that, for some forest communities, their precise spectral signatures, or locations in spectral space, can be predicted. It is that understanding that we will discuss in the next section.

Forest Information Classes and Spectral Signatures

A spectrometer is a device that measures the amount of radiation reflected or emitted from an object. By measuring the ratio of incident-to-reflected light on a leaf or a landscape patch as a function of the wavelength or frequency of the reflected light, the *reflectance* of the leaf or patch can be calculated. The spectrometer can also be used to measure the ratio of incident light to light transmitted through the leaf, and leaf *transmittance* can be calculated. Measurements of leaf reflectance and transmittance, as shown in Figure 17.2, give rise to "spectral signatures" (i.e., variations in reflectance with wavelength associated with a land-cover type or biophysical property). In Figure 17.2 is displayed the spectral reflectance of aspen leaves (*Populus tremuloides*) and black spruce needles (*Picea mariana*).

Common to the reflectance signatures of vegetation is the absorption of light at approximately 0.4 and 0.69 micron where leaf chlorophyll acquires light for photosynthesis, and the "green" peak at about 0.552 micron where reduced absorption and scattering of light by leaf pigments give rise to the characteristic green appearance of vegetation.

Aspen and black spruce differ from each other in the near-infrared portion of the spectrum beyond 0.7 micron where leaf and needle cell structure plays a dominant role in reflectance characteristics. In the near-infrared region, there is little absorption of light, and scattering from the cell walls is the dominant mechanism governing radiation interactions. As additional layers of leaves are added, multiple scattering in the near-infrared causes reflectance of a canopy to increase.

In the mid-infrared portion of the electromagnetic spectrum, in the neighborhood of 1.6 to 2.0 microns, absorption of light by leaf water is the key feature determining reflectance. In the far-, or thermal, infrared region (not shown in Figure 17.2) at approximately 3.0 to 12.0 microns, thermal emissions of radiation, rather than reflected radiation, are dominant. The thermal infrared region is used primarily to infer radiative or kinetic temperatures of forest stands.

Even the highest-resolution civilian satellite sensors, with spatial resolutions as small as 10 meters, view not only leaves but groups of canopies. Thus, pixel-level reflectance is not necessarily the same as the leaf-level reflectance. Each pixel views leaves, needles, and a variety of other scene elements—twigs and branches, understory, and canopy shadows (see Figure 17.3). The ratio of the total light reflected from a pixel to that incident on the pixel is defined as the pixel reflectance. Thus pixel reflectance is determined not only by the canopy leaf reflectance, but also by branches, twigs, and understory, as well as the degree of shadowing by the canopy of itself, its neighbors, and the understory. Thus, characteristic reflectances of forested landscape elements are determined by the type of leaves, number of layers of leaves (leaf area index), the ratio of leafy to woody materials, and the spatial arrangements of the leaves and tree canopies—in other words, pixel-level reflectance is a function of the total community composition and structure of the forest canopy, and can potentially be used to infer these characteristics using computer-implemented algorithms.

Figure 17.4 demonstrates these effects for a variety of wetland black spruce stands of varying density. The predominant understory species for all stands is sphagnum moss. The reflectance of pure sphagnum moss, shadows, and pure sunlit canopy with no shadow (a unique point commonly known as the "hot spot") are plotted. The reflectance values for the remainder of the pixels plotted were taken (as described in the Figure 17.4 caption) over the black spruce/sphagnum stands. For very sparse black

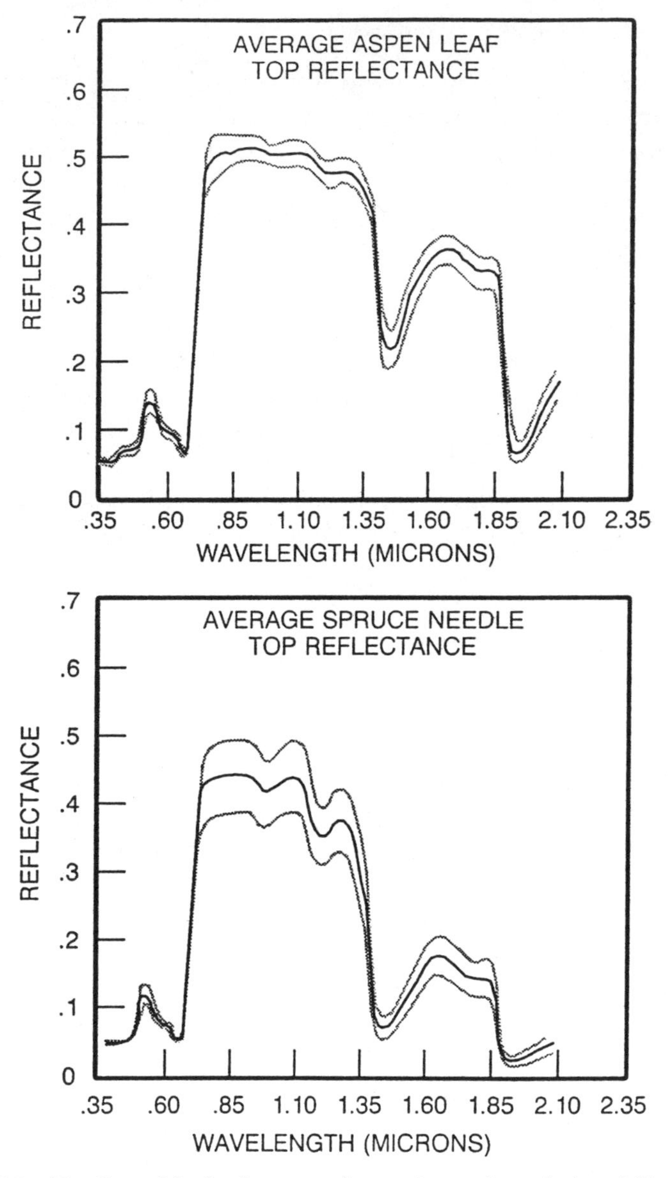

Figure 17.2 Needle and leaf reflectance of aspen leaves (top plot) and black spruce needles (bottom plot). The absorption troughs to either side of 0.55 micron are a result of leaf chlorophyll, highly efficient absorbers of light at 0.40 and 0.69 micron. The peak at 0.55 micron is a result of other plant leaf pigments. The large values of reflectance observed between 0.7 and 1.37 microns (near-infrared region) is a result of light scattering by the cell walls of the leaves. The absorption troughs at 1.37 and 2.0 microns are absorption of light by leaf water. Differences in these leaf-level features are reponsible for much of the spectral separability between different species. Within species, sensitivity to chlorophyll amount, leafy biomass, and leaf water content, provide information related to plant biomass and condition. These data were collected as part of a field experiment in the Superior National Forest in Minnesota (Hall et al. 1991)

Figure 17.3 Vertical photograph of dense black spruce canopy in the Superior National Forest. The photograph is approximately 10 meters on each side. Satellite pixels view collections of canopies like these and thus the pixel-level reflectance is determined by sunlit canopy reflectance, shadows, and sunlit background reflectance. At a pixel level, reflectance is determined not only by leaf-level reflectance (see Figure 17.2) but also by individual canopy morphology (which affects shadowing), crown closure, sun and view angle, and background reflectance.

spruce stand densities (upper right of triangle), the sphagnum understory dominates pixel reflectance. As stand density increases, sunlit canopy area and associated shadow area increasingly obscure the sphagnum background; shadows increase faster than sunlit canopy area, particularly for low sun angles, driving the pixel reflectance more strongly toward the shadow reflectance value than the pure sunlit canopy value. For denser stands, shadow and sunlit canopy dominates the sphagnum background, and the pixel-level reflectance is roughly an equal mixture of shadow, sunlit sphagnum, and canopy.

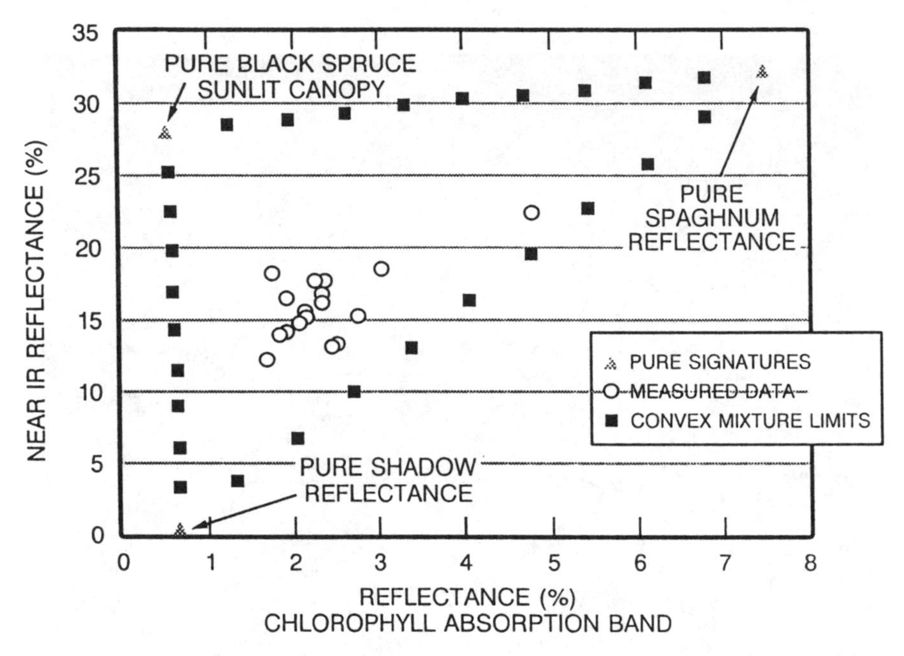

Figure 17.4 Chlorophyll absorption band versus near-infrared reflectance for wetland black spruce canopies growing on a sphagnum moss background in the Superior National Forest (Hall et al. 1991). Reflectance data were acquired from approximately 30 × 30-meter stands with a helicopter hovering at 200 meters above ground level using a radiometer simulating thematic mapper bands. Reflectance values for pure sphagnum were obtained by hand-held measurements; sunlit canopy reflectance obtained from measurements near "hot spot" (sun directly behind sensor to minimize visible shadows); shadow reflectance is the ratio of light levels reflected from shadows incident sunlight to the incident radiation above canopy.

The reflectance of aspen canopies behaves quite differently with varying canopy density from that of black spruce. Figure 17.5 displays the reflectance of aspen canopies, at a number of phenological stages, stand densities, and ages, overlayed with the black spruce reflectance. Shown also are reflectance values for red pine and water. Leaf litter from the previous year is the predominant aspen understory in early spring, which has a quite different reflectance than sphagnum moss. At full canopy, there is little difference between one aspen canopy and the next, even though for the canopies shown there is a range in overstory leaf area index from one to three, and quite a range in stem density and age. The major difference in reflectance lies in seasonal phenological change as the aspen canopies leaf out. What can be seen from this figure is that at certain times of the year, aspen canopies appear identical to black spruce canopies.

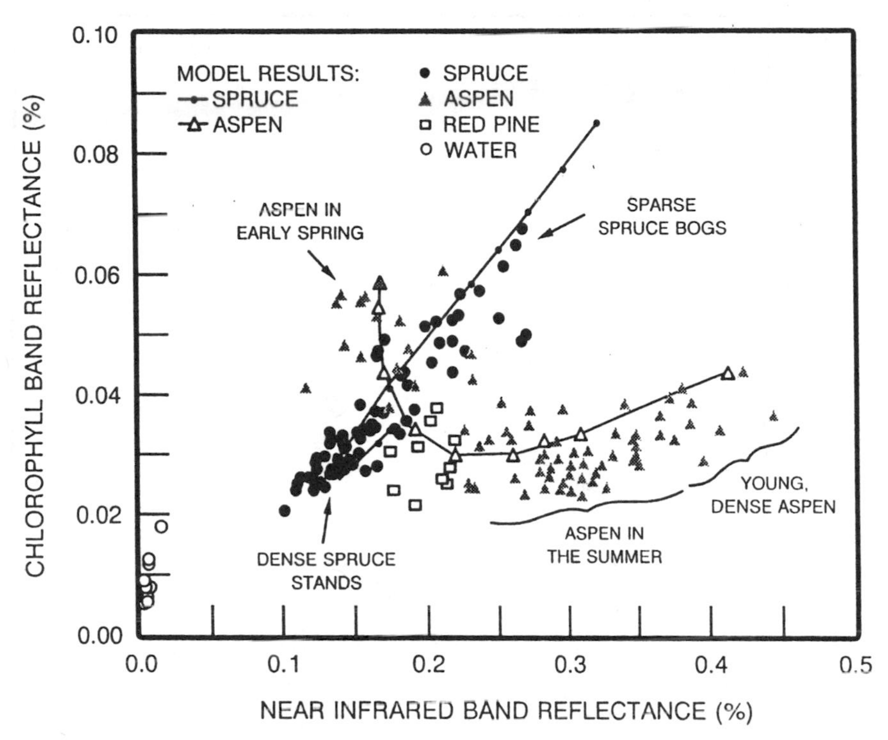

Figure 17.5 Near-infrared (bandwidth) versus chlorophyll absorption band (bandwidth) reflectance for aspen, black spruce, red pine canopies, and water for the Superior National Forest (Hall et al. 1991). Data were acquired from approximately 30 × 30-meter stands with a helicopter hovering at 200 meters above ground level using a radiometer simulating thematic mapper bands. Solid lines are reflectance values predicted by a canopy radiative transfer model that accounts explicitly for the effects of canopy shadowing of the background or understory.

Thus, care must be taken in selection of sensor acquisition dates in separating these two species.

In addition to the visible and near-infrared bands, satellite optical imagers currently available image the ground in the mid-infrared region from 1.35 to 2.0 microns and the thermal-infrared at approximately 12.0 microns. Although many spectral bands are possible, the independent information available to discriminate among vegetation types is contained almost entirely in the visible band (chlorophyll amount), the near-infrared band (leaf area index), and the thermal-infrared band (canopy temperature). For live vegetation, the green band is highly correlated to the remaining visible bands because they all respond to leaf chlorophyll or leaf pigment. The visible bands are in turn highly correlated to the mid-

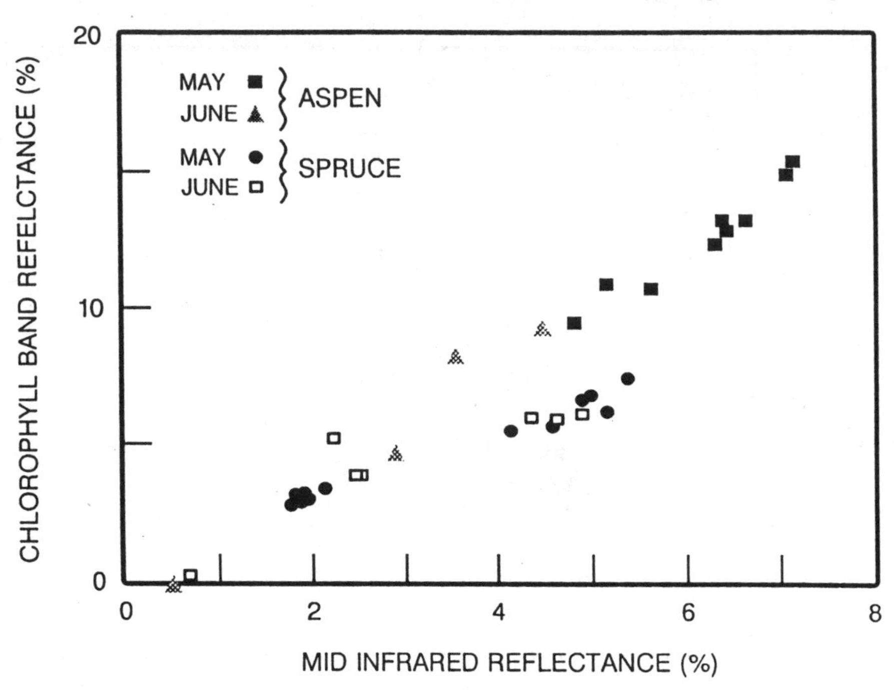

Figure 17.6 Mid-infrared (bandwidth) versus chlorophyll absorption band (bandwidth) reflectance for aspen and black spruce canopies of the Superior National Forest (Hall et al. 1991). Data were acquired as described in Figure 17.8. Data show that as total stand chlorophyll content increases, so does total stand water content. However, the slope of the line is clearly different between the two species, indicative perhaps of morphological or physiological differences.

infrared bands (see Figure 17.6), because leaf chlorophyll and leaf pigment are correlated with leaf water content. Figure 17.6 displays stand-level reflectance in the chlorophyll band, around 0.69 micron, versus reflectance in the water absorption band at 2.0 microns (thematic mapper bands) for aspen and black spruce canopies during the months of May and June. Note that even though the stand chlorophyll and water contents are highly correlated, the relationship between the two bands is different for black spruce and aspen. Such differences provide a means for differentiating between the two stand types and possibly can provide information on the differences in relationships between chlorophyll and water.

Finally, we see in Figure 17.7 the near-infrared versus visible reflectance plotted for an entire Landsat MSS scene over the Superior National Forest test site. Overlayed on this plot is a "landscape taxonomy," that is, a division of the landscape into boreal forest successional stages (see Hall et al. 1991 for additional discussion). In addition to the black spruce and deciduous (primarily aspen and birch) classes of Figure 17.7 we see re-

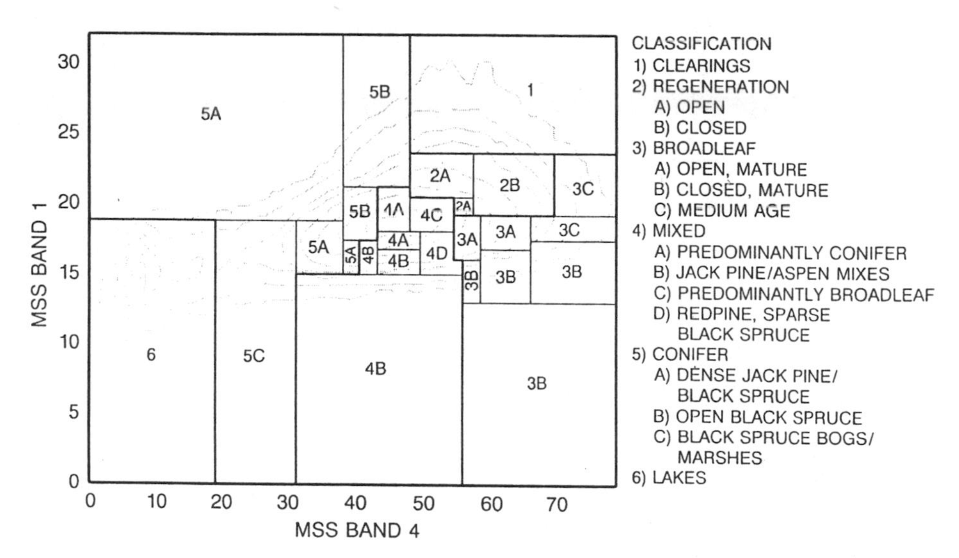

Figure 17.7 Scatterplot showing correspondences between ecological states and reflected light intensity as measured in July 1983 by Landsat-4 MSS in wavelength band 4 (near-infrared, 0.8–1.1 μm) and band 1 (visible light near the green peak, 0.5–0.6 μm). The boxes define correspondences of spectral regions with ecological states. The units on each axis are digital counts (0–256) proportional to radiance as recorded by the Landsat MSS.

generation (shrubs and grasses), clearing (logging and fire), and mixed deciduous/conifer classes. These are cardinal stages of forest succession in a boreal ecosystem. Note the similarity between the location in the band 1–band 4 space of MSS and that of Figure 17.5. The densest black spruce classes have the lowest band 1–band 4 reflectances, and the aspen in this June image have high band 4 reflectances in comparison to the black spruce. In particular, the young aspen have the highest band 4 reflectances. Not shown in Figure 17.5 is the entire range of successional classes shown in Figure 17.7; helicopter data were not available from classes such as regeneration and mixed deciduous and conifer. The decision boundaries were developed using ground-observed plots as described in Hall et al. (1991). Rectangular decision boundaries were used to simplify the classification process and were chosen to minimize classification error between the different spectral classes.

Remote Sensing of Forest Biophysical Characteristics

In addition to the forest information classes discussed above, ecosystem managers also need information on stand biophysical parameters such as

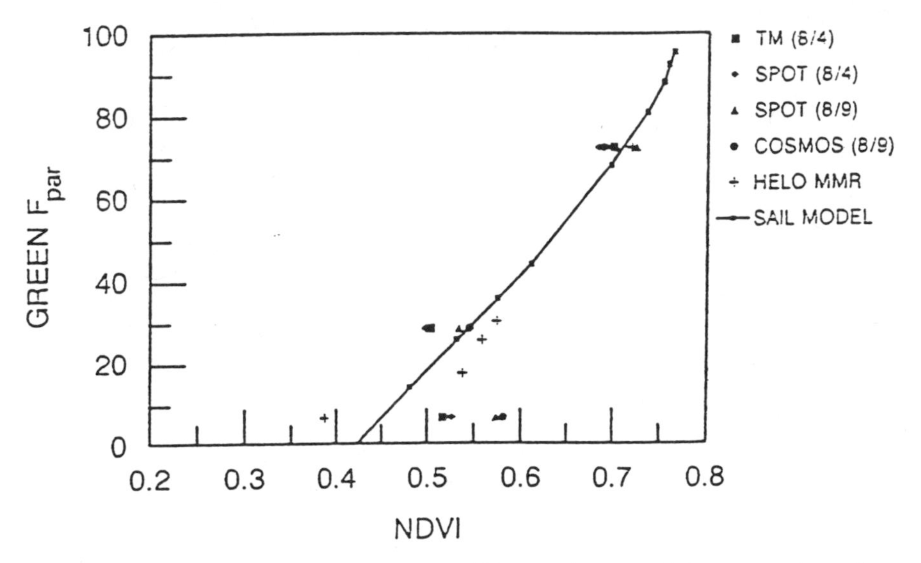

Figure 17.8 Ground measured Fpar versus NDVI as determined from atmospherically corrected and calibrated Landsat TM, SPOT HRV, Cosmos 1939 MSU-E, and helicopter MMR during FIFE. Plotted for comparison are a canopy radiative transfer model prediction.

biomass, leaf area index, and the fraction of photosynthetically active radiation (Fpar) absorbed by the canopy, and energy exchange rates related to growth and photosynthesis (net radiation, evaporation rates, sensible heat flux, and so on).

The relationship between reflectance and the stand biophysical parameters derives from the absorption of visible and mid-infrared light by leaf chlorophyll and water, and the scattering of near-infrared light by leaf cell walls. As leafy material increases in a stand, visible and mid-infrared reflectances decrease while near-infrared reflectance increases. Ratios and differences of the reflectances in these spectral regions also increase with increases in leafy materials and are known as "vegetation indices." One such index is the "normalized difference vegetation index" (NDVI). NDVI is the difference in the near-infrared and chlorophyll reflectance divided by the sum of these reflectances. The NDVI has the property that it increases with stand biomass and leaf area index. NDVI also increases with Fpar, as shown in Figure 17.8 for prairie grassland vegetation.

Calibration and Atmospheric Correction

Many applications using satellite remotely sensed data can be accomplished with "raw" or uncalibrated data, that is, satellite digital count values not calibrated in terms of radiance or corrected for atmospheric effects. But in cases where relative measures of surface radiance or reflectance are required, the data must be corrected for at least relative atmospheric and calibration differences between acquisitions. If absolute measures of surface reflectance are required, the satellite images must be both absolutely calibrated and atmospherically corrected. If training data are available within the scene, classification can proceed without calibration or atmospheric correction. However, if the signatures developed from one image are to be applied to a different image, then the images must be radiometrically "rectified" to each other, that is, the digital count values in one image must relate in exactly the same way to surface reflectance as in the other image.

Radiometric rectification can be accomplished by finding, in each image, surface elements whose reflectance does not change much from one acquisition date to the next. Usually, deep-water bodies, old-growth conifers, rock outcrops, or urban features (parking lots, roof tops, streets, and so on) provide such features. Rectification of two images can then be accomplished by requiring the digital count values of these "reflectance invariant" features to be equal in both images. An algorithm for accomplishing this has been developed and evaluated in Hall et al. (1991). Once radiometric rectification has been accomplished for a series of images, each image will appear as if it were taken with the same sensor through the same atmosphere. That is, the relationship between digital count values and surface reflectance for each image will be identical. If absolute values of surface reflectance are then needed, one image can be calibrated and atmospherically corrected and these same corrections will apply equally to each radiometrically rectified image. This is particularly useful in using long time series of data to evaluate ecosystem change. If training data are available in the current image, then signatures can be developed for that image and applied to an image acquired years earlier. One caution is that vegetative signatures between years must be adjusted for seasonal phenological differences, even if acquired near the same date each year. Such techniques, including the phenological adjustment, have been successfully applied to examine decadal-term ecosystem dynamics over a boreal forest in Hall et al. (1991).

Atmospheric correction and calibration of satellite images requires measurements of both atmospheric composition at the time of a satellite overpass and sensor calibration coefficients relating sensor digital count values to radiance reflected from the surface. On existing satellites, calibration coefficients are measured prelaunch and change as optics age and sensors degrade. Long-term drift in these coefficients are partially accounted for by viewing earth surface elements with known and stable reflectance such as White Sands, New Mexico, and the deep ocean; however, uncertainties of a few percent still exist (see Markham et al. 1987 for a discussion of calibration). Correction of images for atmospheric attenuation and scattering by suspended aerosols and water vapor requires measurements of the column abundance of these constituents, and an atmospheric radiative transfer model to calculate the magnitude of their effects. Kaufman (1988) provides a good discussion of this process. Current techniques for measuring atmospheric constituents combined with the best radiative transfer models were tested during the FIFE experiment and shown to produce estimates of surface reflectance to within approximately 10 percent of the surface reflectance value.

Classification Accuracies Achievable with Current Sensors

As can be seen in Figure 17.7, class boundaries in the band 1–band 4 MSS spectral space partition the space into the various successional stages of interest. Although these rectangular decision boundaries were chosen to minimize misclassification rates, there is, of course, misclassification between successional classes. These misclassification errors are defined in terms of frequencies of correct classification (diagonal elements of Table 17.2) and frequencies of incorrect classification (off-diagonal elements). Table 17.2 shows the frequencies obtained using 1983 MSS data to classify successional stages in the Superior National Forest (Hall et al. 1991).

The frequencies of correct classification range from 0.80 to 0.96. The largest errors result from confusion of conifers with mixed states (0.138) and regeneration with clearing (0.118). While quantitative, these numbers in and of themselves are not the whole story. How do they translate into map and area estimation accuracies?

In terms of mapping, the misclassified pixels can occur in two forms: (1) random, isolated pixels within patches, causing a "moth-eaten" appearance to the patches; or (2) entire patches or stands can be misclassified. The problem of whole misclassified patches can usually be addressed by adjusting decision boundaries to pick up the errant class. Random misclassifications can be reduced by further minor adjustments of decision boundaries, but will remain at some level if the classes of interest are not

TABLE 17.2 Frequencies of correct and incorrect classifications based on satellite remote sensing

		Classified States					
		Clearings	Regener.	Broadleaf	Mixed	Conifer	Other
	Clearings	0.88	0.00	0.00	0.00	0.069	0.00
"Gnd	Regener.	0.118	0.882	0.00	0.095	0.00	0.00
Obs"	Broadleaf	0.00	0.059	0.957	0.048	0.00	0.00
States	Mixed	0.00	0.059	0.044	0.857	0.138	0.00
	Conifer	0.00	0.00	0.00	0.00	0.793	0.00
Other	Other	0.00	0.00	0.00	0.00	0.00	1.00

Matrix elements are the estimated frequencies (based on ground-observed patches) of classifying a ground-observed patch (pixel) into information class a if ground observations indicated that it was in information class b. The diagonals of the matrix are the estimated frequencies of correct classification. The off-diagonal elements are the estimated frequencies of incorrect classification.

completely distinguishable in the spectral data. For mapping, error rates of 10 to 20 percent are generally tolerable, producing usable maps that can then be processed with boundary-finding algorithms and smoothed to reduce the "speckle" left by random misclassifications.

For inventory purposes, where area estimates are important, error rates as high as 10 to 20 percent would be intolerably high were it not for canceling errors. To some extent, errors of commission balance errors of omission, reducing the absolute effect on area estimation accuracies. Furthermore, given good estimates of the errors of omission and commission from a representative sample of ground-mapped sites, independent of the training data sets, the area estimates can be adjusted for these errors, reducing their impact even further. While the effect of classification error rates on area estimation accuracies depend on omission and commission rates between classes and patch size distributions, as a general rule, classification error rates of 20 percent can yield areal estimation errors of less than 10 percent.

Remote Sensing and Ecosystem Management: Future Directions

While satellite data analysis algorithms and data analysis and display systems are adequate to serve a number of current applications needs, a number of advancements should improve even further the utility of satellite remote sensing for ecosystem management. These advancements should be seen in several areas: (1) improved satellite sensors and platforms, (2) improved data analysis and display systems, (3) cheaper and more easily accessible satellite data, (4) improved ecosystem management models, and (5) improved remote sensing algorithms.

Improved satellite platforms and sensors are on their way. Landsat 7 is under development for launch in 1998. Landsat 7 will carry an Enhanced Thematic Mapper (ETM) with improvements over Landsat 5, and possibly a pointable sensor with improved spatial resolution and signal-to-noise ratio (HRMSI). The Earth Observing System, consisting of a morning (AM) and an afternoon (PM) platform and currently scheduled for launch beginning in 1998 (the AM platform), will carry five separate sensors: CERES to measure radiant energy flux at the top of the atmosphere; MOPPIT to measure methane and carbon monoxide concentrations throughout the total atmospheric column; MISR, to provide multi-angle views of the earth's surface and information needed to correct satellite data for atmospheric aerosols; MODIS, a 36-narrowband imager with 1-kilometer spatial resolution; and ASTER, with bands in the visible and near-infrared similar to Landsat MSS, and with additional thermal bands to measure thermal emissions from the earth. These sensors should provide significant improvements in capability to map ecosystem information classes, as well as support the global applications for which they are intended.

A vision of improved data analysis and display systems has already been discussed. As the applications community has grown in the past few years, the stimulus has spawned a number of commercial developers and vendors of software who have packaged existing image-analysis programs and coupled them to GIS software. In the next few years as users become more numerous and sophisticated, we should see integration of these capabilities with existing graphical user interfaces, statistical analysis packages, and more; and we hope to see the applications community linked electronically through the "information highways" already in place and to be improved upon greatly in the future.

With the Remote Sensing Act of 1992 now law, Landsat 7 data will be made available to users at the cost of reproduction. This should make possible many uses of Landsat data that were heretofore restricted by data costs.

Finally, a number of activities currently underway in the global change program should result in improved understanding of ecosystem-level processes, including hydrology, ecosystem productivity, mesoscale meteorology, biogeochemical cycling, and pollutant cycling. Such understanding will be captured in ecosystem simulation models that should provide much badly needed information relevant to ecosystem management.

These advances in sensors, computational capabilities, data availability, and ecosystem process models will bring new requirements for satellite data processing algorithms. Added to the image analysis systems currently in place will be remote sensing algorithms for estimating biophysical parameters, soil moisture, and surface temperature necessary to drive and validate the ecosystem process models.

Conclusions

Research funded by NASA and other government agencies over the past 40 years has led to the development of a satellite remote sensing technology that, combined with GIS and other data system components, is finally beginning to match some of the early high expectations and claims for its performance. (1) Current satellites produce high-quality, calibrated multiband imagery, with radiometric and geometric precision adequate for many of the quantitative mapping tasks required by forest ecosystem managers. (2) Atmospheric correction algorithms can remove atmospheric effects to a precision approximately 5 percent to 20 percent of the signal strength. Radiometric rectification algorithms can render multiyear and spatially disparate satellite data series to a common radiometry with similar precision. (3) Image processing algorithms exist that permit a user to infer and map a large variety of land-cover classes and biophysical parameters required by ecosystem managers and ecosystem process models. (4) Geographic information systems and graphic user interfaces, integrated with satellite images and image-processing subsystems, greatly facilitate analysis and display of disparate data sources. (5) Database management systems render possible the storage and retrieval of complex, disparate data sources. (6) Electronic networks now permit users to access databases around the world to exchange not only data, but results, ideas, and technology. Additional funding to increase network access, bulletin boards, and electronic data archives should accelerate the number and sophistication of remote sensing applications. (7) Satellite data continuity should no longer be an issue with better and more abundant earth resources satellite coverage available in the near future from Landsat 7, and the EOS AM platform. (8) The Remote Sensing Act of 1992 should make data less expensive to the user and thus more widely used. Many applications that were heretofore not cost-beneficial will be so with the anticipated pricing structures.

References

Hall, F.G., P.J. Sellers, M. Apps, D. Baldocchi, J. Cihlar, B. Goodison, H. Margolis, and A. Nelson. 1993. BOREAS: Boreal Ecosystem-Atmosphere Study. IEEE Geoscience and Remote Sensing Society newsletter, 86, pp. 9–18, ISSN 0161–7869.

Hall, F.G., D.B. Botkin, D.E. Strebel, K.D. Woods, and S.J. Goetz. 1991. Large-scale patterns of forest succession as determined by remote sensing. Ecology 72(2):628–640.

Journal of Geophysical Research, FIFE Special Issue. 1992. P.J. Sellers and F.G. Hall (eds.). vol. 97, D17.

Kaufman, Y.J. 1988. Atmospheric effect of spectral signature. IEEE Transactions on Geoscience and Remote Sensing 26(4):441–451.

MacDonald, R.B., and F.G. Hall. 1980. "Global Crop Forecasting." Science 208:670–679.

Markham, B.L., and J.L. Barker. 1987. Radiometric properties of U.S. processed Landsat data. Remote Sensing of Environment 22:30–71.

Remote Sensing of Environment, AgRISTARS Special Issue, April 1979.

Chapter 18

Adaptation of Military Space-Based Remote Sensing Technology to Ecological Information Needs

Donald R. Artis, Jr.

This chapter describes the potential for the Department of Defense (DoD) to adapt space-based remote sensing technology to support environmental monitoring and to support other government agencies and private organizations in environmental protection and natural resource conservation. The DoD space-based remote sensing program must complement other programs, rather than replace or duplicate those programs. This may include:

- Application of defense-related remote sensing technologies to civil environmental functions and ecological information needs
- Use of DoD sensors for civil applications
- Support for the Environmental Task Force (ETF)

Civil applications, in the context of this discussion, are those functions normally associated with the use of space-based sensors to support the civil works mission of the U.S. Army as performed by the Army Corps of Engineers. The evolution of the national security role of DoD will be framed in the future by concerns about the environment, by the need to maintain technological excellence at minimal cost, and by the trend toward joint and cooperative programs with other government agencies. DoD's traditional role as the military arm of the American people could expand to resemble more closely national considerations that are the current purview of the Departments of State and Commerce, such as economic stability; clean, safe work environments; and the free interchange of goods and services. DoD will have to address its evolving role in those areas through traditional as well as nontraditional partnerships.

DoD has traditionally supported the following capabilities in its use of space:

- Seeing a threat
- Exploiting a favorable balance of time with respect to the threat's decision execution cycle
- Using satellites for communication and navigation
- Evaluating the degree of success of inflicting damage
- Monitoring meteorological conditions

DoD's traditional capabilities in space will continue, but those capabilities will, in all likelihood, expand to include new and more demanding requirements on our resources (people, facilities, and funding). This may include providing assistance to the civil sector for functions such as law enforcement, providing humanitarian relief for natural (e.g., hurricanes) and human-made disasters (e.g., Somalia), and providing support for environmental programs ranging from pollution monitoring to compliance.

Strategy: Cooperative Exploitation of Technology

DoD must consider adopting a strategy that would involve contributing to, using, and exploiting civil space-based remote sensing to satisfy DoD needs as well as to contribute to other agencies' needs. DoD's strategy for the use of space should include:

- Capitalizing on emerging space systems' capabilities
- Exploiting space activities that contribute to the successful execution of all DoD missions
- Assuring access to space and the use of space capabilities for United Nations supported missions

The DoD space program must maximize cooperative exploitation of non-DoD space programs such as NASA's Mission To Planet Earth (MTPE) program. The MTPE program will provide the scientific basis for understanding global change and is NASA's contribution to the U.S. Global Change Research Program. NASA proposes to accomplish this by orbiting satellites to study the Earth from space on a global scale and on a continuing basis. In addition to its traditional national security related activities, DoD may have an increased role in fostering cooperative relationships with agencies such as NASA and in supporting civilian government interests in the United States and abroad. The future role of DoD in space must emphasize two types of activities to meet future needs:

- Participation in established programs to ensure continuation where there is a potential benefit to DoD or for DoD support of other government agencies
- Initiation of new technology programs with a high potential for use by DoD to include technology to support the evolutionary development of new DoD missions or mission support activities provided to other government agencies

DoD Applications of Space-Based Instruments

Table 18.1 shows the relationship between DoD applications of space-based instruments and the instruments existing or planned to be on orbit. While these instruments are not inclusive of all possible space-based sensors, they are the ones most likely to exist and are the most likely to have potential utility for DoD. Engineering and Housing Support (EHS) and Training/Training Lands are distinct functions performed by DoD in support of its nontactical military missions and include such activities as postconstruction repair and training operations support. Table 18.2 lists civil application areas that DoD, through the Army Corps of Engineers (USACE), participates in routinely.

Current DoD space-based remote sensing programs have concentrated on achieving traditional DoD national security objectives; future DoD programs must strive to minimize duplication and maximize integration, interoperability, and cooperation with other government agencies' space programs. This may be more difficult to achieve than a clearly and singularly military application. In the future, DoD space remote sensing programs may be required to expand their sphere of influence to include providing environmental support because of an increasing emphasis by the Congress on multiple uses of this capability. A report entitled "A Post Cold War Assessment of U.S. Space Policy," published by the Vice President's Space Policy Advisory Board in December 1992, states that while there remains a need to maintain distinct civil and national security space sectors, planning should be centralized across sectors and its execution streamlined within the respective sectors. This must be done to foster synergism among the civil, military, intelligence, and commercial space-

TABLE 18.1 Space-based instruments and DoD applications

	DoD potential application	
Space-based instrument's platform or program	Civil Applications Support (Civil Works)	Engineering and Housing Support & Training/Training Lands Applications
MTPE (11)	Yes	Yes
Landsat (2)	Yes	Yes
French SPOT (1)	Yes	Yes
European Space Agency Earth Resource Satellite (ERS)-1 (1)	Yes	Probably
Japanese Earth Resource Satelite (JERS)-1	Yes	Probably
Canadian RADARSAT	Yes	Probably
NOAA Polar-Orbiting Environmental Satellite (POES) (1)	Yes	Probably
NOAA Geostationary Operational Environmental Satellite (GOES) (1)	Probably	Unlikely
NAVSTAR (GPS)	Yes	Yes
DMSP	Yes	Yes

All on-board instruments unless number noted

based remote sensing programs. As noted in the Space Policy report, delivering mature, critical technologies to enable space-based remote sensors to meet future requirements, whether for tactical military or for civil works, will also require comprehensive and favorable cost and operational effectiveness analyses to support future operational requirements.

Potential DoD Participation in Civil Applications of Space-Based Instruments

Analyses must be conducted by DoD to define its role in the post–Cold War uses of space for other than tactical considerations. In that vein, DoD

TABLE 18.2 U.S. Army Corps of Engineers Civil Application Areas[1]

Aquatic vegetation
Archeological and cultural resources
Assistance to other nations
Construction monitoring
Dam repair
Desertification
Dredging
Drought studies
Emergency operations
Energy usage monitoring
Environmental impact assessments
Environmental monitoring
Erosion studies
Flood control projects
Flood damage assessment
Floodplain
Forestry management
Fuel and chemical spills
Geology and soils studies
Hazardous, toxic, and radioactive wastes
Infrastructure development and relocation
Land cover
Mapping
Natural resource management
Navigation
Pollution monitoring
River/coastal engineering functions
Sedimentation analyses
Sensitive species
Snow cover assessment
Urban area studies
Wetlands delineation

[1]The Corps of Engineers performs these functions as a DoD mission.

has initiated several efforts to contribute to some of its space exploitation requirements including:

- Exploiting existing and planned sensors on civil Earth sensing satellites such as Landsat, SPOT, and JERS-1
- Contributing to and focusing research and development (R&D) to support the Space-based Global Change Observation System (S-GCOS) program
- Collaborating and cooperating with the U.S. Navy, NASA, NOAA, EPA, and USGS in requirement definitions for space sensors of joint interest

Table 18.3 is a list of space-based remote sensing instruments (MTPE as well as non-MTPE) that may support the DoD civil application area of forestry management on DoD lands. Similar lists have been compiled for each of the application areas in Table 18.2 and are described in the USACE Mission To Planet Earth Task Force Report, 14 August 1992. DoD is in the process of identifying how to assist NASA in its MTPE program. A USACE MTPE task force was formed in March 1991 to determine if and how the USACE, as the Army's and DoD's civil works proponent, should become involved in the MTPE program. The USACE's interest in the MTPE program is based on the possibility that the MTPE program will measure parameters needed by the USACE to support its civil and environmental protection missions. The USACE's districts, divisions, and engineering and housing support directorates have a need for remotely sensed imagery, data, and/or information that is currently both expensive and time consuming to acquire. If space-based, remotely sensed data from the MTPE program can be used to satisfy some of those data needs, the potential for large cost and time savings makes USACE participation in the MTPE program very attractive. The USACE is in the process of preparing a program implementation plan to be completed at the end of April 1993 that will (1) describe how to support the MTPE program and use it for USACE/Army/DoD purposes and (2) identify potential joint or collaborative research and development with NASA and/or NOAA.

Table 18.4 shows global change variables currently being, or planned to be, analyzed by the Army for its own purposes. The results of those ongoing or planned analyses may contribute to other organizations' or agencies' programs for global change measurements, not only in forestry management but in many other application areas as well. This table also

TABLE 18.3 S-GCOS Space-based instruments of potential use to the U.S. Army for forestry management on DoD lands

Mission-to-Planet-Earth (MTPE) planned instruments	Non-MTPE planned or existing instruments
NASA/NASDA[1] Advanced Spaceborne Thermal Emission and Reflection Radiometer (ASTER)	NASDA JERS-1's Visible Near IR/Shortwave IR Instrument
NASDA Advanced Visible and Near-Infrared Radiometer (AVNIR)	Landsat 7's Enhanced Thematic Mapper (ETM)+ and the High-Resolution, Multispectral, Stereo Imager (HRMSI)
NASA High-Resolution Imaging Spectrometer (HIRIS)	NOAA's Polar Orbiting Environmental Satellite (POES) Advanced Very High-Resolution Radiometer (AVHRR)
NASDA Investigator of Micro-Biosphere (IMB)	NOAA's Geostationary Environmental Satellite (GOES) Imager French SPOT's High-Resolution Visible Imaging System (HRVIS)

[1]NASDA is Japanese Space Agency

shows the current U.S. space-based instrument platforms used to analyze these particular variables. They include the following satellites:

- DoD Defense Meteorological Satellite Program (DMSP)
- NOAA's Polar-Orbiting Environmental Satellites (POES)
- NOAA's Geostationary Operational Environmental Satellites (GOES)
- Landsat

Data sources other than those listed will also be exploited, but space-based sensors on the satellites listed above are currently being analyzed, or planned to be analyzed, by the Army for the potential contributions that could be made toward the measuring and understanding of the global change variables listed on Table 18.4.

Potential new customer relationships may require DoD to assume new and expanded responsibilities, albeit as a supporting partner, to another

TABLE 18.4 Potential Army uses of or contributions to global change variables using space-based instruments

Variables	DMSP (D); POES (P); Landsat (L); GOES (G)
Trace Species	
$H_2 0$	D, P, G
0_3	P
Ocean Variables	
Sea surface temperature	D, P, G, L
Sea ice extent	D, P, L
Sea ice type	D, P, L
Sea ice motion	D, P, L
Sea level	D, P, L*
Atmospheric Response Variables	
Surface air temperature	D, P, G
Tropospheric temperature	D, P, G
Stratospheric temperature	D, P, G
Pressure (surface)	D, P, G*
Tropical winds	D, P, G*
Tropospheric water vapor	D, P, G
Components of earth radiation budget	D, P, G*
Cloud amount, type, height	D, P, G, L
Tropospheric aerosols	D, P, G*
Land Surface Properties	
Surface radiating temperature	D, P, G, L
Snow cover	D, P, G, L
Snow water equivalent	D, P, L
River runoff (volume)	P, L
River runoff (sediment loading)	P, L
Surface characteristics (for albedo, roughness, infrared, and microwave emittance)	D, P, G, L
Index of land use changes (broad classification of vegetation types)	D, P, G, L*
Index of vegetation cover	D, P, L*
Index of surface wetness	D, P, L*
Soil moisture	D, P, L, G
Biome extent, productivity, and nutrient cycling	D, P, L*

*derived potential utility but not directly measured or observed. Actual usage would be through an all-source analysis combining DoD satellite data such as from DMSP with other satellite data such as from the GOES, POES, and Landsat satellites.

government agency. These governmental partnerships will evolve over time but are likely to increase in both number and type.

To exploit space, DoD must take advantage of leveraging opportunities when they exist and technology breakthroughs as they occur. This includes adopting commercial techniques, processes, and systems where appropriate. The true measure of success for technology development is the transition of the technology into an operational capability, assuming cost and operational effectiveness analyses warrant the transition.

DoD must continue to monitor and support the work of the Environmental Task Force (ETF) in supporting the ecological information needs of the scientific and engineering communities. The report entitled "A Post Cold War Assessment of U.S. Space Policy," mentioned above, noted the need to reduce security constraints associated with national security space program data. The ETF will support that need by determining the applicability of data, products, images, and other outcomes from classified sources that could be used to assist the civil community's need for environmental science data. Environmental stewardship for civil applications is not necessarily a function of the ETF, but it is a strong concern of the Army Corps of Engineers and is being pursued as aggressively as possible by the Corps in each of its civil, as well as its military, responsibilities.

DoD must complement other agencies' space programs, rather than replace or duplicate existing programs. Therefore, two types of activities should be pursued to meet future DoD requirements for other than tactical military applications:

- More active and visible participation in established, non-DoD programs to enhance DoD capabilities, such as the MTPE program
- Direct participation in other programs to ensure continuation where there is a potential benefit to DoD, such as the DoD/NASA Landsat 7 program

Conclusions

To support the ecological information needs of the civil sector for global change measurements, the Department of Defense should:

- Continue to support NASA's Mission To Planet Earth by preparing an Earth Observation Program Implementation Plan to help define how to implement DoD's, and in particular the Army's, participation in the Space-based Global Change Observation System program

- Participate as an active partner with civil applications working groups to coordinate civil and DoD needs for environmental data
- Continue to support the work of the ETF
- Consider expanding its role in established, non-DoD space-based remote sensing programs to enhance and improve DoD capabilities and to offer DoD expertise in the conduct of those space programs, as appropriate and consistent with national security and technology transfer considerations

Acknowledgments

The author wishes to acknowledge the contributions and advice of the following people in the preparation of this chapter: Dr. Richard B. Gomez, Office of the Undersecretary of Defense for Acquisition, Office of the Secretary of Defense; Dr. Harlen L. McKim, Chief, Remote Sensing/Geographic Information System Center, U.S. Cold Regions Research and Engineering Laboratory, Hanover, New Hampshire; Richard J. Szymber, Battlefield Environment Directorate, U.S. Army Research Laboratory, White Sands Missile Range, New Mexico.

References

U.S. Army Corps of Engineers. 1992. Mission-To-Planet-Earth Task Force Report.

U.S. Army Corps of Engineers. 1990. Report of Joint NASA-USACE Task Force on Planet Surface Systems Partnership.

Vice President's Space Policy Advisory Board. 1992. A post cold war assessment of U.S. space policy.

Part IV

Synthesis

Chapter 19

The Potential and Limitations of Remote Sensing and GIS in Providing Ecological Information

Kass Green

Introduction

The purpose of this book is to (1) identify the needs of ecologists and resource managers engaged in forest ecosystem management and policy analysis, and (2) explore current and potential applications of remote sensing and GIS to meet those needs. In summary, a number of unifying themes are worth exploring.

Almost every author identified conflict and competition over land use as the driving force behind the exploding need for spatial information of ecosystem interrelationships. The need to balance the use of ecosystem products while maintaining ecosystem function and aesthetic values drives the integration of GIS and image processing into ecosystem management. As the world's population has grown, so have demands on forests and rangelands to provide multiple resources for diverse client groups. While the earth's population continues to expand, its land base is static. Rapid land cover and land use change has intensified conflicts over land management, resulting in an immediate need for accurate information about land status and resource relationships. Conflict over land allocations can only continue to intensify. As a result, the management of forest ecosystems has become increasingly complex, requiring sophisticated information management tools and timely data.

From harvesting and land use conversion-induced fragmentation of spotted owl and neotropical bird habitat, to increased tree density resulting from fire exclusion in the southwest, to habitat eradication in the midwest, people have significantly affected the forest ecosystems of the world. In the United States, the closing of the American frontier at the turn of the century precipitated a growing concern for the wise use of forest resources. Today, any landscape change is often viewed with alarm.

Over the last 10 years, GIS and remote sensing have emerged as promising tools for prioritizing and analyzing resource management alternatives. Because land use conflicts are spatial, any rigorous analysis of land management alternatives must include consideration of the spatial interactions between people and their environments. When coupled with field data, the combination of GIS and remote-sensing technologies may offer the analytical tools needed by ecosystem managers in assessing and monitoring ecosystem condition and management alternatives over time.

A number of policy and management needs have been identified, along with several current and potential means by which GIS and remote sensing might meet those needs. In addition, potential users of this emerging technology should be aware of its promise, but also its pitfalls. Finally, there is a need for better communication between technology experts and land managers when applying GIS and remote sensing to ecosystem management.

The Need: Sustained Ecosystems Versus Sustained Yield

While the precise definition of ecosystem management is still emerging, there is an overriding shift in management emphasis away from a goal of sustainable yield and toward a goal of sustainable ecosystems (Aplet et al. 1993). Several ecosystem management concepts continually resurface:

- Minimizing habitat fragmentation
- Maintaining biodiversity
- Identifying and managing cumulative effects regardless of ownership boundaries
- Analyzing, predicting, and monitoring impacts across various scales of time and space

As William Gregg observes (Chapter 1), ecosystem management requires "complex tradeoffs that enable use of ecosystems to meet human needs in ways that sustain natural ecosystem function and components in a changing environment."

Figures 19.1a and 19.1b illustrate both the change in management emphasis and the power of remote sensing and GIS as tools for ecosystem management. Figure 19.1a is a Landsat TM scene of the Olympic Peninsula in Washington. The pattern of vegetation is a function of ownership. The cities of Seattle, Tacoma, and Olympia are easily identified. Olympic

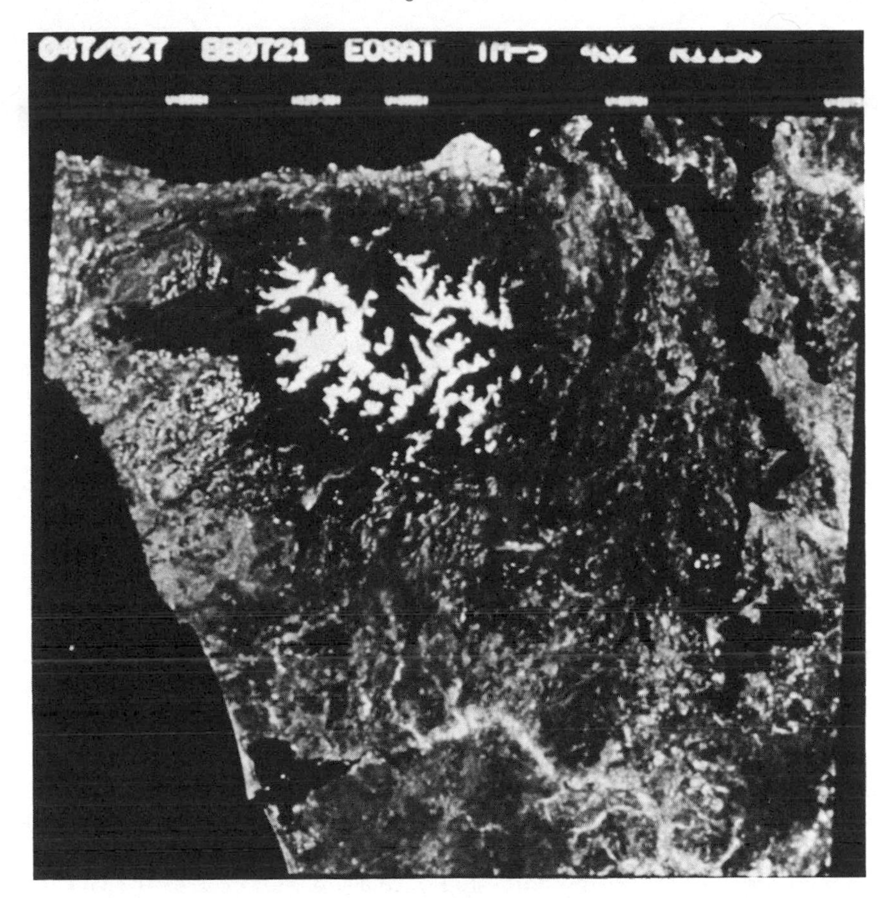

Figure 19.1a Landsat TM scene of the Olympic peninsula

National Park reflects the homogeneous expanses of late-successional forest lands. Olympic National Forest and the State of Washington lands show their fragmented pattern of harvesting over the last 40 years. The pattern of harvesting and regeneration of industry lands shows little fragmentation, but also few areas of larger dense stands of trees.

One of the most interesting areas on the Olympic Peninsula is the Shelton Sustained-Yield Unit, established by Congress in the 1940s to combine the lands of the Olympic National Forest and Simpson Timber Company into a sustained-yield management unit. The purpose was to sustain the economic viability of the town of Shelton. At the time of the agreement, Simpson's lands had been harvested, while the lands of the national forest had not. The unit combined both ownerships for the calculation of allowable harvest levels, allowing for the harvest of the

Figure 19.1b SPOT image of National Forest lands

national forest while the Simpson lands regenerated. As Figure 19.1a shows, the establishment of the unit was a success in terms of sustained yield. Simpson lands now support relatively large second-growth forests, which are presently being harvested once again. However, as the SPOT image of Figure 19.1b shows, the National Forest lands are largely harvested, and while regenerating, the landscape is fragmented. A quick look at the remotely sensed data shows that the Shelton unit was a successful demonstration of sustained-yield forestry, but a questionable application of sustainable ecosystem management.

Tools for Ecosystem Management

GIS and remote sensing are means, not ends. To identify and analyze the tradeoffs implicit in ecosystem management, Jerry Franklin (Chapter 2) challenges GIS and remote sensing to provide ecosystem managers with the ability to (1) inventory and monitor resources, (2) plan both site-specific and regional management, and (3) analyze policy alternatives. However, Roger Hoffer (Chapter 3) notes that the prerequisite to providing these abilities is the careful linking of information needs with technology capability. The biggest task facing users of GIS and remote-sensing technologies is matching the tools to the problem at hand. The spectrum of GIS and remote-sensing technologies is broad, requiring a careful tailoring of response to problem.

Historical failures associated with the use of GIS and remote sensing are abundant and well known (Meyer and Werth 1990). However, recent technological and organizational innovations have dramatically altered the methods available for ecosystem management. The integration of GIS and image processing into ecosystem management directly results from five recent technological and organizational advances (Green 1992).

- First, remotely sensed data has improved. Digital imagery includes Landsat thematic mapper (TM) data (with a spatial resolution of 30 meters), both multispectral and black-and-white SPOT data (with spatial resolutions of 20 and 10 meters, respectively), video, digital photography, and airborne scanner data.
- Second, computer hardware has become more powerful and less expensive, creating the ability for interactive processing, analysis, and evaluation. In addition, the decrease in the cost of peripherals, such as scanners and electrostatic plotters, allows the input and output of completely integrated image and GIS data.
- Third, GIS and image-processing software have become more fully integrated and sophisticated. The integration makes possible the analysis of relationships among spatial location, spectral variation in the image, and land-cover variation on the ground.
- Fourth, procedures for the assessment of map accuracy have been fully implemented and accepted as critical elements in the production of any GIS layer derived from remotely sensed data (Story and Congalton 1986, Congalton and Green 1993), resulting in the development of unbiased, quantitative measures of map accuracy.
- Finally, training in GIS and remote sensing is now offered in many natural resource university courses, resulting in professionals who understand GIS, remote sensing, and ecosystem dynamics.

The result of these advancements is the integration of remote sensing with GIS for ecosystem management. As David Mladenoff and George Host (Chapter 14) explain, "Although the usefulness of both GIS and remote sensing in resource management have not fulfilled their promise in the past, it now appears that technology, cost, need, and the conceptual framework are converging in a way that will continue to prove practical and effective." Accompanying the integration is a decrease in the cost to produce traditional maps required for ecosystem management, a decrease in the amount of time required to produce the GIS layers, an increase in the detail of information linked to the layers, a decrease in the cost to monitor change over time, and an increase in the variety of maps and outputs available from the GIS database.

The workshop participants demonstrated that Jerry Franklin's challenges of (1) inventorying and monitoring of resources, (2) planning both on the ground and regionally, and (3) analyzing policy alternatives are already being met. John Steffenson explains how the Forest Service was able to use the technologies to map old growth in the Pacific Northwest. Forrest Hall (Chapter 17) illustrates the employment of the technologies to monitor global change. Jessica Gonzales (Chapter 10) discusses the inventory of old-growth forests in New Mexico. Mladenoff and Host present numerous examples of the use of the technologies for inventory and monitoring ecosystems. John Sessions, Richard Flamm, and Craig Allen (Chapters 5, 13, and 8) demonstrate that GIS and remote-sensing tools also exist for management planning and analyzing policy alternatives. Jim Rochelle's three-dimensional representation of the incongruity between the location of spotted owl telemetry sites versus the location of spotted owl reserves displayed the power of GIS to instantly reveal complex spatial relationships. Numerous examples of the promise of the technologies were presented throughout the workshop.

However, this new panacea can result in frustration and disappointment if the technologies are applied carelessly. The experiences noted in the symposium papers pointed to several requirements for successful integration of GIS and remote sensing into ecosystem management:

- The classification system for the data must be well designed. Jessica Gonzales mentioned that there are as many classification systems as there are people building GIS data sets. There is no one universal classification system. However, implementation of three simple criteria for classification systems will help to ensure their viability. Each classification system must be
    - totally exhaustive; all things must have a name.

- mutually exclusive; different things must not have the same name.
- hierarchical; all classes can be collapsed upward to more general classes.

Database design is the easiest and most expensive step to ignore. Flawed database design will haunt a project to its completion and beyond.

- The temporal and spatial scale of the project must be determined. As Margaret Moore stressed, ". . . we need to recognize and emphasize the relationships between spatial patterns and ecological processes, over time, and at several hierarchical scales." While GIS and remote sensing can accommodate multiple scales, the cost, flexibility, and quality of a project will all be directly impacted by the choice of scale.

- The appropriate data source must be identified. The only reason that remotely sensed data can be used to build GIS data layers is the high correlation between spectral response and land use and land cover. Video, aerial photography, satellite imagery, and airborne scanner data all vary, to some extent, with variation in land cover. The task of the GIS/remote sensing analyst is to discover when the data source varies with the components of the classification system, and when it does not. The strength of the correlation will determined the choice of data source.

- The accuracy of the data needs to be assessed and reported. A data set without accuracy assessment is useless because the reliability of the data cannot be conveyed to a land manager, a constituent, a judge, or Congress.

- Budget and schedule must limit the scope of a project. Tradeoffs among cost, schedule, detail, and sophistication of analysis always are possible. A project must be completed within the constraints of available resources. A late project may have no value regardless of its quality. Table 19.1 compares two projects with seemingly similar final products—a GIS database of land cover type—but with very different scales, classification systems, time frames, and accuracy assessment procedures.

- Finally, as Mark MacKenzie's presentation highlighted, standardized formats are needed for the exchange of information across projects. The promise of GIS and remote sensing have led to the proliferation of data capture projects and data sets. The implementation of standardized formats can help to eliminate duplication of effort.

Table 19.1 Comparison of two projects with similar products but different parameters

	Project I	Project II
Goal	Regionwide analysis of cumulative effects	Day-to-day land management *and* regionwide policy analysis
Scale	1:48,000	1:24,000
Land cover classes	5	over 1,500
Area covered	70 million acres	20 million acres
Time frame	9 months	4 years
Cost	$120,000	$2,700,000
Number of accuracy assessment sites	400	4,000

Communication

Perhaps the greatest task facing the integration of GIS and remote sensing into ecosystem management is the honest communication of the promise and pitfalls of the technologies to the users of the information. As Douglas Schleusner pointed out, a technically excellent project will fail without a committed constituency for the information. The enthusiasm and expectations for GIS and remote sensing are high. Building and maintaining a constituency requires tempering the enthusiasm with straightforward evaluation of technological limitations.

Roger Hoffer counseled that the use of GIS and remote sensing often stagnate because of incomplete knowledge and a lack of communication between those who need the information (the stakeholders) and the GIS/remote sensing analysts. Work on ecosystem management need not stop because we lack a definitive definition of an ecosystem or complete knowledge of the applications of GIS and remote sensing to ecosystem management. GIS and remote sensing allow ecosystem managers to simulate possible future conditions and test the sensitivity of assumptions prior to action. Thus, the technologies can be used as tools to enhance communication and discovery. Conditions can be initially stipulated, assumptions detailed, and results presented and qualified. Stakeholders can then conceptualize new assumptions and conditions, and new results can be generated. Through this iterative process the stagnation circle can be transformed into an information spiral toward increasingly better information and analysis.

However, information ownership and commitment must be established without the creation of territory. Several of the workshop participants expressed the need to continually share GIS data sets among organizations. Turf wars over data, models, or processes will maim the potential and the power of the technologies.

Conclusions

The shift in management emphasis from sustaining yields to sustaining ecosystems requires the analysis of complex ecosystem functions across space and time. The tools of GIS and remote sensing can offer support to this analysis. GIS and remote sensing are already supporting ecosystem inventory, planning, and analysis of management decisions. The preceding chapters present an enlightening array of applications where the technologies are being used to predict the location and magnitude of impacts and how those impacts will affect land cover, land use, wildlife populations, county incomes, jobs, recreation opportunities, and infrastructure. Many tools already exist, and the data sets are being built.

However, GIS and remote sensing are both a panacea and a Pandora's box. The panacea exists in the promise of the technologies to meet the challenges of ecosystem inventory, monitoring, planning, and policy analysis. The Pandora's box contains the pitfalls of using the tools wrongly, capturing data poorly, miscommunication of information, conveying incorrect results, and overselling the capabilities. The Pandora's box can be avoided through solid project design, honest information portrayal, and constant communication.

Because increasing demands on the land are intensifying conflicts over land use, the need for integrating GIS and remote sensing into ecosystem management will expand. GIS and remote sensing will continue to contribute to ecosystem management and the resolution of land-use conflicts because, as many of the presenters acknowledged, we are presently tapping a minimal amount of the information that exists in the remotely sensed data and in the relationships between GIS layers. The potential for these technologies to enhance wise care of the earth is inspiring.

References

Aplet, G., N. Johnson, J. Olson, and V.A. Sample. 1993. Defining Sustainable Forestry. Island Press: Washington, DC.
Green, K. 1992. Spatial imagery and GIS. Journal of Forestry 90(11):32–36.

Meyer, M., and L. Werth. 1990. Satellite data: Management panacea or potential problem? Journal of Forestry 88(9):10–13.

Story, M., and R. Congalton. 1986. Accuracy assessment: A user's perspective. Photogrammetric Engineering and Remote Sensing. 52(3):397–399.

Congalton, R., K. Green, and J. Teply. 1993. Mapping old-growth forests on national forest lands in the Pacific Northwest from remotely sensed data. Photogrammetric Engineering and Remote Sensing. 57(6):677–687.

20

GIS as a Catalyst for Effective Public Involvement in Ecosystem Management Decision-Making

Zane J. Cornett

The opinions expressed are those of the author and are not intended to represent those of the USDA Forest Service.

Introduction

John Naisbitt, author of Megatrends (Naisbitt 1982), has commented that we mass-produce information like we mass-produce cars. Our capacity to produce data has exceeded the human capacity to absorb and understand, creating a psychological malady called "information anxiety." It is the "black hole" between data and knowledge, and happens when information doesn't tell us what we want or need to know. GIS is an information management tool that has profound potential for bridging that black hole, not only for reaching decisions in natural resource management, but also in communicating complicated data in a form that becomes usable information for scientists and the public alike.

In the complex and litigious society we live in, decision-makers frequently try to reduce risk and uncertainty by demanding more and more information. While economic constraints are themselves an impediment to meeting that demand, it has become apparent that no amount of data will resolve the conflicts between divergent interests—each with a legitimate voice in how public lands are managed. There are evolving approaches to public involvement that show promise of bringing those divergent interests if not together, then at least into an arena where consensus can be reached (Sirmon 1991). Using GIS in a participatory decision-making process can be extremely effective, even in its more basic forms. With the advent of high-speed workstation technology, ever more powerful software, and integrated databases, the opportunities for interactive data analysis and display in a public forum have arrived.

Ecosystem management builds on the principles of the Brundtland Report (World Commission on Environment and Development 1987) and New Perspectives (see, for example, Kessler et al. 1992) in its approach to sustainable natural resource management. Three components to management practices must occur simultaneously: they must be ecologically sound, economically feasible, and *socially acceptable*. Ecologically sound pertains to ecosystem resiliency, health, and thresholds. Economically feasible requires new ways of valuing noncommodity benefits along with marketable goods. For natural resource management to be socially acceptable, decision-making must be a public and participatory process.

The complexities of ecosystem management virtually require the use of GIS technology. The need to view and analyze ecosystems at a landscape level demands the spatial capabilities that only GIS can provide. The same capabilities will assist in modeling conditions and attributes that feed into the new economic equations, as well as providing data for those currently in use. Social acceptability may prove to be the most difficult aspect of ecosystem management to implement—therefore defining the most critical role for GIS. Visualization is a very powerful form of communication. Using GIS as a communications tool in public settings has potential that has been relatively untapped. Let's explore that potential.

Public Involvement and GIS: a Land-Management Planning Case Study

The community of Cooper Landing, Alaska, is completely surrounded by the Chugach National Forest and sits in the middle of a forest that is dominated by Lutz spruce (a hybrid of white spruce [*Picea glauca*] and Sitka spruce [*P. sitchensis*]). Forest pest management surveys detected a spruce bark beetle infestation in the early 1960s, though it was considered a minor problem at the time. However, beginning in the late 1970s and throughout the 1980s the outbreak grew into an epidemic. There were a number of proposals to deal with the infestation by chemical treatment and/or sanitation harvests, but these were resisted by the local community and antilogging interests. By 1990, the epidemic of spruce bark beetles had infested more than 28,000 acres around Cooper Landing, with two-thirds of the spruce stands suffering an average mortality of greater than 90 percent. The threat of catastrophic wildfire was real.

The Cooper Landing Cooperative Project was initiated by the Chugach National Forest and the Alaska Region State and Private Forestry Forest Health Management Group as a joint effort of the USDA Forest Service, citizens of Cooper Landing, timber industry and environmental community representatives, Alaska Departments of Fish and Game and Natural Resources, the USDI Fish and Wildlife Service, and the Kenai Peninsula Borough. The initial gatherings of this extremely diverse public working group could politely be called contentious—the state troopers were called in to break up a fight at the conclusion of the first meeting. Even so, the project leaders were committed to a participatory decision-making process. They adhered to this philosophy to the point of refusing to either define the problem or the scope of the project for the group. With dead trees surrounding the meeting site, it did not take long for the group to decide that the project objectives were to reduce the risk of catastrophic fire in the Cooper Landing area and restore treated lands to a forested condition.

From the beginning, GIS played a prominent role in the Cooper Landing Cooperative Project and was utilized in a way that was fairly unique at that time. (The Chugach has a cooperative agreement with the USGS EROS Alaska Field Office [in Anchorage], where the Forest's GIS database and product development occurs. Additional GIS analysis on community property valuations from tax records was conducted by the Kenai Peninsula Borough in Soldotna.) At every working-group meeting there were GIS plots and corresponding tabular information pertinent to the current meeting agenda. In addition, plots were displayed at local post offices, libraries, and government offices throughout the planning process. The following summary of GIS products and how they were used tracks the progress of the working group throughout the project:

- The process of defining the problem and the scope of the project was facilitated by the first two plots presented to the group. On the first, the general vegetation types within the project area were displayed, identifying the forested lands susceptible to spruce bark beetles. For the second plot, information aggregated from aerial surveys of infestations were superimposed on the vegetation types, using increasing color intensities to indicate the age of infestations. Extensive field surveys had shown a direct correlation between age of infestation and both percent mortality and fuel loading. Thus, the color intensities on the second plot also gave some indications of fire potential.

- The next phase was to develop more precise information on potential fire dangers around Cooper Landing. Working closely with state fire behavior specialists, a fuel model was developed for the local area. GIS layers used in the model were vegetation types, the age of spruce bark beetle infestation and correlated fuel types and loading, elevation, and aspect; slope was found to play an insignificant role in this ecosystem. Using the fuel model and local wind and precipitation data, six different fire behavior models were developed, representing historic ranges of fuel moisture and wind conditions. With the visualization of these models, the working group was able to see where potential fire danger threatened key community, ecological, and cultural sites, as well as strategic locations for fire-suppression initiatives.
- Development of alternative treatments that would meet project objectives was the next step. Two additional plots were produced to initiate this phase. One displayed a "minimum-action" alternative—the need for strategic fire breaks around the community was identified in a previous action. The other plot represented the maximum treatment alternative, which was based on biological, environmental, and regulatory constraints. After some initial hesitation at "messing up pretty plots" (one of the project leaders had to scribble on one of the plots before anyone else would apply a felt pen), the working group developed six additional alternatives in four hours! Their resources were the complete set of GIS plots to date, computer printouts of associated information referenced by specific GIS polygons, and familiarity with the local landscape.
- At the next meeting, the working group was presented with GIS plots of "their" alternatives. The prescriptive treatments within each alternative were analyzed and refined with input from working group members and Chugach National Forest resource specialists. As those refinements occurred (which resulted in a total of five alternatives in the final Environmental Assessment), the working group had fresh plots representing the current status for each meeting. If questions came up that could not be answered from computer printouts or resource specialists' reports, they were researched and/or modeled and the results offered at the subsequent meeting.

The selected alternative from the Cooper Landing Cooperative Project was implemented without appeal. The district ranger and the project leaders have stated many times that even faced with an urgent timeline due to the fire situation, they never would have gotten through the planning phase of the project without the use of GIS.

The Future Is Today

While we were using what I would call a "near-interactive" GIS process in the Cooper Landing effort, current technology would allow us to conduct analysis and display the results in real time. With a wholly integrated database, a high-speed workstation, and a large screen or projected display, alternatives and questions can be explored immediately. Models can be run repetitively with varying assumptions to pursue "what if?" scenarios. The response to using these GIS capabilities while conducting land-management planning in a public and participatory process should be profound and rewarding.

Humans are visual creatures. It is much easier for most people to relate to "pictures" of information than to text or numbers. Even in the Cooper Landing project, where there was a time lag between "public input" and "government" response, the participants *knew* they were being listened to because when they came to the next meeting they saw graphic products that represented *their* ideas. When they made changes, the results were visible not only on the plots, but also in the numbers that were used in calculations and reports. In a truly interactive process, everyone will be looking at the same data or derived information, rather than someone's interpretation (which some fear as a censoring process). No one will "go away" and then come back with an answer—the answers will be produced in a public forum. While the validity of models and the quality of the data itself may still be questioned, the public can have much more confidence that critical information is not being "kept" from them.

Imagine the trust that can be developed where public working-group members can pose their own assumptions and "what if?" questions, watch the results be generated and displayed, and then immediately be able to discuss the implications of those results with resource managers and decision-makers. Think of the intent of NFMA and NEPA, and the philosophy of ecosystem management—all of which place great value on public involvement. Reflect on that concept and intent in the following scenario.

A public working group has struggled diligently to define the conditions they desire in the landscape surrounding their community. They have nearly reached consensus, when one of the members points to a polygon on a plot and states that she is satisfied with everything they've decided except that one area. She then offers an alternative desired condition and a slightly modified polygon boundary. As the group ponders her suggestion, they explore what they know about that part of the ecosystem by querying the databases and displaying the various resource components of the subject area, and those on adjacent lands. They bounce the idea around awhile and look at changes in the various acreage reports that would result from the boundary adjustment. A few questions are posed to the wildlife biologist and hydrologist who are part of their support team, and a couple of other alternatives are displayed on the screen. One by one the members of the working group (including the line officer) indicate that they can live with her suggestion, and consensus is reached.

What is astounding about this scenario is that the process, from suggestion to conclusion, could transpire in 45 minutes. The potential ramifications were explored with all members present and participating, alleviating concerns of behind-the-scenes manipulations or unseen political pressures. What is frustrating about this scenario is that the only real impediment to making it a widespread reality is the paucity of sufficient, integrated databases, and too few land managers skilled in bringing diverse interests together to discuss landscape-level issues.

There is another benefit of GIS that is critical to public acceptance of land management planning and other aspects of ecosystem management—it was key to the Cooper Landing Project and to the scenario described above. With GIS, the communication inherent in visualization goes beyond just having a picture—the "pictures" represent real places on the ground. For example, when the Cooper Landing working group looked at a plot of one of their alternatives, they were also able to picture in their minds the actual landscape the plot represented, because they were familiar with the area. Thus, when a polygon treatment was identified as "summer tractor logging with natural regeneration and fill-in planting as needed," working group members knew specifically where the proposed treatment was scheduled to take place.

In the interactions described in the theoretical scenario, the ability to deal with the suggestion quickly relies significantly on the fact that the group is discussing the desired condition of a specific piece of ground, not the number of acres allocated to a certain land use. The GIS would also generate reports of acres and projected outputs, but those become secondary to addressing issues of landscape patterns and concerns about specific locations of special interest.

Beyond Planning

Neither public involvement with GIS nor ecosystem management culminates in the world of planning. For example, I believe that one of the greatest opportunities for meaningful public involvement with GIS technology lies unexplored at the front counters of land management agencies across the country. Recreation users who have questions about where to view wildflowers, catch fish, or cross-country ski would certainly appreciate the latest information and most up-to-date maps available. Why can't they design their own maps (using point and click technology) to access the same databases that the resource managers use? Customers could include on their map all of the information pertinent to their interests, and leave off everything extraneous—and have fun doing it! Their map could be drawn to a scale that reflected their needs and interests. The price of the map would be based on the costs of the specific product they created, rather than buying a map that may be out of date and probably contains more information than the customer really wants. Those same databases could display photographs, interpretive information, satellite images, current weather and road/trail/stream conditions, fire danger levels and restrictions, or projections of how full the campgrounds and picnic areas are likely to be. Talk about being responsive to public needs!

James Burke (1993) recently posed the question "What is the point of knowledge if not to run alternative futures from which to make choices?" Indeed! GIS provides wonderful opportunities for exploring alternative futures in natural resource management. But how do we know if we've made the right choice? What if the ecosystem responds to management in ways that are different from those we predicted and wanted?

The philosophy of ecosystem management (see, for example, Society of American Foresters 1993), acknowledging the limitations of our understanding of ecosystem functions, realistically anticipates that such a scenario will occur. Thus, ecosystem management is also adaptive management and requires two reiterative steps: (1) monitoring system response to manipulation, and (2) adjusting management practices to refine the course toward desired conditions. GIS and public involvement both play a role in each of these steps, as they did in the planning and initial decision-making phases.

The same recreation users who are utilizing GIS to create their own maps should also be considered an excellent resource for "monitoring" as they spend time exploring the landscape—so do the mushroom collectors, ranchers, researchers, loggers, and others who work and play in the outdoors. In this world of declining budgets, these individuals augment agency personnel not only in the number of eyes that are "monitoring" the landscape, many of them also bring historical knowledge and experiential wisdom that land managers may not have. GIS provides a platform to capture this informal monitoring, as well as the information generated through detailed studies.

After the results of informal and analytical monitoring are entered into the ecosystem management database, they can be displayed and compared to those that were predicted in the planning phase. The spatial and visualizing capabilities of GIS become the windows for both the resource managers and the public to "see" what management changes (adaptations) might be necessary. Comparisons of actual versus predicted ecosystem responses become site-specific rather than a summary of acres treated, widgets produced, and user-days tallied:

> "We had planned on doing a controlled burn of 15 acres in the slope above Jane's Creek—that was this area outlined on the photo on the left side of the display, and represented by the light gold polygon on the right side of the screen. The management intent of the burn was to reduce the percent ground cover of exotic invading forbs in the understory, and to rejuvenate the bunch grasses that historically dominated this area. Well, the fire burned a little hotter than we had planned, and as the video image and this light green polygon indicate, we ended up treating 23 acres instead of 15. The bad news is that the fire intensity killed a few pole-sized trees we hadn't intended on losing; the good news is that we seem to have eradicated the invading forbs from the site, and forage production is about 40 percent higher than our models predicted. We're still waiting on the nutritional analysis of this season's grass to see if there's any change in forage quality."

Using GIS to analyze, visualize, and communicate monitoring results essentially starts the process over again. Even if monitoring confirms the predicted course toward desired conditions, that assessment (and the supporting information) should be presented to the public along with the request for confirmation that expectations are being met by reality. In the Forest Service, this cycle would be reflected in forest plan revision or amendment processes. However, rather than treating those processes as cyclic, they should be viewed as dynamic. Ecosystem management, supported by thorough monitoring, the analytical and communication potentials of GIS, and continuous public involvement, should allow us to make natural resource management decisions frequently and in relatively small increments—reflecting adjustments in course, rather than the significant changes in management direction we are faced with today.

References

Burke, J. 1993. The complexity effect. Keynote address to the 13th Annual ESRI User Conference. Palm Springs, California. May 24–28, 1993.

Kessler, W.B., H. Salwasser, C.W. Cartwright, Jr., and J.A. Caplan. 1992. New perspectives for sustainable natural resources management. Ecological Applications 2(3):221–225

Naisbitt, J. 1982. MEGATRENDS: Ten new directions transforming our lives. Warner Books. New York.

Sirmon, J.M. 1991. Evolving concepts of leadership: Toward a sustainable future. Forest Perspectives 1(2):8–9

Society of American Foresters. 1993. Task force report on sustaining long-term forest health and productivity. Bethesda, Maryland.

World Commission on Environment and Development. 1987. Our common future. Oxford University Press, New York.

21

Realizing the Potential of Remote Sensing and GIS in Ecosystem Management Planning, Analysis, and Policymaking

V. Alaric Sample

Remote sensing/GIS promises to become a central tool in land and resource management planning and decision-making, not just for analysis but for consensus-building and for public involvement in public resource management. GIS has significantly improved past capabilities in resource inventory, modeling, and monitoring. In resource inventory it offers a cost-efficient way to assess ecological characteristics over landscape-scale area and to periodically update basic data. Combined with our growing understanding of the functioning and response of natural ecosystems, GIS technology is improving our ability to model changes over time in the interrelationships among ecosystem components at the landscape scale. In monitoring, remote sensing and GIS provide important new capabilities in change analysis and in providing feedback to ecosystem models.

As a spatial analytical tool, GIS has inherent advantages over earlier multiresource planning models, such as those based on linear programming. The recent experience with FORPLAN, a linear programming model used in national forest planning, serves as one example. Many of the difficulties with linear programming planning models such as FORPLAN, in terms of public involvement as well as plan implementation, stem from the fact that there is no spatial representation of the optimal solutions developed with the model. The optimal solution might indicate the timber volume to be harvested or the fish and wildlife habitat capacity to be maintained, but these are averages over the entire analysis area, often a national forest of a million or more acres. The models give no indication as to where on the ground these activities should take place in order to actually implement the solution.

In public involvement in national forest planning, linear programming planning models were as frustrating as they were informative, during both plan development and plan implementation. Many of the individuals and organizations that sought involvement in national forest planning did so out of a concern for a special place—a watershed important for high-quality water for municipal needs, a riparian area along a quiet trout stream, a picturesque mountainside in view of a popular hiking and recreation area, a favorite elk-hunting spot visited year after year. What many people were anxious to know about a proposed plan was what it would mean for their special place. Despite the millions of dollars spent on national forest planning—and on the development of FORPLAN in particular—such seemingly simple questions could not easily be answered. People were asked to support the plan despite their continuing uncertainty over the very things they cared most deeply about.

When the time came to implement the plans, what actually took place on the ground was inevitably different from what at least some segments of the public had anticipated. Accusations of bad faith were exchanged and many specific projects were delayed or halted by legal or administrative appeals. Resource managers felt that the public's commitment to the forest plan should translate to public support for the specific projects necessary to carry out the plan. Segments of the public, on the other hand, felt no commitment whatsoever to management activities that they had inquired about repeatedly during the planning stage and were unable to obtain any specific information on.

Spatial displays using GIS will not eliminate the basic challenge of balancing the tradeoffs among competing or conflicting resource users, but they can make it easier for all the stakeholders to see and understand what activities are being proposed and where it is anticipated they will take place on the landscape. Use conflicts over particular geographic locations are more likely to be recognized and resolved during plan development, rather than cropping up during implementation—causing frustrating delays and potentially necessitating significant revisions to the plan.

But there are other potential pitfalls that GIS shares with nonspatial planning models that have gone before. One of the difficulties with FORPLAN and other planning models is that they are a "black box" to many policymakers and other segments of the interested public. With FORPLAN, there was little understanding beyond the scientists and resource managers themselves of how the model developed the solutions it provided, and how the solutions were influenced by the mathematical constraints and coefficients put into the model. Rather than helping resource managers to communicate the opportunities and tradeoffs associated with alternative management options, the models served to alienate nontechnical participants, engendering mistrust and a sense that decisionmakers were hiding behind the technology.

GIS and resource planning models can also be highly complex, with practitioners speaking in an intimidating stream of acronyms. The basic concept of a geographic information system—collecting resource information on a series of overlays and then displaying maps on the basis of criteria that group important characteristics such as slope, aspect, and vegetation type—is simple and intuitive. As resource managers become more familiar with the technology they should keep in mind how intimidating they themselves found it at first. They must work to keep the use of GIS in public involvement simple and intuitive while still communicating essential information about the resource characteristics being displayed and about basic assumptions such as resolution and accuracy. They must be careful to not let their ability to create marvelous, multicolored displays get in the way of the need to communicate effectively with people about what is expected to happen in their special places.

GIS will be a crucially important tool in the next major step toward implementing ecosystem management at the landscape scale: the cooperative establishment and accomplishment of ecosystem management goals on lands in mixed public and private ownership. GIS can provide all landowners within the ecosystem with a comprehensive view at the landscape scale, one in which they can readily see their land as a single piece in a larger puzzle.

This ecosystem-level view can also serve as a basis for facilitating a greater understanding of the influence their management decisions have on surrounding lands and what implications their neighbors' land and resource management decisions might have for the resources and values on their land. Several states are currently considering forest-practice regulations that require watershed-level analysis, aimed at minimizing the negative cumulative effects of individual actions on water quality. Major corporate and nonindustrial forest landowners around the United States are recognizing that a cooperative approach to ecosystem management that considers the role of private forest lands in the overall landscape matrix may be the key to protecting habitat and avoiding acute endangered-species controversies in the future. This kind of understanding, and newly available capabilities for landscape-scale analysis using GIS, can form the basis for a far higher level of coordination, cooperation, and collaboration than has existed before among adjacent public and private forest landowners.

What is still needed in many areas of the United States is the resource information to go into the GIS. The federal government has had authority to conduct inventories on all forest lands, public and private, since the 1930s, but inventories generally have been limited to statewide surveys focused primarily on timber resources; good spatial information on broader ecosystem values and resources has generally been lacking. This information would be of value not just in providing a current inventory of ecological and economic values, but in monitoring the change and redistribution of those values across the landscape over time. Efforts to develop such information are under way currently in many parts of the United States. For example, the Forest Service's Satellite Vegetation Mapping project in the Pacific Northwest is intended to classify, map and inventory all vegetation on forest and range lands in Oregon and Washington, not just the vegetation classifiable as late-successional forest. An inventory of "old-growth" forests is limited to portraying one portion of the landscape—under a particular definition that continues to evolve—at one particular point in time. Ecologists, resource managers, and planners need to be able to project changes in vegetation patterns at the landscape scale as they might vary under different management regimes and land-use allocations. In time, vegetative mapping needs to incorporate historical information as well, in order to gain a better understanding of the natural range of variation and how this compares with landscape characteristics, fire regimes, and climate variations that have developed since European settlement.

GIS is a highly flexible analytical tool, capable of adapting to new information needs, new definitions of ecologically important characteristics, and even changing social values over time. Recent efforts at late-successional forest ecosystem classification in the Pacific Northwest show the value of having digital databases from which maps can be drawn and redrawn as our understanding of the critical ecological characteristics that define high-priority old-growth forests continues to improve. Early efforts at late-successional forest ecosystem classification directly using remotely sensed imagery were poorly suited for change analysis and could not be adapted easily to reflect an increasingly precise definition of the ecological characteristics of late-successional forest ecosystems. Current efforts using digitized imagery in a GIS database allow the mapping and remapping of the area using a variety of different assumptions specified by ecologists. A digital GIS database permits ecologists to test the effect of changing assumptions on buffer widths and minimum critical habitat size on assessments of the extent and location of late-successional forests and connectivity across the landscape. This capability also allows for the prioritization of remaining old-growth to separate the most critical areas from the least critical. If the political compromise reached for the Pacific Northwest old-growth forests calls for a gradual stepping-down of old-growth harvest levels during an economic transition period, such an analysis can help assure that harvesting is directed to the least critical areas first, leaving most valuable areas intact.

Digital GIS databases also provide the flexibility to adapt inventory, planning, and monitoring to changing social values. Ten years ago few individuals outside the scientific community were even using the term biodiversity—now it has become a broad social concern that has fundamentally reoriented our approach to managing forest ecosystems, whether they be in wilderness areas or in the larger "landscape matrix" of managed forestlands. This has required the gathering of additional resource information, through both remotely sensed and ground-based inventories. But this information can easily be added into the digital GIS database so that it can be viewed in relation to other important attributes and can serve as a basis for new maps with greater sensitivity to biodiversity considerations.

Much closer cooperation is needed in the future, particularly between resource managers and GIS applications specialists. Although both ecologists and resource managers are being challenged by the need to focus their analysis at the landscape scale, ecologists seem to be further along in using the potential of GIS to meet current and emerging information needs. Resource managers are only beginning to fully grasp the capabilities of GIS and to think creatively about how it can be used to address not only analytical needs but the need for informed, interactive involvement of policymakers and the public in resource-management planning and decision-making. Resource managers must be careful to not get caught up in a "stagnation cycle," described earlier by Roger Hoffer (Chapter 3)—managers unwilling to work with researchers because researchers are not tailoring information and new techniques to managers' needs, and researchers unwilling to work with managers because managers won't take the time and effort to articulate their specific needs to researchers.

GIS can also foster the closer coordination and cooperation among adjacent landowners that is crucial to ecosystem management at the landscape scale. There are few areas of the United States in which the delineation of landscape-scale areas along ecological (rather than political or property) boundaries does not encompass a patchwork of public and privately owned lands. The sustainable management of forest ecosystems at the landscape scale is being recognized increasingly as critical to the protection of biological diversity, water quality, and other ecological values—and to avoiding many of the social and economic costs associated with the current crisis-driven, species-by-species approach to conserving habitat for sensitive or threatened plant and animal communities. A common GIS database developed for the entire landscape-scale area can play a central role in bringing adjacent landowners together to assess existing ecological values and to establish commonly agreed upon ecosystem management objectives.

Resource management planning of the future may entail spatial displays of resource information, and stakeholders' expression of their needs and concerns through interactive, real-time development of alternative overlays of land and resource use. Zane Cornett's description (Chapter 20) of the use of GIS to allow concerned citizens to truly participate in decision-making on the Cooper Landing Cooperative Project in Alaska is but a foretaste of the central role that GIS is certain to play in the future in forest planning. A visual spatial display in which citizens can see the areas most important to them, their physical and ecological characteristics, and how these can be expected to change under alternative management scenarios can convey more information more effectively than hundreds of pages of narrative descriptions and statistical tables. It also facilitates the recognition of the role that a particular area plays in the larger landscape and thus promotes a clearer understanding of the need for an integrated approach to resource management.

Interactive, real-time analysis of resource management options may become a key component in such acute, high-stakes political issues as the development of an ecologically sound, economically viable, and socially responsible solution to the old-growth/spotted owl controversy in the Pacific Northwest. The analysis of 14 planning alternatives for the old-growth forests of the Pacific Northwest by the Scientific Panel on Late-Successional Forest Ecosystems (the "Gang of Four," see Chapter 2) relied on the development of hand-drawn maps, analyzed by dozens of scientists over a period of several months. The development of an actual plan for the region will have the benefit of a digitized database and GIS and is likely to be developed through the active interplay among ecologists, resource managers, policymakers, and the concerned public. Land allocation and management prescriptions can be proposed and displayed, then adjusted and displayed again, repeating until the best possible compromise is found. In the end, we may find that the availability of this new analytical tool has facilitated an entirely new kind of interactive role for scientists in the development of political solutions to public resource management controversies.

The accessibility of GIS technology enables the public not just to understand analyses by resource managing agencies, but to conduct their own proactive analysis for comparison. GIS technology and hardware are becoming increasingly accessible and affordable to individuals and organizations that can develop independent analytical capabilities. These groups can play a key role in developing a broader public understanding of what information the GIS data is conveying to them and how to communicate needs, concerns, and perhaps even unforeseen opportunities to resource decision-makers.

Alternative GIS techniques and interpretations of remotely sensed data keep policymakers cognizant of the fact that the technology does have its limitations and drawbacks, that there is art as well as science involved in its interpretation, and that there is no single "right" answer. GIS—no matter how sophisticated—is a powerful tool to be used by decision-makers, but it is incapable in itself of making a single decision. This is, and will remain, the province of the informed and sensitive judgment of scientists and resource managers, in consultation with the full diversity of individuals and communities whose interests are at stake in the decision.

This general accessibility of GIS technology is perhaps most important simply from the standpoint of the value of an informed, engaged citizenry able to converse intelligently and debate constructively with scientists and resource managers on increasingly complex questions over the protection and use of valuable public resources.

Continuing rapid advances in remote sensing/GIS technology mean that even the quantum leap that has already taken place in its use by ecologists and resource managers has only scratched the surface of what is possible. As with any new technology, the challenge will be to assure that GIS technology continues to serve the practitioners and users, rather than the other way around. Ecologists and resource managers must continue to improve their ability to articulate new and emerging information and analysis needs. GIS applications specialists must continue to improve their ability to understand these needs and to craft custom solutions from their rapidly expanding array of technical capabilities. Such improvements in coordination and communication could not help but increase the likelihood that remote sensing/GIS technology will live up to its fullest promise and potential in ecosystem management planning, analysis, and decision-making.

About the Contributors

CRAIG D. ALLEN is an Ecologist with the National Biological Survey at the Jemez Mountains Field Station, Los Alamos, New Mexico.

DONALD R. ARTIS, JR. is Chief of Environmental Sciences, U.S. Army Topographic Engineering Center, Alexandria, Virginia.

DAVID CAWRSE was District Ranger, USDA Forest Service, Nantahala National Forest, Highlands Ranger District, Highlands, North Carolina, at the time his chapter was prepared and is now Ecosystem Management Specialist for the Chatooga Project located at Clemson University, South Carolina.

DAVID T. CLELAND is the Research Liaison between the Lake States National Forests and the North Central Forest Experiment Station, USDA Forest Service, Rhinelander, Wisconsin.

WARREN B. COHEN is a Research Forester with the USDA Forest Service, Pacific Northwest Research Station in Corvallis, Oregon.

ZANE J CORNETT is Manager of Resource Information with the USDA Forest Service, Chugach National Forest, Anchorage, Alaska.

SARAH CRIM is a Timber Management Analyst with the USDA Forest Service, Pacific Northwest Region, Portland, Oregon.

THOMAS R. CROW is Project Leader with the Landscape Ecology Research Unit, North Central Forest Experiment Station, USDA Forest Service, Rhinelander, Wisconsin.

RICHARD O. FLAMM was Research Associate with Oak Ridge National Laboratory, Environmental Sciences Division in Oak Ridge, Tennessee, at the time his chapter was prepared and is now with the Florida Marine Research Institute in St. Petersburg, Florida.

JERRY F. FRANKLIN is Professor of Ecosystem Analysis at the College of Forest Resources, University of Washington, Seattle, Washington.

JESSICA GONZALES is a Remote Sensing Analyst and Vegetation Inventory Coordinator for the San Juan–Rio Grande National Forest Integrated Resource Inventory Center, Dolores, Colorado.

KASS GREEN is President, Pacific Meridian Resources, Emeryville, California.

WILLIAM P. GREGG, JR. is the Chief of the International Affairs Office for the National Biological Survey in Arlington, Virginia.

FORREST G. HALL is Research Scientist with NASA at the Goddard Space Flight Center in Greenbelt, Maryland.

JAMES B. HART is Associate Professor in Michigan State University's Forestry Department.

ROGER M. HOFFER is Professor, College of Natural Resources, Colorado State University, Fort Collins, Colorado.

GEORGE E. HOST is an Ecologist with the Natural Resources Research Institute, University of Minnesota, Duluth, Minnesota.

K. NORMAN JOHNSON is Professor, Department of Forest Resources, Oregon State University, Corvallis, Oregon.

MARK D. MACKENZIE is an Ecologist with the Environmental Remote Sensing Center, University of Wisconsin-Madison, Madison, Wisconsin.

DAVID J. MLADENOFF was with the Natural Resources Research Institute, University of Minnesota, Duluth, Minnesota, at the time his chapter was prepared, and is now Forest Ecologist, Wisconsin Department of Natural Resources, Bureau of Research, and Assistant Professor, Department of Forestry, University of Wisconsin-Madison, Wisconsin.

MARGARET M. MOORE is Associate Professor, School of Forestry, Northern Arizona University, Flagstaff, Arizona.

PETER H. MORRISON is the Research Director of the Sierra Biodiversity Institute in North San Juan, California.

EUNICE A. PADLEY is a Forest Ecologist with the Huron–Manistee National Forests in Cadillac, Michigan.

SCOTT M. PEARSON is Research Associate with Oak Ridge National Laboratory, Environmental Sciences Division in Oak Ridge, Tennessee.

JAMES A. ROCHELLE is Senior Wildlife Biologist with Weyerhaeuser Company in Tacoma, Washington.

V. ALARIC SAMPLE is Vice President for Research and Director of the Forest Policy Center at AMERICAN FORESTS in Washington, D.C.

DOUGLAS P. SCHLEUSNER is Forest Planner for the USDA Forest Service, Santa Fe National Forest, Santa Fe, New Mexico.

JOHN SESSIONS is Professor, Department of Forest Engineering, Oregon State University, Corvallis, Oregon.

THOMAS A. SPIES is a Research Forester with the USDA Forest Service, Pacific Northwest Research Station in Corvallis, Oregon.

MONICA G. TURNER is a Research Scientist with Oak Ridge National Laboratory, Environmental Sciences Division in Oak Ridge, Tennessee.

CHARLES C. VAN SICKLE is Assistant Station Director with the USDA Forest Service, Southeastern Forest Experiment Station, Asheville, North Carolina.

Index

Island Press Board of Directors